SINGAPORE
BUSINESS

**World Trade Press
Country Business Guides**

CHINA Business
HONG KONG Business
JAPAN Business
KOREA Business
MEXICO Business
SINGAPORE Business
TAIWAN Business

SINGAPORE BUSINESS

The Portable Encyclopedia For Doing Business With Singapore

Christine A. Genzberger
David E. Horovitz
Jonathan W. Libbey
James L. Nolan
Karla C. Shippey, J.D.
Chansonette Buck Wedemeyer

Edward G. Hinkelman
William T. LeGro
Charles Smithson Mills
Stacey S. Padrick
Kelly X. Wang
Alexandra Woznick

Auerbach International • Baker & McKenzie
CIGNA Property and Casualty • Ernst & Young
Far Eastern Economic Review • Foreign Trade
Singapore Trade Development Board

Series Editor: Edward G. Hinkelman

WORLD TRADE PRESS ®
Resources for International Trade
1505 Fifth Avenue
San Rafael, California 94901
USA

Published by World Trade Press
1505 Fifth Avenue
San Rafael, CA 94901
USA

Cover and book design: Brad Greene
Illustrations: Eli Africa
Color Maps: Gracie Artemis
B&W maps: David Baker
Desktop Publishing: Kelly R. Krill and Gail R. Weisman
Charts and Graphs: David Baker and Kelly R. Krill
Prepublication Review: Aanel Victoria

Copyright © 1994 by Edward G. Hinkelman. All rights reserved.
No part of this book may be reproduced, transmitted, or copied in any manner whatsoever without written permission, except in the case of brief quotations embodied in articles and reviews. For information contact the publisher.

The name *World Trade Press* and the representation of the world map imprinted on an open book appearing in this publication are trademarks and the property of Edward G. Hinkleman.

Permission to reprint copyrighted materials has been given as follows: Articles from *Asia Pacific Legal Developments Bulletin* (vol 8., no. 3, Sept. 1993), copyright © 1993 by Baker & McKenzie, reprinted with permission of Baker & McKenzie, Singapore, and the author Edmund H.M. Leow. Excerpts from *Ports of the World* (14th ed.), copyright © 1993 CIGNA Property and Casualty Co., reprinted with permission from CIGNA Property and Casualty Companies. Excerpts from *1993 Worldwide Corporate Tax Guide* and *1993 Worldwide Personal Tax Guide*, copyright © 1993 by Ernst & Young, reprinted with permission of Ernst & Young. Articles and cartoons from *Far Eastern Economic Review,* copyright © Far Eastern Economic Review, reproduced with permission of the publisher. Export and import statistics from *Foreign Trade*, copyright © 1993 by Defense & Diplomacy, Inc., reprinted with permission of the publisher. City maps originally sourced from the *HSBC Group's Business Profile Series,* with permission of The Hongkong and Shanghai Banking Corporation Limited, copyright © 1991 by The Hongkong and Shanghai Banking Corporation Ltd. Excerpts from *Singapore Trade Connection,* copyright © 1993 by Singapore Trade Development Board, reprinted with permission of Singapore Trade Development Board.

Library of Congress Cataloging-in-Publication Data
Singapore business : the portable encyclopedia for doing business with
   Singapore / Christine Genzberger . . . [et al.].
      p. cm. – (World Trade Press country business guides)
   Includes bibliographical references and index.
   ISBN 0-9631864-6-9 : $24.95
    1. Singapore—Economic conditions. 2. Singapore—Economic
policy. 3. Investments, Foreign—Government policy—Singapore.
4. International business enterprises–Singapore. I. Genzberger,
Christine. II. Series.
HC445.8.S543 1994   93-45979
658.8'48'095957–dc20   CIP

Printed in the United States of America

ACKNOWLEDGMENTS

Contributions of hundreds of trade and reference experts have made possible the extensive coverage of this book.

We are indebted to numerous international business consultants, reference librarians, travel advisors, consulate, embassy, and trade mission officers, bank officers, attorneys, global shippers and insurers, and multinational investment brokers who answered our incessant inquiries and volunteered facts, figures, and expert opinions.

A special note of gratitude is due to those at the U.S. Department of Commerce and the Singapore Trade Development Board.

We relied heavily on the reference librarians and resources available at the Marin County Civic Center Library, Marin County Law Library, San Rafael Public Library, San Francisco Public Library, University of California at Berkeley libraries, and U.S. Department of Commerce Library in San Francisco.

Our thanks to attorneys Robert T. Yahng and Anne M. Kelleher, with Baker & McKenzie, San Francisco, who spent precious time in assisting us with the law section. We also extend our sincere appreciation to Barry Tarnef, with CIGNA Property and Casualty Co., who graciously supplied information on world ports.

We also acknowledge the valuable contributions of Philip B. Auerbach of Auerbach International, San Francisco, for translations; all the patient folks at Desktop Publishing of Larkspur, California; and Leslie Endicott and Susan August for reviewing, proofing, and correcting each draft.

Thanks to Elizabeth Karolczak for establishing the World Trade Press Intern Program, and to the Monterey Institute of International Studies for its assistance.

To Jerry and Kathleen Fletcher, we express our deep appreciation for their immeasurable support during this project.

Very special thanks to Mela Hinkelman whose patience, understanding, generosity, and support made this project possible.

DISCLAIMER

We have diligently tried to ensure the accuracy of all of the information in this publication and to present as comprehensive a reference work as space would permit. In determining the contents, we were guided by many experts in the field, extensive hours of research, and our own experience. We did have to make choices in coverage, however, because the inclusion of everything one could ever want to know about international trade would be impossible. The fluidity and fast pace of today's business world makes the task of keeping data current and accurate an extremely difficult one. This publication is intended to give you the information that you need in order to discover the information that is most useful for your particular business. As you contact the resources within this book, you will no doubt learn of new and exciting business opportunities and of additional international trading requirements that have arisen even within the short time since we published this edition. If errors are found, we will strive to correct them in preparing future editions. The publishers take no responsibility for inaccurate or incomplete information that may have been submitted to them in the course of research for this publication. The facts published indicate the result of those inquiries and no warranty as to their accuracy is given.

Contents

Chapter 1	Introduction	1
Chapter 2	Economy	3
Chapter 3	Current Issues	17
Chapter 4	Opportunities	23
Chapter 5	Foreign Investment	41
Chapter 6	Foreign Trade	47
Chapter 7	Import Policy & Procedures	53
Chapter 8	Export Policy & Procedures	61
Chapter 9	Industry Reviews	65
Chapter 10	Trade Fairs	79
Chapter 11	Business Travel	135
Chapter 12	Business Culture	145
Chapter 13	Demographics	159
Chapter 14	Marketing	163
Chapter 15	Business Entities & Formation	171
Chapter 16	Labor	181
Chapter 17	Business Law	189
Chapter 18	Financial Institutions	207
Chapter 19	Currency & Foreign Exchange	219
Chapter 20	International Payments	223
Chapter 21	Corporate Taxation	237
Chapter 22	Personal Taxation	243
Chapter 23	Ports & Airports	247
Chapter 24	Business Dictionary	251
Chapter 25	Important Addresses	267
	Index	299

Introduction

Singapore is one of the world's most dynamic economies. A tiny island city-state with no resources other than a fine harbor and a strategic geographic location on the Straits of Malacca between the Indian Ocean and the South China Sea, Singapore is not only a thriving commercial and industrial center in its own right but also a regional hub and gateway to Southeast Asia's rapidly growing economies.

Since the mid-1960s Singapore has built a modern, internationally oriented trade, manufacturing, and service economy largely from scratch. This vibrant economy is based on low-cost, high-quality export production, trade, and state-of-the-art business and financial services. It grew at an average rate of 12.2 percent per year between 1970 and 1992. Despite its small size, Singapore is the world's seventeenth-largest exporter and fifteenth-largest importer. In 1992 its total trade was three times the size of its economy, the highest ratio in the world. Singapore's trade has grown at an annual rate of about 13.9 percent since 1970.

Ranked first in the world among newly industrialized countries in the 1993 *World Competitiveness Report*, Singapore is a market well worth investigating from a number of perspectives. For buyers Singapore can provide a wide range of competitive goods at virtually any level of sophistication. It is a leading producer of office and data-processing machines, electrical machinery, petroleum products, telecommunications equipment, industrial machinery, apparel, and organic chemicals, among many other items. Its businesses can handle anything from the smallest to the largest orders.

From the seller's standpoint Singapore needs a wide range of agricultural and industrial raw materials, intermediate components, and specialty items to feed its own active industries and those of the regional economies it services. Both the upgrading of its industrial base and its large public-sector development projects require materials, capital goods, and services. And the rising demands of Singapore's affluent and sophisticated consumers offer opportunities to place goods in the colony's avid consumer market. Singapore has relied on Japan for many of its goods, but the strength of the yen has made Japanese products less competitive, opening up opportunities for new suppliers. For manufacturers Singapore has a pool of well-educated semi-skilled to highly skilled labor experienced in the areas already noted as well as in many others. Singapore has up-to-date plant and its business infrastructure is second to none. These conditions provide a competitive base for a variety of outsourcing needs. Increasingly, Singapore is managing production for some of its less-developed neighbors, particularly in the Growth Triangle which it forms with Malaysia and Indonesia.

For investors Singapore is one the world's most attractive business venues. Restrictions are few and procedures are well established in a stable system presided over by a pro-business government. Singapore's industrial, trade, service, and financial markets are all well developed and highly active, providing a variety of investment opportunities. It particularly encourages foreign participation in its emerging high-technology industries, trade, and business and financial services areas, offering substantial tax breaks, operating exemptions, and other incentives. Singapore is refocusing its economy on high-technologies and high-value-added, clean, capital-intensive products and sophisticated business and financial services as it gives up some of its traditional edge in low-cost, low-technology production and assembly. The proactive stance of Singapore's technocratically skilled and staunchly pro-business government and the adaptability of its small- and medium-sized businesses argue that it will successfully make the transition to a new economy.

A sophisticated, world-class business center, Singapore moves at a fast pace. The overall tempo and level of change is expected to accelerate, making Singapore one of the most complex and challenging—as well as one of the most compelling—places on the globe in which to do business.

SINGAPORE Business was designed by businesspeople experienced in international markets to give you an overview of how things actually work and what current conditions are in Singapore. It will give you the head start you need as a buyer, seller, manufacturer, or investor to be able to evaluate and operate in Singaporean markets. Further, it tells you where to go to get more specific information in greater depth.

The first chapter discusses the main elements of the country's **Economy,** including its development, present situation, and the forces determining its future prospects. **Current Issues** explains the top three concerns affecting Singapore now and during its next stage of development. The **Opportunities** chapter presents 11 major areas of interest to importers plus 10 additional hot prospects, 14 major areas for exporters plus 14 more hot opportunities. Discussions of 11 major sectoral growth areas and a section on Singapore's special free trade zones, industrial estates, science park, international business park, and teletech park follow. The chapter also clarifies the nature of the government procurement process that drives Singapore's multibillion dollar development plans in construction, telecommunications, utilities, land development, roads, ports, aerospace, environment, biotechnology, petrochemicals, and tourism. **Foreign Investment** details policies, regulations, procedures, and restrictions, with particular reference to incentives and Singapore's developing high-technology and service sectors.

Although Singapore is banking on services and high technology as the wave of the future, it remains a diversified export-oriented economy with many thriving low- and medium-technology operations. The **Foreign Trade, Import Policy & Procedures,** and **Export Policy & Procedures** chapters delineate the nature of Singapore's trade: what and with whom it trades, trade policy, and the practical information, including nuts-and-bolts procedural requirements, necessary to trade with it. The **Industry Reviews** chapter outlines Singapore's 11 most prominent industries and their competitive position from the standpoint of a businessperson interested in taking advantage of these industries' strengths or in exploiting their competitive weaknesses. **Trade Fairs** provides a comprehensive listing of trade fairs in Singapore, complete with contact information, and spells out the best ways to maximize the benefits offered by these chances to see and be seen.

Business Travel offers practical information on how to travel to Singapore including travel requirements, resources, getting around, local customs, and ambiance, as well as comparative information on accommodations and dining in the country. **Business Culture** provides a user-friendly primer on local business style, mind-set, negotiating practices, and numerous other tips designed to improve your effectiveness, avoid inadvertent gaffes, and generally smooth the way in doing business in Singapore. **Demographics** presents the basic statistical data needed to assess the Singaporean market, while **Marketing** outlines resources, approaches, and specific markets in the colony.

Business Entities & Formation discusses recognized business entities and registration procedures for setting up operations in Singapore. **Labor** assembles information on the availability, capabilities, and costs of labor in Singapore, as well as terms of employment and business-labor relations. **Business Law** interprets the structure of the Singaporean legal system, giving a digest of substantive points of commercial law with additional material from the international law firm of Baker & McKenzie. **Financial Institutions** outlines the workings of the financial system, including banking and financial markets, and the availability of financing and services needed by foreign businesses. **Currency & Foreign Exchange** explains the workings of Singapore's complex foreign exchange market operations. **International Payments** is an illustrated step-by-step guide to using documentary collections and letters of credit in trade with Singapore. Ernst & Young's **Corporate Taxation** and **Personal Taxation** provide the information on tax rates, provisions, and status of foreign operations and individuals needed to evaluate a venture in the country.

Ports & Airports, prepared with the help of CIGNA Property and Casualty Company, gives current information on how to physically access the country. The **Business Dictionary,** a unique resource prepared especially for this volume in conjunction with Auerbach International, consists of more than 425 entries focusing specifically on Singaporean business and idiomatic usages to provide the businessperson with the basic means for conducting business in Singapore. **Important Addresses** lists more than 550 Singaporean government agencies and international and foreign official representatives; local and international business associations; trade and industry associations; financial, professional, and service firms; transportation and shipping agencies; media outlets; and sources of additional information to enable businesspeople to locate the offices and the help they need to operate in Singapore. Full-color, detailed, up-to-date **Maps** aid the business traveler in getting around the major business districts in Singapore.

SINGAPORE Business gives you the information you need both to evaluate the prospect of doing business in Singapore and to actually begin doing it. It is your invitation to this fascinating society and market. Welcome.

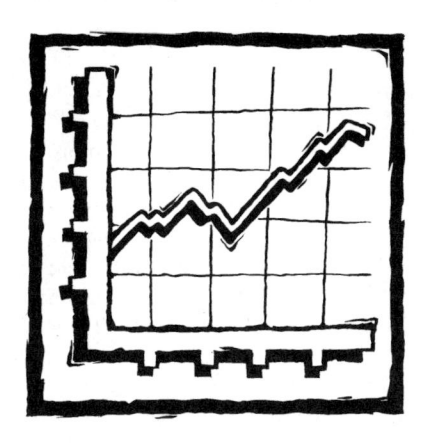

Economy

In less than 30 years, Singapore, a small former colony with no natural resources or real independent economic base, wracked by ethnic tensions and labor and political unrest, and threatened by its larger neighbors, has become the most modern, successful, stable, and sophisticated economy in Southeast Asia. This multiethnic city-state consists of a main island and some 58 islets located at the tip of the Malay Peninsula between Malaysia and Indonesia. With a total area of 639 square km (247 square miles) three-and-a-half times the size of Washington, DC, but slightly smaller than the five boroughs of New York City—and a population of about 3.1 million, the average population density in Singapore is more than 4,400 square km (12,500 per square mile), which makes it one of the most crowded places on earth.

The tropical island is mostly low and swampy. Some 49 percent of its land area is urban, while 0.1 percent is officially classed as agricultural, 4.5 percent as forest, 2.5 percent as tidal marsh, and 43.9 percent as "other"—a classification that includes inland waters, certain developed and undeveloped public spaces, and coconut and rubber plantations. Singapore has no natural resources except its deepwater natural harbor and its strategic location at the opening of the Straits of Malacca between the Indian Ocean and the South China Sea which makes it a gateway to Southeast Asia.

Singapore is about 78 percent ethnic Chinese—the largest concentration in Southeast Asia—14 percent Malay, 7 percent Indian, and 1 percent other. Dependent on trade since its founding, Singapore has parlayed its role as a business broker into its current dominant position.

HISTORY OF THE ECONOMY

Premodern Singapore was home to tigers and villages of fishermen and pirates under the nominal control of a succession of local states. After falling under the influence of the Portuguese in the 1500s and the Dutch in the 1600s, its modern history began in 1819 when Sir Thomas Stamford Raffles established a treaty port on the island under the auspices of the quasi-official British East India Company in order to compete with the Dutch. The company bought the island outright from the local overlord in 1824 and combined it with other like entities into the semi-official Straits Settlements. It was not declared an official colonial possession of Great Britain until 1867.

Colonial Growth

Singapore grew rapidly from a village of a few hundred prior to 1819 to a town of 5,000 in 1821 and a city of more than 80,000 in 1860. Between the 1820s and the 1870s, Singapore was a fast-growing, wide open free port catering to trade, outfitting, and transshipment as well as to such unsavory practices as slave trading, drug running, prostitution, and piracy.

The opening of the Suez Canal in 1869 increased Singapore's strategic importance, and its garrison was developed until it became a major British military base. Rubber plants smuggled in from Brazil and disseminated into surrounding areas in the 1870s established Singapore as the hub of the Southeast Asian rubber industry. The greater official British presence and the increasing legitimatization of its economy made Singapore increasingly respectable and prosperous. Trade expanded 700 percent between 1873 and 1913. This was also the period during which waves of Chinese immigrants began to arrive; they would, along with the British, set Singapore's cultural tone and ultimately become the inheritors of the colony.

This complacent prosperity lasted until the Japanese conquered the supposedly impregnable Singapore in short order in 1942. Japan exercised a harsh occupation of the island from 1942 to 1945. The chastened British returned to face a regional insurrection led by Communist Malays to the north. This tense situation led the British to declare a state of emergency in 1948 that lasted until they withdrew 12 years later. Between 1953 and 1959, negotiations were held to establish terms for Singapore's independence, the inevitability of which became increasingly apparent. The first local governing body was elected in 1955, and formal independence finally came in 1959.

Post-Colonial Development

In 1954 a core of Communists and leftists founded the People's Action Party (PAP). The party's patriarch and the architect of modern Singapore, Lee Kuan Yew, won a substantial victory in the 1959 elections. Over the years, the PAP consolidated its hold on power, gradually shifting its orientation in stages from far left to far right. Using the far-reaching Internal Security Act both to purge itself internally and to eliminate external opposition, the authoritarian PAP has become not merely the dominant political party but also the only real legitimate political and social force in Singapore.

In 1963 Singapore joined with three other former British colonies—Malaya, Sabah, and Sarawak—to form the Federation of Malaysia. This led to conflict with neighboring Indonesia. There was internal tension with Singapore's new partner, Malaya, over the fact that Singapore was primarily Chinese and that its Malay population held a subordinate place in its society and economy. In 1965 the Federation broke up, leaving Singapore to fend for itself with no natural resources or independent economy amidst relatively hostile neighbors.

The Modern Economic Base

During the 1950s Singapore's developing economic base consisted primarily of such low-level resource-based activities as lumbering, rubber cultivation, fishing, shipbuilding, tin smelting, and brick manufacture. Its domestic economy was predominately rural; fruit and vegetable cultivation and pig and poultry raising were the main activities.

During the 1960s the government took steps to upgrade the local economy, focusing on international trade and industrial production of light and medium industrial goods, particularly in such labor-intensive functional areas as assembly. The government sponsored a variety of public works and other programs, built industrial parks, and opened the economy up to foreign participation on favorable terms. In 1968 the government sponsored a development bank and a trading company to provide mechanisms and marketing operations to service the investment that the preceding measures were designed to attract.

The strategy began to bear fruit in the 1970s as Singapore's economy became increasingly export oriented. As the 1980s progressed, Singapore established a regional lead in such areas as electronics, transportation, machinery, and petroleum refining and began its shift into upmarket areas with such tangibles as semiconductors, components, and high-tech assembly of precision instruments and such services as trading, telecommunications, and international finance. Petroleum and electronics have been Singapore's two major product categories, and as world prices and demand for them ebb and flow, so does the country's prosperity.

In the 1990s, while concentrating on maintaining its edge, Singapore is looking for the next wave to ride. The government has focused on high-value-added, clean, advanced technologies and on state-of-the-art services. It is also attempting to exploit such lower-end opportunities as the growing labor and consumer markets in China and Vietnam. Officials fear that Singapore's past successes could make it complacent and thereby close off future success. The government wants to see growth continue, but not at a pace that prices Singapore out of the market by allowing wage rates and costs to get out of hand. This desire conflicts to a certain extent with Singapore's stated goal of becoming the Switzerland of the Far East, which entails becoming more affluent and continually raising the standard of living.

SIZE OF THE ECONOMY

Singapore's economy is about half the size of Hong Kong's. And although its population and territory are also about half the size of the colony's, Singapore and Hong Kong have similar per capita incomes. As the only developed nation in Southeast Asia, Singapore's economy is roughly on a par in size with that of its territorially much larger neighbor, Malaysia, and more than twice that of the even larger Vietnam. Singapore's gross domestic product (GDP) was US$44.6 billion in 1992. Its GDP grew from US$1.7 billion in 1960 to US$40.8 billion in 1991, an average annual increase of 16.3 percent in nominal terms.

After a sharp recession in 1985, when the economy contracted by 1.6 percent, the real growth rate slowed to an average of 8.3 percent from 1986 through 1991. Real growth peaked in 1988 at 11.1 percent and has been slowing ever since—to 9.2 percent in 1989, 8.4 percent in 1990, 6.7 percent in 1991, and 5.6 percent in 1992. As of mid-1993, the economy was growing at a 5 percent annual rate. However, preliminary figures for all of 1993 show a scorching overall growth of 9.8 percent to about US$49 billion. It remains to be seen whether this renewed growth will be sustainable.

Standard of Living

Singapore's per capita gross national product (GNP), which was only US$435 in 1960, had grown to US$14,731 in 1992, an increase of 11.3 percent over 1991. The performance of the economy in 1993 boosts the figure above US$15,000 and rapidly approaching the US$16,000 mark. Average annual

Singapore's Gross National Product (GNP)

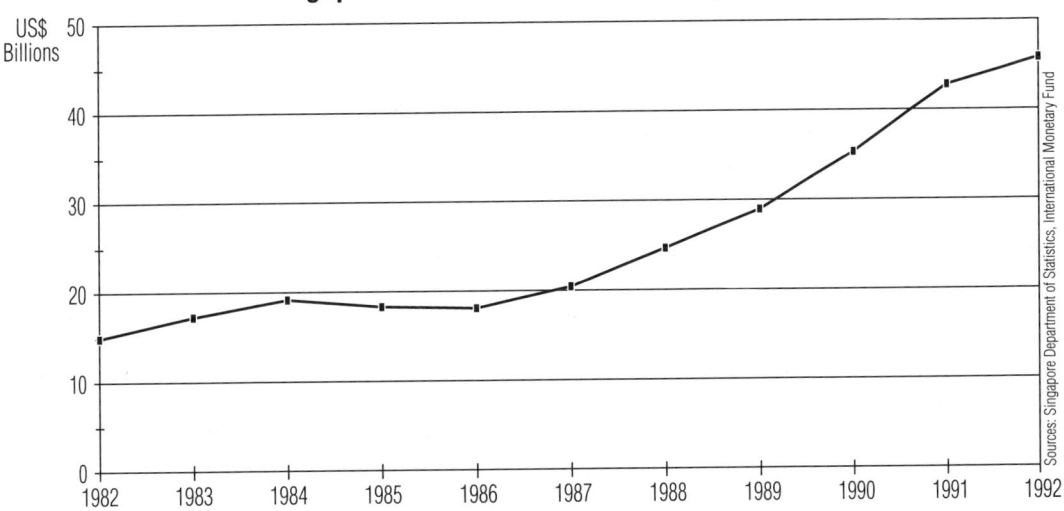

Sources: Singapore Department of Statistics, International Monetary Fund

growth was 10.2 percent from 1970 through 1990. The current figure places Singapore fourth in per capita income among major Pacific Rim nations, behind Japan, Australia, and Hong Kong and within striking distance of incomes in the affluent nations of Europe and Scandinavia.

CONTEXT OF THE ECONOMY

Like Hong Kong, Singapore had no real economy before it became a European treaty port. However, its situation is even more dramatic than that of Hong Kong, because in 1965, when it became independent as the Republic of Singapore, it was little more than a transshipment point. It had little in the way of a coherent political or economic structure, and it was at odds with its much larger neighbors. Its government, which has been described as "visionary, hands-on, and popular, leaving nothing to chance," can claim the vast bulk of the credit for Singapore's rapid rise.

Government Control

The picture in Singapore is one of sharp contrasts between what is at once one of the most wide open economies and at the same time one of most heavily managed and strictly paternalistically controlled societies in the world. Through regulation or actual participation, the government is involved in the management of virtually every aspect of business, although its commitment to free enterprise, to relaxing rules that hamper free wheeling business, and to privatizing its interest in state-run operations seems genuine. It has instituted a compulsory savings program, the Central Provident Fund (CPF) but the resulting funds are allocated according to the most ruthlessly capitalistic principles of the highest possible return. Official Singapore is generally composed of talented and scrupulously honest technocrats, and it is highly responsive to business needs. It provides the business climate, support, and infrastructure that have consistently led Singapore to rank among the top four business locations in the world.

Although the government does not plan the economy centrally, it does target industries for development, usually through tax and investment incentives, and it reserves the right to intervene directly in the affairs of individual businesses. There are few import and capital restrictions, and tariffs are low. However, the government's social planning is so heavy-handed and its punishments for infractions of social order are so draconian—indefinite incarceration without trial, administrative trial without jury, heavy fines for what most of the rest of the world considers minor infractions, and corporal and capital punishment—that some observers are voicing concern about a future brain drain due to a lack of personal and intellectual freedom.

Although the government controls student and labor groups and the press and has even attempted to provide its own hand-picked opposition candidates, while disqualifying true opposition candidates, locals generally view its policies as benignly paternalistic. It offers stability, security, full employment, high-grade yet low-cost services, and a rising standard of living within the context of an accepted Confucian-based ethic that highlights the ideals of hard work and obedience to authority. In fact, the greatest expressed fear is that of creeping Westernization, which locals worry could cause Singapore to lose its disciplined edge. This concern is well founded, as the older generation is replaced by a generation accustomed to a high standard of living that has no experience of underdevelopment.

The Lack of Local Giants

In contrast to many other Asian systems, Singapore's economy is not organized around dominant, integrated local firms, a fact that can be explained by its relatively short history; its diminutive size, which makes the domestic market a distant second to the international market; the generally even-handed treatment that the government accords domestic firms; and its emphasis on attracting foreign firms rather than protecting the locals by artificial means. The realistic official philosophy seems to be that if the outsiders can be kept happy, there will be plenty to go around, while if Singapore kills the goose that lays the golden egg, it could revert to a swampy backwater.

THE UNDERGROUND ECONOMY

Singapore's economic ambiance is cutthroat, but basically honest. Although the economy is wide open, there is a great deal of regulation. Moreover, those who enforce it are reasonably efficient and generally honest bureaucrats—a rare breed in Asia or elsewhere, for that matter. Virtually everybody who has the ambition and ability to take advantage of the system's opportunities benefits from it, so there is little incentive to try to get around regulations. And the fact that things are so tightly controlled, penalties are so harsh (although in practice they are usually less harsh than the law provides for), and Singaporeans have bought so heavily into the system are all substantial disincentives to illicit activity. In a country where selling chewing gum is illegal and there are heavy fines for jaywalking, street crime is virtually unknown, as is organized crime. Such practices as bid-rigging, oligopolistic market manipulation, special relationship-based deals, bribery, and the granting of favors are rare as well, and although gift giving is practiced as part of the Chinese ethic, people take great pains to avoid the appearance of impropriety. However, sharp practice and tough negotiating are the order of the day, and business operators definitely do not want to run afoul of the government.

In fall 1993 a scandal surfaced involving allegations that the head of the influential Singapore Trade Development Board (STDB) had used his position to gain special treatment. According to the charges, the highly placed executive, a collector, had arranged the sale of Chinese art and antiques to government-controlled institutions under conditions favorable to himself. In comparison with what is often seen as standard business practice elsewhere, the improprieties may seem relatively minor. However, the level of outcry that the incident provoked and the rarity of such corruption charges are cause for comment.

Intellectual Property Rights

Singapore is virtually unique among Asian nations in its stance on intellectual property. It not only professes to protect such property rights, it actually does so, having criminalized the possession of counterfeit items for use as well as for sale. Moreover, the laws on the books are actually enforced, especially for hard goods, such as audio and video products and such items as clothing and watches. Enforcement is less effective for harder-to-catch problems, such as software piracy. Trademark protection also has been strengthened, although patent protection needs additional work. Singapore follows British practice in that patents of outsiders are generally unprotected unless they have been registered in the United Kingdom.

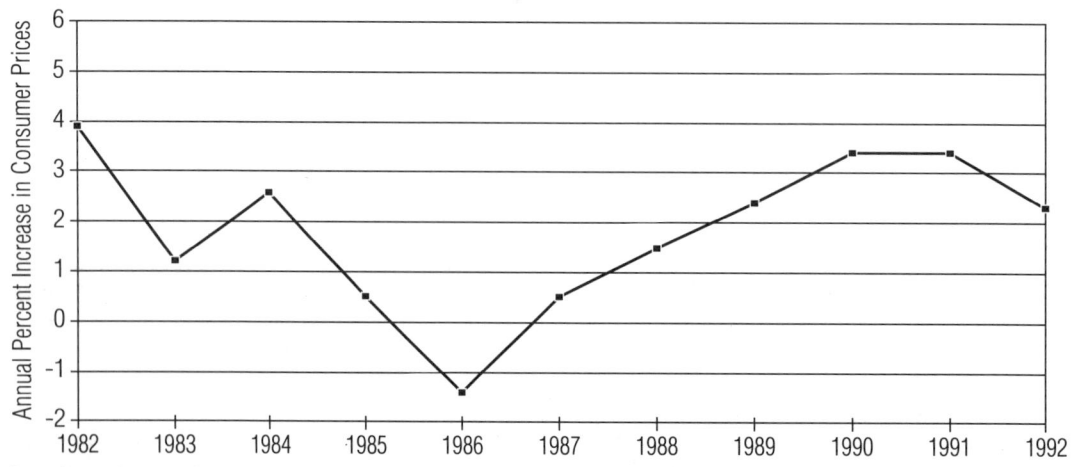

**Singapore
Inflation: 1982-1992**

Sources: Singapore Department of Statistics, International Monetary Fund

INFLATION

Despite rapid growth in Singapore since the 1960s, inflation has largely been kept in check not only by government controls but also by policies which, while mandating high savings, allow fairly liberal use of those savings as well as private consumption. Given the high wage increases of recent years and the continually rising standard of living, it seems likely that official figures understate actual inflation.

Prior to the early 1980s, annual inflation ran in the neighborhood of 8 percent. But between 1985 and 1992, as measured by consumer prices, the average annual increase dropped to 1.5 percent. Recession kept growth in prices to a –1.4 percent in 1986, but growth subsequently rose to a high of 3.4 percent in 1990 and 1991 before easing to 2.3 percent in 1992. Midyear 1993 inflation was running at an annual rate of 2.4 percent; the consensus among forecasters is that it will rise to 3 percent for all of 1993 and remain there well into 1994.

Another measure of expectations on inflation is provided by the prime rate, which in late 1993 stood at 4.8 percent, down from its recent high of 7.4 percent in 1990. Observers expect to see it climb to 5.6 percent by the middle of 1994, as expected demand picks up late in 1993.

LABOR

Singapore has a well-educated, hardworking labor force. School attendance is mandatory for 10 years, and attendance beyond that minimum is encouraged. Training has received fairly high priority: employers are expected to provide a substantial amount on their own, and they must also contribute to a government training fund.

The Labor Shortage

With a total labor force of only about 1.6 million in 1992, Singapore has fallen victim to its own success, and the current labor shortage is growing. Unskilled and semi-skilled workers are in especially short supply. Job hopping is causing employers concern with some job categories show a 25 percent annual rate of turnover as employers bid for scarce labor. To date, the government has taken a liberal view of imported labor, which makes up as much as 20 percent of the work force by some estimates. For some low-skill jobs, such as construction, employers have been allowed to import as many as five foreign workers for each local worker. However, concern that Singapore will not be able to provide services for guest workers or control them is growing, and the government has served notice that it will not accommodate business by easing access any further.

To underwrite services and discourage unnecessary labor imports, the government currently requires employers to pay monthly fees that can amount to as much as 30 percent of wages for each foreign worker. Some efforts are being made to encourage women to enter the work force. Because Singapore's business environment has traditionally been male dominated and because this conflicts to some extent with Singapore's traditional family values policy, these efforts have as yet been tentative.

Unemployment

Singapore's growth rate and chronic labor shortage made unemployment a minor concern during the 1980s and early 1990s. In 1960, roughly 14 percent of the labor force was unemployed, but unemployment had fallen to 3.3 percent by 1988, and since then it has fluctuated around the 2 percent mark. It was 2.2 percent in 1989, 1.7 percent in 1990, 1.9 percent in

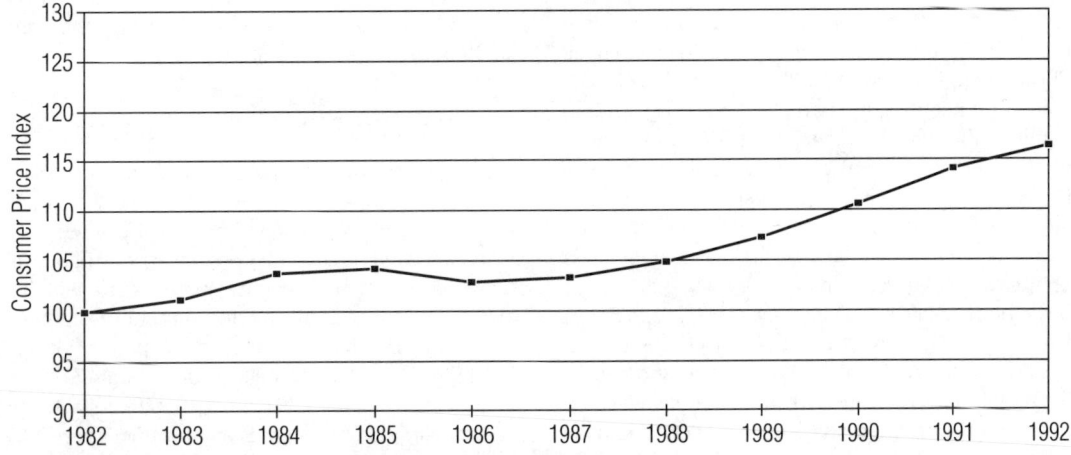

Singapore Consumer Price Index (CPI)

Sources: Singapore Department of Statistics, International Monetary Fund, International Financial Statistics

Structure of the Singaporean Economy - 1991

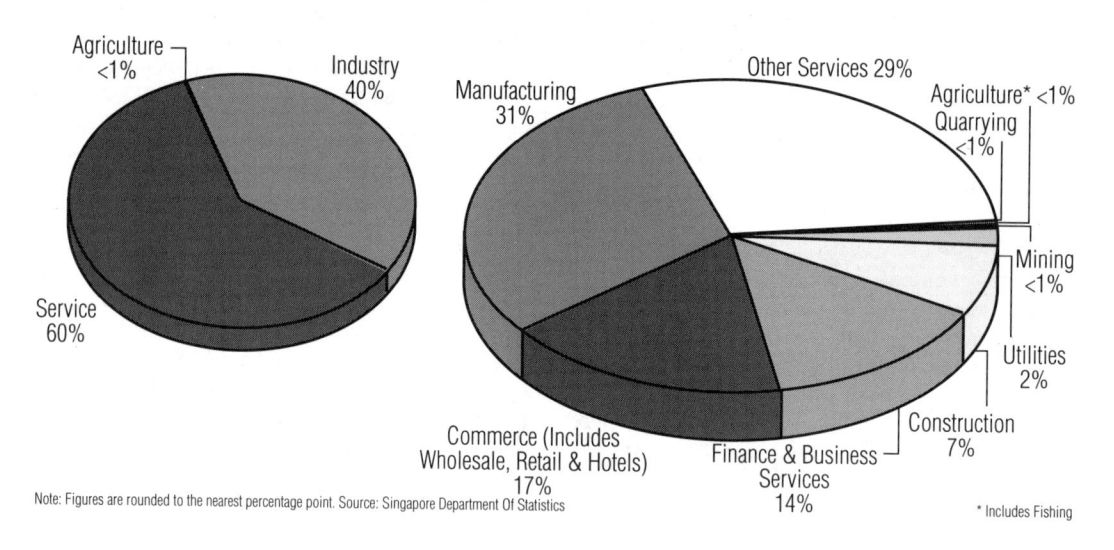

- Agriculture <1%
- Industry 40%
- Service 60%

- Manufacturing 31%
- Other Services 29%
- Agriculture* <1%
- Quarrying <1%
- Mining <1%
- Utilities 2%
- Construction 7%
- Finance & Business Services 14%
- Commerce (Includes Wholesale, Retail & Hotels) 17%

Note: Figures are rounded to the nearest percentage point. Source: Singapore Department Of Statistics

*Includes Fishing

1991, and 2 percent in 1992. Singapore essentially has a full-employment economy.

Unionization

Singapore is known for labor peace and good labor-management relationships, which are overseen by a decidedly probusiness government. Labor disputes and strikes were endemic during the early 1960s, an extension of the political and ethnic tensions that wracked the body politic. Today labor unrest is effectively nonexistent, the result in part of laws and government policies restricting strikes. A right-to-work country, Singapore has seen union membership drop steadily since the 1970s. Union participation was 21.9 percent in 1980, but it had fallen to 14 percent by 1991. Despite a 1992 campaign by the National Trade Union Congress (NTUC) to encourage workers to sign up, the trend toward less unionization is likely to continue.

Labor Costs

Singapore's high standard of living and continuing labor shortages have helped drive up wage costs. The average weekly wage in 1991 was US$206.40, plus mandated benefits of 9 percent (US$18.58), and an employer Central Provident Fund (CPF) contribution of 17.5 percent (US$36.12), for an average weekly total of US$261.10. Employers also customarily pay an annual bonus amounting to between 8 percent and 33 percent of base salary. Singapore's labor costs compare with nominal average weekly wages plus basic benefits of US$245 in Taiwan, US$295 in South Korea, US$705 in Japan, and US$7.37 in China.

Wage rates vary widely by skill level and supply in the labor market. In 1991 low-level, semi-skilled production workers made an average of US$146.20 without benefits, less than three-quarters of the national average, while managers made US$661.30, three-and-a-quarter times the national average.

The government has called for restraint in wage settlements, noting that wages increased by 7.6 percent in 1991, while productivity grew by only 1.5 percent, a trend which dates back to 1988. Although the government exercises some control over wage increases through the National Wages Council (NWC), it has done little more than suggest levels in order not to interfere with the free operation of labor markets. The labor shortage and the ever-increasing standard of living are sure to mean that wages will continue to rise.

Workweek

Singapore's official workweek is 44 hours long, but the average actual numbers of hours worked was 46.7 in 1991, less than the average 47.9 hours in South Korea, but somewhat more than Japan's 46.1 hours, and Taiwan's 45.5 hours, and much more than the average workweek of 43.6 hours in the United Kingdom, 39.9 hours in Germany, 39 hours in France, and 34.3 hours in the United States.

SECTORS OF THE ECONOMY

Singapore still relies heavily on its manufacturing sector, the largest single contributor to GDP, accounting as it did for 30.2 percent in 1991. However, total manufacturing and industry, with 39.6 percent of GDP, contributes far less than the services sector which is at 60.1 percent and still growing. Agriculture and fisheries contributed an insignificant 0.3 percent of GDP in 1991, and its share is falling rapidly. Some 34.8 percent of the work force is employed in manufacturing and industry; 64.4 percent is employed in the service sector; and a dwindling 0.8 percent is employed in agriculture and related enterprises.

The domestic economy consists of local firms that the government defines as small and medium enterprises (SMEs). These make up 88 percent of total business establishments and employ 40 percent of the work force. However, SMEs account for only 23 percent of Singapore's added value, an indication of its dependence on international operations and foreign investment.

The average annual fixed investment in Singapore between 1988 and 1992 was around 35 percent of GDP. This figure has grown steadily and continues to do so, albeit at a lower rate than before. By comparison, annual fixed investment in Japan was 33 percent of GDP for the same period, and in the United States it was only 15 percent. National savings have averaged 45 percent of GNP. The high savings rates, due primarily to mandatory employee contributions to the CPF, have financed government spending on social programs and infrastructure improvements and enabled Singapore to minimize its borrowing. External debt has fallen from an already minimal US$71 million in 1989 to US$19.8 million in 1992.

AGRICULTURE

Singapore is an island and its inhabitants historically have been involved in fishing and subsistence pursuits, but it lacks the agricultural and fishing tradition of other Asian nations. There is no large agricultural lobby influencing economic and political decisions. Agriculture and fisheries provide only 0.3 percent of GDP but employ a disproportionate 0.8 percent of the work force—largely holdovers from the more rural past. The situation is changing rapidly, as the government eliminates unproductive activities, such as pig raising, which it has discouraged since 1990. Employment in the agricultural sector shrank by 22.4 percent in 1991 alone.

Attempts in the 19th century at plantation agriculture generally failed; coconut and rubber trees do grow in Singapore, but they never have been economically significant and they do not occupy land with a higher potential use value. Singapore is self-sufficient in the production of poultry and eggs, and it can fill about one-third of the domestic demand for vegetables and 25 percent of demand for fish, but all other foodstuffs must be imported.

Future Directions

Singapore is directing agricultural efforts into high-value-added areas that have export potential instead of chasing the unachievable goal of self-sufficiency. The government has set up agrotechnology parks to encourage the development of such products. Prawn and fish farming, the breeding of ornamental fish and aquarium plants, the growing of orchids, and the production of fruit tree nursery stock are some examples of Singapore's recent agribusiness efforts.

MANUFACTURING AND INDUSTRY

Manufacturing and industry, including mining, utilities, and construction, provided the initial impetus for Singapore's independent economic growth, and they continue to add heft to its economic might. But the situation is changing rapidly as high labor costs force producers to shift lower-end manufacturing and assembly functions to lower-cost venues.

Manufacturing experienced a slowdown in 1991, paced by a decline in the demand for electronics, which posted a growth rate of only 0.4 percent due to the worldwide recession. The rate of growth in the index of industrial production fell from 9.9 percent in 1990 to 5.3 in 1991, and the index turned negative on a quarterly basis in 1992, although on a full-year basis it showed positive but anemic overall growth. However, growth rebounded sharply in 1993.

Manufacturing Base

Singapore has a diversified manufacturing base, although its small size has historically led it to avoid investment in heavy industry and focus instead on medium to light industry. Exceptions include shipbuilding, oil rig fabrication, and oil refining, all areas in which Singapore leads in Southeast Asia. Singapore has strong positions in medium-range manufactures, such as industrial chemicals, machinery, appliances and other electrical machinery, specialized transport equipment, and textiles and apparel.

Singapore is strong in the manufacture of electronic components, including electronic valves, semiconductors, and printed circuit boards, and in fabrication of metal components of electronic parts. Output of computer equipment, primarily disk drives, and telecommunications equipment declined somewhat due to the global recession in the early 1990's. However, in 1993 output in its electronics sector grew by a minimum annualized rate of 25 percent during each quarter. Pharmaceuticals and related chemical products represented the fastest-growing market in 1991; this sector grew by 30 percent in that year on top of a 31 percent jump in 1990. Singapore also has a significant presence in printing, food processing, the manufacture of furniture and jewelry, and the manufacture of aerospace and precision scientific equipment.

Future Directions

Singapore is targeting high-value-added, engineering-intensive areas for future production investment. Although the actual contribution of such areas is still small, the country is banking on the development of its products to fuel the next level of high-tech manufacturing growth. Focus areas include information technology, biotechnology, microelectronics, automation, process controls and artificial intelligence equipment, lasers and optical applications equipment, and communications technology. (Refer to "Opportunities" and "Industry Reviews" chapters.)

Construction: The Fallback Position

Construction, which in 1991 accounted for 7.1 percent of GDP, exploded in the early 1990s. Sector output grew by 21 percent in 1991, accounting for a full percentage point in growth of GDP. Half of the construction activity in 1991 involved public sector projects, and this share increased to almost 60 percent in 1992, indicating government pump priming in an attempt to jump-start the economy.

SERVICES

The service sector, which includes financial and business support services, commerce and trade, transportation, health care, and government services, produced 60.1 percent of the GDP and employed 64.4 percent of the work force in 1991. This sector appears to be headed for a two-thirds share of the economy by decade's end.

Services Structure

Commerce, including wholesale and retail trade and hotels and restuarants, accounted for 17.3 percent of GDP in 1991, the largest contributor in the services sector. Financial and business services, the next largest and most dynamic area of the service sector, accounted for 14 percent of total GDP in 1991. This subsector led growth in services, with double-digit advances in output annually during the late 1980s and early 1990s, although the worldwide slowdown brought the rate of growth in such services down to 5.4 percent in 1991. Growth in the finance component slipped to 2.9 percent in 1991, while growth in business services grew at a 7.6 percent annual rate. Transport and communications follow with 13.5 percent. Commerce grew at a 6.4 percent annual rate, driven more by international trade activity than by the relatively minor domestic consumer market. Transport and communications grew by 8 percent in 1991, down only slightly from the previous year.

Finance

Singapore is already the financial hub of Southeast Asia, and roughly 25 percent of government revenues come from levies on financial transactions. Banking and financial services grew at an average annual rate of more than 10 percent between 1960 and 1991, a rate greater than overall growth in the GDP. Singapore is the center for trading in the Asian dollar market—the regional equivalent of the Eurodollar market—the assets of which grew at a 20 percent annual rate between 1985 and 1990. Singapore is also the major banking center in the region, and it has designs on Hong Kong's role as a regional and world financial market after 1997. However, the level of government control that prevails in Singapore will probably scare away some operators used to Hong Kong's no-holds-barred style. Nevertheless, Singapore is already the premier foreign exchange market in Asia.

Singapore's stock exchange, which in early 1993 had a market value of around US$61 billion, is stable and well established. However, despite a 20 percent rise in value in the first half of 1993, it is still less dynamic than some other regional exchanges. For example, Hong Kong's exchange has a market value of around US$133 billion, while Malaysia's less developed market is valued at US$63 billion, and both are growing at a much faster rate. Foreigners are allowed to hold Singaporean stocks, but restrictions and the relatively small size of the market have limited the ability of foreigners to operate freely on the exchange.

The government is planning to offer a 20 to 25 percent stake in Singapore Telecom, the communications monopoly, with a view to increasing the number of shares in the hands of Singaporeans and boosting the market value of the exchange by an estimated 20 percent. The government is also considering offering shares in state-owned utility, airport, port, and mass transit companies. In 1992 Singapore fund managers ran a record US$22.9 billion in investment funds, a 300 percent increase from 1989. Some 50 percent of this amount was in equity accounts, and 80 percent represented foreign money.

Transportation and Communications

Singapore's strategic location and its natural harbor has helped it to reach the level of development that it enjoys today, and the government has invested heavily in improving on these advantages. Singapore is one of the busiest ports in the world in terms of total tonnage handled as well as the world's second busiest container port, and it services more than 700 shipping lines. Another important element in Singapore's competitive edge is its advanced telecommunications system, which it is continuously upgrading. This sector has grown at an average annual rate of just under 10 percent since 1960.

The Internal Service Economy

Singapore's small size means there are few wholesalers and distributors and relatively few retail outlets. Larger firms are vertically integrated and handle both wholesale and retail operations. While there are many traditional mom-and-pop stores, there are also large, Western-style outlets using Western merchandising techniques. To date, the government has kept these functions largely in local hands.

Future Directions

The Singaporean government is actively trying to attract new service industries to strengthen its position as a regional and international service provider. Targeted areas include virtually all aspects of finance and banking but particularly such advanced fee-based services as funds management, risk management, capital markets operations, securities underwriting, futures markets operations, trade financing, and reinsurance.

The government is also interested in increasing Singapore's expertise in air and sea transport and distribution; telecommunications and information services; health care; leisure and recreational services; education and training; agrotechnology services; engineering consulting; management consulting; legal and accounting services; laboratory, research, and testing services; and advertising and public relations.

However, the level to which the government micromanages specifics can be seen in a recent edict limiting the number of physicians through sharp reductions in the number of foreign physicians to be licensed, a freeze in admissions to the national medical school, and radical reductions in the number of foreign medical schools whose graduates would be eligible to practice in Singapore, all in the name of holding down health care costs by limiting excess supply.

TRADE

Trade is Singapore's reason for being, and more than 100 international trading firms operate in Singapore. In 1992 Singapore's total trade was about two-and-three quarters times its total GDP—US$130.4 billion, an all-time high and a 4.3 percent increase over the US$125 billion registered in 1991. Recession shrunk trade by 3.2 percent in 1986, but the growth rate rebounded to a postrecessionary high of 30 percent in 1988. It slowed to 10 percent in 1989 and 11.4 percent in 1990, and the worldwide economic slowdown brought it down to 5.4 percent in 1991. Midyear figures for 1993 show total trade growing at a brisk annualized rate of 15.6 percent.

Growth of Trade

Between 1960 and 1991 total trade grew at an average annual rate of 11.5 percent. Reexports, which accounted for 94 percent of all trade in 1960, shrank to 35 percent in 1991 as other countries established direct trading relationships and reduced the traffic passing through third-party Singapore. Most of the growth in trade has resulted from export-oriented manufacturing activity, although entrepôt commodity trading continues to be strong. Singapore is the third largest oil refining center worldwide and the largest in Southeast Asia, giving it an important regional role in petroleum processing and distribution.

Exports have more than doubled since 1985, growing at an average annual rate of 10.6 percent, while imports have not quite doubled, growing as they have at an average annual rate of 10.2 percent during the same period. Exports reached a high of US$61.1 billion in 1992, up 3.6 percent from 1991. Imports were US$69.3 billion in 1992, up 5 percent over 1991. Midyear figures indicate that exports have grown at an annualized rate of 13.4 percent in 1993, while imports have surged to 17.4 percent.

Balance of Trade and International Reserves

Singapore has maintained a negative balance of trade averaging a steady 6.2 percent of total trade since 1985. Imports have represented an average 53.1

Singapore's Foreign Trade

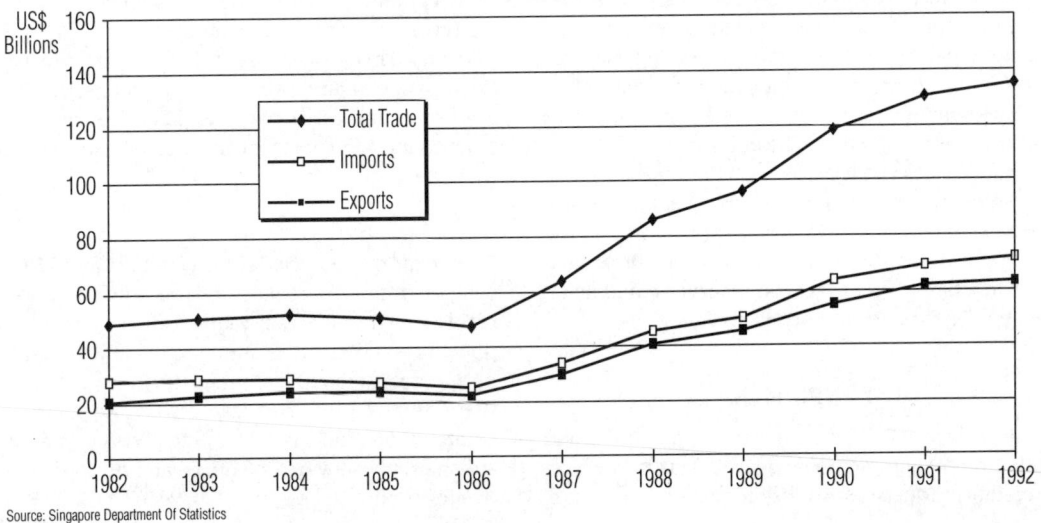

Source: Singapore Department Of Statistics

Singapore's Leading Exports By Commodity - 1992

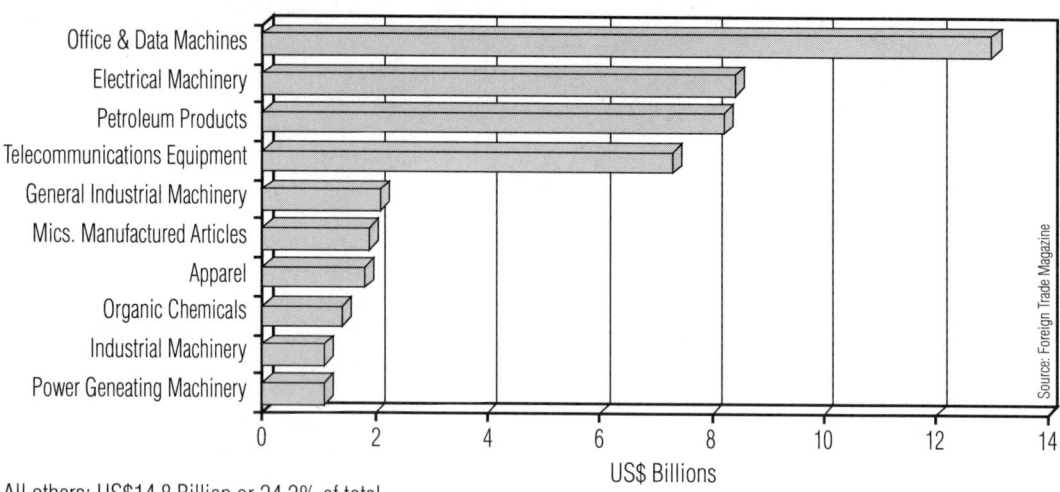

Commodity	US$ Billions (approx)
Office & Data Machines	~13
Electrical Machinery	~8.5
Petroleum Products	~8
Telecommunications Equipment	~7
General Industrial Machinery	~2
Mics. Manufactured Articles	~2
Apparel	~1.8
Organic Chemicals	~1.5
Industrial Machinery	~1.3
Power Geneating Machinery	~1

Source: Foreign Trade Magazine

All others: US$14.8 Billion or 24.2% of total
Total 1992 Exports: US$61.1 Billion

percent share of trade and exports a 46.9 percent share since 1985; the variation has been less than 1 percentage point. Singapore's trade deficit grew from US$2.4 billion in 1989 to US$6 billion in 1992, and that figure could double to US$12 billion in 1993 if trends in the first half hold for the rest of the year.

Despite the mounting trade deficit, which appears to be structural, Singapore's international reserves remain high: US$41.4 billion in March 1993, equal to 93 percent of the country's 1992 GDP and up almost 20 percent from the level of US$34.6 billion a year earlier. This figure is double China's international reserves and half the level of Japanese reserves, which continue to be an international marvel. Singapore's international reserves have more than doubled since 1985, growing at an average annual rate of 10.9 percent.

Singapore's trade imbalance can be accounted for by its need to import inputs both for local consumption and for its export manufactures. Its high reserves can be accounted for by its negligible international borrowing position, its astronomical internal savings rate, and its increasingly important financial services sector. The financial sector brings in commissions and fees based on the large and growing volume of transactions occurring in Singapore's financial markets. The government has primed the pump by funding infrastructure projects. Prolonged and substantial draws against reserves could materially alter the current financial balance.

EXPORTS

In 1992 Singapore's main exports were office and data processing machines, electrical machinery, petroleum products, and telecommunications equipment. Together, these four categories represent about 60 percent of Singapore's total exports. The remaining categories each make up less than 2 percent of the total. Processed petroleum products account for 13.4 percent of total exports by value. However, at least 46.9 percent of Singapore's exports are manufactures tied to the electronics industry either as manufactured or assembled end user products or as components, such as valves, integrated circuits, semiconductors, and replacement and original equipment parts.

In 1991 nonoil exports rose by 7 percent, down from the 8.5 percent increase recorded in 1990 and resulting largely from the worldwide economic slump. The slowed growth was due mainly to reduced demand for disk drives, computer peripherals, and consumer electronics, such as television sets and cassette recorders. Growth came primarily in microcomputers (up 118 percent), computer printers (up 28 percent), and computer monitors (up 32 percent). This growth had a downstream effect by increasing demand for integrated circuits, cathode ray tubes, and other parts. Regional demand kept petrochemical exports strong, and oil exports rose by 12 percent.

Reexports

Reexports increased sharply in 1991 by 10.5 percent, up from 1.8 percent in the preceding year. Most of this growth came from demand for electronic components, such as integrated circuits and computer peripherals and parts, although the reexport of synthetic fiber fabrics was also strong. Reexports of commodities, such as rubber and plywood, were off substantially as were consumer and business electronics products.

Singapore's Leading Imports By Commodity - 1992

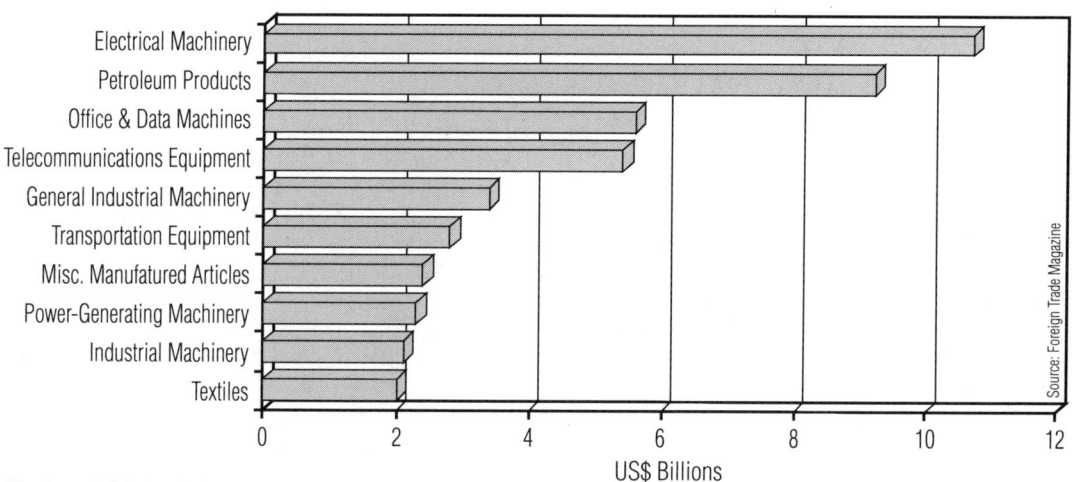

Commodity	US$ Billions (approx.)
Electrical Machinery	~10.8
Petroleum Products	~9.2
Office & Data Machines	~5.5
Telecommunications Equipment	~5.3
General Industrial Machinery	~3.5
Transportation Equipment	~3.0
Misc. Manufatured Articles	~2.5
Power-Generating Machinery	~2.3
Industrial Machinery	~2.2
Textiles	~2.0

All others: US$23.4 Billion or 33.8% of total
Total 1992 Imports: US$69.3 Billion

Source: Foreign Trade Magazine

Rising local labor and other overhead costs and increased competition from offshore low-cost sites is eroding Singapore's competitive advantage. The government, which is focusing on high-technology, high-value-added areas, is trying to turn away from lower-end commodity manufactures and increase Singapore's role as a regional and global service center.

IMPORTS

Singapore's main imports in 1992 were electrical machinery, petroleum products, office and data processing machines, and telecommunications equipment. Together, these four categories account for 44.6 percent of its imports. The remaining categories account for less than 5 percent of all imports. Petroleum products account for a 13.3 percent share. Electrical products represent 31.3 percent of total imports.

The drop in oil prices in 1991 following the price runup during the Gulf War in 1990 kept the value of Singapore's imports from surging. Nonoil imports rose by 6.2 percent in 1991. The economic slowdown reduced imports of consumer items. Imports of electronic components, largely for reexport, increased, as did purchases for Singapore's air and tanker fleets. Midyear 1993 figures suggest that, while Singapore continues to benefit from oil prices that remain relatively low, it suffers from the strong yen, which makes the goods of Japan, its primary import supplier, especially costly. If the worldwide recovery continues and strengthens, Singapore will undoubtedly be able to find alternative, lower cost suppliers. However, in the near term, it must pay the going rate for inputs to keep its all-important export industry going.

TRADING PARTNERS

Export Partners

The share of Singapore's trade with its various export partners has remained relatively stable over the past few years. The United States, neighboring Malaysia, the European Community (EC), Hong Kong, and Japan were Singapore's largest export markets in 1992, accounting for more than 60 percent of total exports. Singapore's regional markets, which include Thailand, Taiwan, and Australia, take another 20 percent of its exports. The remaining 20 percent are destined for countries that each buy less than 2.5 percent of Singapore's total exports.

The United States is Singapore's largest buyer, although in 1991 US purchases fell slightly to 19.7 percent of the total owing to the worldwide economic slowdown. Exports to the EC grew by 3.3 percent; those to Japan grew by 6.4 percent, and to Malaysia by a whopping 22 percent, up from modest 4.5 percent growth recorded in 1990. This surge was due to increased purchases of electronic components, construction equipment, and petrochemical products.

Import Partners

In 1992 Singapore bought 55 percent of all imports from Japan, the United States, and Malaysia, Japan alone supplied about 23 percent of total imports. Oil purchased from Saudi Arabia accounted for an additional 5 percent of Singapore's imports, and goods from countries in Asia, such as China, Taiwan, Thailand, Hong Kong, and South Korea, accounted for an additional 17 percent. The remaining 23 percent of Singapore's imports came from countries each of which accounted for less than 3 percent of the total. Imports from Japan rose 10 percent, while those from Malaysia, the United States, and the EC have eased slightly.

FOREIGN PARTICIPATION IN THE ECONOMY

Singapore, which has based its development on a foundation of international trade and foreign investment, has worked to make itself as user-friendly as possible to overseas businesses. It has always welcomed foreign investment, and it currently is especially interested in attracting capital-intensive, high-technology, high-value-added manufacturing and highly skilled, sophisticated service industries. In the early 1990s more than 3,000 foreign firms were doing business in Singapore.

Restrictions on Foreign Investment

Singapore is a free port. There are no import or export duties on raw materials, equipment, or most products. In fact, there are no tariffs on 90 percent of all items. Currency controls were abolished in 1978, and there are few restrictions on foreign ownership, although foreigners are barred from certain strategic industries, such as telecommunications, utilities, and defense-related manufactures. The government limits foreign ownership of national banks, the print media, and some other firms deemed to be strategic, and there are restrictions on foreign ownership of real estate. The government also supports local majority ownership of firms involved in shipping; tourist-related activities, such as travel agencies and hotels; and retailing. All operations must be registered with the government.

Incentives for Foreign Investors

Generous incentives sometimes mean that foreign businesses get an even better break than local firms. Incentives include tax concessions, credits, and holidays; research and development incentives; expansion incentives; investment allowances; investment offsets to cancel out certain losses; and special loan and grant programs to fund expansion and the purchase of capital goods. There are also incentives for foreign firms that provide the expertise to enable small- to medium-sized local firms to produce locally goods that formerly had to be imported. The government does not discourage any allowed investments, although it denies incentives to less-favored projects.

Singapore recently revised its tax codes with an eye toward attracting foreign investment. The legislature approved a 3 percent value-added tax (VAT), to go into effect on April 1, 1994. This new tax will broaden the tax base and offset drops in corporate and personal income taxes, which are relatively high compared to those of such regional competitors as Hong Kong and Taiwan, and are thought to have held back some foreign investment.

Size of Foreign Participation

Foreign direct investment in Singapore was US$3.6 billion in 1991, down 6.5 percent from the peak of US$3.85 billion reached in 1990. Cumulative foreign investment in the manufacturing sector expanded 20 times between 1965 and 1990, with 45 percent of the increase coming between 1985 and 1990. In 1990 89 percent of all investment in the manufacturing sector came from foreign sources, and 70 percent of that total represented expansion of existing facilities.

The United States is Singapore's largest foreign investor, with cumulative investment of US$9 billion as of 1991. In 1990 United States sources accounted for 48 percent of all foreign investment and for 33 percent of all cumulative investment. Japan ranks second. In recent years, it has overtaken the United States in amounts invested, although its annual commitment is not substantially more than that of the United States and it will not surpass US cumulative investment anytime soon. The countries of the EC rank third. The Scandinavian countries and Australia have also made substantial investments.

GOVERNMENT ECONOMIC DEVELOPMENT STRATEGY

Singapore has a free market capitalist economy, but its government plays a major role in setting policy and directing the activities of participants. Although the government has little direct, day-to-day involvement in the ownership or operations of economic entities, it maintains strict oversight. While the government denies that it actually manages the economy, it does acknowledge that it provides the conditions for its operation, and it emphasizes that it would react rapidly and firmly to ensure that things not go astray.

Basic Policies

Government policies call for a basic reliance on a free market economic system; the use of government investment in infrastructure, a high-quality work force, and a stable environment to make Singapore an easy place to do business; the attraction of outside investment; the competitive servicing of external markets; the institution of a meritocracy to reward inputs; and a concern for the physical and social environment.

The government is moving away from the low-cost producer strategy that it embraced during the 1970s and 1980s and toward a strategy of upgrading skills to enable it to compete in high-end, value-added products. This strategy involves upgrading the education and training of personnel; moving labor-intensive operations offshore and promoting capital-intensive operations locally; developing a local research and development capability; and positioning itself as the premier high-tech manufacturing and high-touch service center in the region. In short, Singapore wants to be known for its expertise.

Strategic Plans

In 1984 the government announced Vision 1999, a plan that called for Singapore to become a developed country within 15 years. Success in achieving that goal was to be measured by matching Switzerland's then current per capita GNP. In 1991 the government set the Strategic Economic Plan, which called for Singapore to match the level of development of the Netherlands by 2020 and that of the United States by 2030; the measures used to assess progress are to be more flexible than just the per capita GNP measures established in the earlier plan.

According to the official statement of Singapore's Economic Development Board (EDB), these goals are to be achieved by "enhancing human resources, promoting national teamwork to achieve common economic goals, adopting an international orientation, creating a climate which supports innovation, developing core capabilities of industry and service clusters, promoting economic redevelopment, tracking Singapore's international competitiveness, and reducing the country's vulnerability by encouraging multinationals to treat Singapore as a home base and by developing local enterprises to world-class standards."

Branching Out

Singapore's progress toward its goal of becoming the Switzerland of the Far East has been hampered by its lack of diversification through outward foreign investment. In 1991 it invested only US$701 million in overseas projects. Although Singapore is credited with involvement in more than 700 projects in China valued at US$1 billion, in 1992 Singapore's cumulative total investment in China between 1979 and 1991 was only US$897 million. In contrast, Taiwan and Hong Kong together have invested an estimated US$30 billion in China over the past 10 years.

Singapore has a venture with Malaysia and Indonesia to move some of its low-end production offshore to adjacent areas of Johore and Batam. In this Growth Triangle, Singapore will provide the expertise, while the less-developed partners provide the labor and real estate for such labor-intensive operations as electronic assembly and the manufacture of apparel and plastics. Singapore's major project in China is the start-up of an industrial park in Suzhou, near Shanghai, to be run by Singapore. The city-state is contributing expertise and initial funds of US$75 million; the hope is that Singapore's managerial reputation will draw an additional US$20 billion in outside investment. Similar deals are under negotiation with Vietnam and Myanmar.

Singapore has also set up a venture capital subsidiary of the government-controlled firm Singapore Technologies to buy into small high-tech overseas firms with a view toward bringing the technologies that they generate to Singapore for development and production. This venture has invested in firms in the United States, and it is looking in Europe, Hong Kong, Australia, and New Zealand for additional opportunities. Such ventures are critical to Singapore's strategy of shifting from a production-oriented to an innovation-oriented economy.

The absence in Singapore of large, powerful indigenous companies can be expected to hold it back in its own overseas ventures. The overseas projects announced to date have been spearheaded by state-controlled firms that are more control-oriented than entrepreneurial, and even they may not control the resources necessary to duplicate the home country's pattern of success, which has been largely accomplished by the state rather than by public or private enterprises.

Singapore's success also has been closely linked with its ability to exert almost total control over the environment, including labor, and its ability to provide supporting infrastructure, communications, and industrial development. These elements require control and financing that are accessible only on a governmental level. It is difficult to imagine a sovereign nation that would allow an outside entity to exercise that degree of power in the host country, or that any outside entity could bring to bear the level of funding necessary to bring it off even if it had full authority and cooperation. All in all, Singapore is unique and will have difficulty franchising itself.

POLITICAL OUTLOOK FOR THE ECONOMY

As Singapore's economy crawls out from under the effects of a three-year worldwide economic downturn, it is faced with several potential problems. Singapore's authoritarian government is still very much in command, although it must decide how best to stimulate and transform the city-state's economy while maintaining the level of commitment among its people that has sustained its economic miracle to date.

Singapore appears to be priming the pump through investments in public works that channel funds into the economy via the construction industry, which lives and dies by government expenditure. Singapore's government has announced additional projects to upgrade its general infrastructure, particularly its telecommunications and transportation fa-

:ilities, which are already among the best in the world. Another area of perennial activity is urban renewal and housing construction. Critics argue that Singapore has denatured itself by tearing down everything old and then tearing down everything new in order to make it newer. Some 87 percent of Singaporeans already live in government constructed apartment housing, and 80 percent own or are buying their apartments by drawing on their CPF accounts.

The question is whether such short-term local allocations will detract from other investments that Singapore needs to make to improve its new high-tech competitive position. Such investments do not have an immediate stimulative effect on the economy, but they are more important to Singapore's efforts to secure long-term development along the lines that it has mapped out for itself. However, Singapore's dependence on the global economy, which is remains weak, may give the government little choice but to prop up the domestic economy over the near term.

Changing the Guard?

Prime Minister Lee Kuan Yew, the architect of Singapore's rise and the only leader the country had for 25 years, officially stepped down in 1990. Lee remains in the cabinet, and he is still considered the power behind his handpicked successor, Goh Chok Tong. Lee has stated that he will continue to be involved in policy decisions and that, if the government fouls up, he will even come back from the grave to set things straight.

Singapore's president, a largely symbolic position, was elected by the PAP-controlled parliament until 1991, when it was decreed that the presidency would be filled through popular election. In August 1993 Ong Teng Cheong, a cabinet minister, stood for president. The government set standards that limited the number of potential eligible candidates to fewer than 500 Singaporeans and proceeded to disqualify opposition candidates on various grounds. In the end, it ran one of its own as an opposition candidate. This official opponent went on record as saying that Ong was by far the better candidate and that he himself was running simply because he had been asked to provide another name on the ballot.

The PAP's voter share eroded during the 1980s, and it even lost a couple of parliamentary seats, but its hold on power has never been seriously questioned. The party—and Singapore—fear that a new generation will stray from the party line, which is based on hard work and obedience. There is concern that Singapore's economy will be unable to change from low-cost production to innovation unless the reins are loosened considerably and that it then will no longer be Singapore. Although government technocrats are honest, talented, and dedicated, the younger generation does not seem to have the force of personality and single-mindedness that characterized the older generation. This could be a blessing if it enables them to open up the country's rigid ethos and facilitate change. But if they hew to the old standards and cannot adjust to new realities, it could also prove to be a curse.

SINGAPORE'S INTERNATIONAL ROLE

Singapore maintains diplomatic relations with more than 60 foreign nations, and it is represented at the United Nations, and in most of its major agencies, and before the European Community (EC). Singapore is taking a lead role in improving regional integration as a way of making Southeast Asia more stable and attractive to outside investors. A strengthened regional economy also would help to reduce Singapore's dependence on exports to the developed world. Singapore has been particularly active in the Association of Southeast Asian Nations (ASEAN), which has agreed to set up offices in Singapore and to work toward an ASEAN Free Trade Area, which is to be in effect by 2008.

Current Issues

LEE'S LEGACY: AUTHORITARIAN RULE AND ECONOMIC PROSPERITY IN SINGAPORE

The Legacy Begins

Former Prime Minister Lee Kuan Yew is largely credited with spearheading Singapore's transformation from a backward, colonial jungle swamp to a prosperous urban industrialized city-state following its separation from the Federation of Malaysia in 1965. As head of the People's Action Party (PAP), the only real party to operate in Singapore since Singapore gained its independence from Britain in 1959, he has accomplished this feat by the use of what has been termed "soft authoritarianism." And unlike the reigns of many notorious authoritarian regimes in other countries, Lee's rule has enjoyed strong support from Singaporeans. The government and its people formed an unwritten social contract: soft authoritarianism in exchange for economic prosperity.

Give and Take (The Unwritten Contract)

Life in this rather strict and centralized system has required that Singaporeans give up certain personal freedoms that people in some other, mostly Western, societies take for granted. The authorities have maintained a monopoly on power and have kept opposition and dissent to a minimum. Such elements have been described by the leadership as either too unintelligent to comprehend the problems at hand, too selfish to sacrifice for the common good, or maliciously if not traitorously intent on destroying the Nation.

From the early days of Singapore's independence the state has owned, controlled, or strictly monitored the allocation of capital, labor, and land. The government still owns radio and television stations and exercises tight controls over both domestic and foreign periodicals, including such respected international organs as the *Far Eastern Economic Review* and the *Asian Wall Street Journal*. Both of these have run afoul of the Singaporean government by printing items that were not congruent with the image that the government sought to project. Singapore has gone so far as to seek to exert control over the editorial content of materials sent to it for production by its up-and-coming printing industry even when such items were not intended for distribution in the country.

Basic Western civil liberties, such as the right to a speedy trial and trial by jury, are not a prominent part of the Singaporean system, and the city-state utilizes both corporal and capital punishment. Singaporeans must contribute substantial proportions of their earnings to the Central Provident Fund, a mandatory national social security savings plan. With laws that restrict such mundane things as the hours when residents can drive their cars, jaywalking, gum chewing, spitting, and flushing public toilets, no aspect of daily life in Singapore seems unregulated. The government decreed practices designed to enforce traditional family values, and then, disturbed by the falling birth rate, turned matchmaker, sponsoring mixers that young unmarried Singaporeans were strongly "encouraged" to attend. Lee has described the government's paternalistic political and social philosophy as "We decide what is right. Never mind what the people think."

How do Singaporeans respond to a comment like that? They take it in stride; it sounds appropriate to the vast majority of them. What began as a tiny island with an ethnically and culturally diverse population of nearly two million (half of whom had no formal schooling), no natural resources or adequate water supply, and no defense capability in the face of larger, more powerful, and not overly friendly neighboring countries, has evolved into what is today one of the most technologically advanced, up-to-date, affluent, stable, and secure states in the world.

What other returns do the people of Singapore receive from their participation in the unwritten contract? Rapid economic growth, low inflation, full employment, political stability, an 80 percent rate of home ownership, an average life-expectancy of 76 years, and a mere 0.3 percent poverty rate. Serious crime is also virtually nonexistent. Singapore held foreign exchange reserves of US$41.4 billion (the highest in the world on a per capita basis) in mid-1993. The only industrialized economy in Southeast Asia, Singapore is ranked as the most competitive nation in the developing world according to several authorities and surveys, and is positioning itself to lead the world's information technology sector by the year 2000. In reality, Singapore owes this status to the government's strict and interventionist policies in directing investment and industry.

Is "Soft Authoritarianism" Simply an Asian Phenomenon?

In an interview in 1992 Lee Kuan Yew argued that not only is the Asian community not ready now for Western-style rugged individualism and democracy, it may never be ready. A number of other Asian leaders, such as those of Vietnam, Myanmar, China, and Kazakhstan, seem to agree and are attempting to emulate Lee's political, social, and economic model. Lee differentiates between Western-style democracies (resulting from a more individualistic "Anglo-Saxon temperament") and the more hierarchical, collective societies of East Asia. He contrasts the East Asian consensual approach to decision making using the Western emphasis on anarchic open opposition and debate. Order, discipline, and submission to the collective will, he claims, are key components of traditional Asian culture. Lee has defined his "soft authoritarianism" to mean that there is less opportunity to obstruct what he calls "the majority policy." But the majority policy to date has been the government's policy, which some observers worry may not be adequately innovative or responsive to changing conditions.

Nevertheless, the Asian "soft authoritarian" model is causing economists and other proponents of the "invisible hand" model of economic development to scratch their heads. Strong state management of industry and investment has so far served Singapore well in its rapid advance.

Winds of Change

However, today's increasingly prosperous, well-educated, and well-traveled population may be reaching its limits when it comes to sacrificing personal and intellectual freedom at the behest of a central government, no matter how benign. The younger generation in particular does not remember Singapore's humble origins and early privations and is beginning to question the need for such strict governance. The highly ordered quality of life in Singapore is beginning to be a political issue as Singaporeans begin to feel that they may be hemmed in by too many rules.

In the 1980s, when the PAP witnessed some erosion in its popular support, the government realized that policies and practices devised to deal with a much simpler economy and less-educated and prosperous citizenry in the 1960s were becoming increasingly ineffective. In the 1990s tensions between generations and between advocates of change and those of continuity are mounting.

The recent 1993 presidential election has been interpreted by some observers as an indication that public sentiment is finally heading in a new direction. In 1993 Ong Teng Cheong was elected president, which in Singapore is largely a ceremonial office, in the country's first direct election to fill the position. His victory was not a big surprise, given that Ong, a former party chairman and deputy prime minister, was nominated by the party (as was his opponent, other potential opposition candidates having been disqualified from running). However, his 59 percent win in what was essentially a one-horse race showed somewhat weak support and a high level of passive rebellion on the part of an electorate that had previously acclaimed party candidates with 90 percent plus majorities. Some analysts argue that this "skimpy" margin highlights Singaporeans' growing dissatisfaction with the ruling party and their open skepticism toward Ong's campaign, suggesting that he and other party leaders may begin to look for ways to open up the system to some form of greater popular participation.

As has been the case in several Asian nations, Singapore has become to some extent the victim of its own modernization drive. On the one hand, Singapore seeks to maintain a strong market economy, while on the other, it fears the accompanying Westernization and openness that usually accompanies this model. No longer can the leadership assume that Singaporeans will continue to sacrifice individual interests in exchange for increased economic well-being and stability. The unwritten contract seems to be showing signs of age and seems to be in need of at least some renegotiation.

Goh Goes Beyond Seat-Warming

When Lee Kuan Yew stepped down as prime minister and leader of Singapore in 1990, he appointed Goh Chok Tong to fill the position. Some observers felt that Lee expected Goh to keep his seat warm until he was effectively able to return to proxy power through his son, Hsien Loong. But Goh, rather than hewing religiously to Lee's party line, has been showing more independence. Not in major ways, but enough to cause notice in such a tightly controlled environment. Whereas Lee's approach to increasing Westernization was to stress national ideology and more Confucian indoctrination in the schools, Goh has pragmatically recognized the need to allow for certain freedoms when seeking to enter the "economic big leagues," which involves dealing with Western economies operating with different assumptions.

Goh has already loosened some rules and regulations which had been seen as impeding development to some extent. His style has proved to be more informal, moderate, and open to acknowledging and accommodating the fact as well as the substance of opposition than the more rigid traditional approach. The people of Singapore responded with overwhelming support as shown in Goh's by-election in December 1992. This victory served to stiffen Goh's resolve to continue his less rigid style of government and pull in more allies to support him in this shift. For example, Goh appointed a new cabinet, dropping at least one of Lee's protégés, despite direct pressure from Lee to keep his man in place. And with Lee's presumptive heir Lee Hsien Loong having been recently diagnosed as having cancer, the future of Singapore—so meticulously controlled and structured—is looking as if it might be somewhat more open to change.

INFORMATION TECHNOLOGY: IT 2000

In the 1960s Singapore was a maritime trading port and oil-processing center. In the 1970s and 1980s it developed from a low-cost to a high-tech manufacturing center. Now, in the 1990s Singapore is gearing up to become the "Intelligent Island." Singaporean planners believe that information will be the primary commodity of the 21st century, and their goal is to become one of the predominant future information technology centers in the world to manage and control this coming trade.

In the Beginning

In 1986 the National Computer Board formulated a comprehensive plan to link the Singaporean private and public sectors to fully exploit Information Technology (IT). The comprehensive plan covered all aspects of the creation not only of an industry but of an entire all-encompassing IT system. In 1991, only five years later, Singapore's telecommunications infrastructure was ranked number one in the world by the *World Competitiveness Report*. In the same year Singapore was listed as one of the top ten countries in the world in terms of number of computers per capita by *Computer Industry Almanac*.

The effects of Singapore's IT program have already been astonishing. By computerizing the civil service, the government has watched productivity soar within state bureaucracies. A 1988 audit showed that computerization had reduced or avoided the need for nearly 5,000 government jobs. Moreover, every dollar spent on IT in the civil service reaped a return of S$2.71.

The IT 2000 Report

Such results led the government to announce its IT 2000 Plan in March 1992. The plan calls for the transformation of Singapore into the information-technology center of Asia and a major node in a global network of such centers operating above and outside the traditional antiquated communications systems that have limited and defined international economic and political relations in the past. The plan involves the integration of computer and communication technologies such as the television and the telephone to create a multimedia audio, video, and textual communications and information transfer system. The plan calls for a completely networked society in which all schools, homes, businesses, and government agencies would be linked to an electronic grid. To date Singapore has spent billions of dollars on developing these networking capabilities.

In the IT 2000 report, information is described as the currency of the new age. The report likens information in today's world to precious commodities such as gold and silver. The government's vision is to enable Singaporeans to tap into a vast reservoir of electronically stored information and services. IT will allow the private and public sectors to access, harness, and share information among themselves and the rest of the world, overcoming geographical, political, and temporal boundaries. Through the IT 2000 Plan, Singapore sees itself poised to become "a vast information gateway" and a global hub for headquarters of multinational corporations.

How Does It Work?

Take TradeNet, for example. This network helps manage Singapore's external trade, which was US$135.5 billion in 1992, saving it an estimated US$1 billion annually. Sidestepping the traditional barrage of paperwork, traders can complete one electronic form which can be submitted by modem to the Singapore Trade Development Board (STDB). The information is electronically rerouted to one of eighteen government agencies responsible for issuing trade documents. Within as few as 15 minutes, traders can get approvals in their electronic mailboxes. The traders make any necessary payments, such as application fees and customs duties, through electronic funds transfers from their banks to the STDB. At that point, the necessary permits are automatically routed to the Port and Civil Aviation authorities via TradeNet, to allow immediate physical clearance of goods.

In addition to the revolutionary TradeNet, there are other interactive databases such as RealNet for real estate transactions and information; LawNet for legal research and filings; MediNet for hospital, clinic, and insurance company use; PortNet for surface shipping; StarNet for air cargo; AutoNet for manufacturing automation; InfoLink for government, private, and international databases; $Link for electronic payments; SchoolLink tying all schools to the Ministry of Education; and CoiNet for the construction industry. Clearly, the daily conduct of business is already being transformed. In Singapore's primary industries, such as shipping, these networks gives Singapore a competitive edge by making communications and transactions more timely.

Singapore's civil service, already the most heavily computerized in the world, is constantly being streamlined through added online capabilities. The result is an increasingly efficient government: Singapore is one of the few places in the world where the term "efficient civil service" is not an oxymoron. Whereas it used to take 50 days to register a private company, it now usually takes eight; and a sole proprietorship can be registered with same-day turnaround, a vast improvement over the previous average delay of 30 days. This direct routing and connection of private industry with appropriate government agencies minimizes not only the time and paperwork, but even the number of employees needed in each transaction.

Government Frets Over Internet

The perennial question asked by foreign observers, information technology proponents, and the government is how the government can maintain control in an open information society. For example, one obstacle to the IT 2000 Plan is the government's fondness for information control through censorship. Can the government allow the necessary democratization of information creation without simultaneously allowing the democratization of information access? As former prime minister Lee Kuan Yew was given a demonstration of the nation's latest information technology, his reported comment was, "But how can we control it?"

With the increased usage and spread of the Internet computer network (to which very few Singaporeans currently have access), it will only be a matter of time until Singapore will have an unaccustomed open access to a wide range of potentially contradictory information. To maintain their competitive advantage, Singaporean industries, businesses, and individuals will need to be able to obtain that information with few controls in order to be able to leverage it to realize the IT 2000 goal of becoming the information arbitrageur of the world.

A further aspect to this issue is the question of whether Singapore could become more controlling given enhanced computerized Big-Brother-Is-Watching-You capabilities. Some argue that given its strict control orientation, efficient civil service, and greater computer capabilities, Singapore would be enabled to intrude to an even greater degree into the lives and affairs of its citizenry and businesses.

Will a small equatorial island become the technological world leader in the Information Age? Or will the government fall short of its goal because of its perceived need to control what cannot truly be controlled? Tune in to the Year 2000 for the answer.

HELP WANTED: A SOLUTION NEEDED FOR SINGAPORE'S LABOR SHORTAGE

All looks bright on Singapore's economic horizon, except for one persistent problem: a shortage of workers. While Western countries gaze enviously at Singapore's low, almost nonexistent, unemployment rate (less than 2 percent), Singapore's leaders wring their hands as they seek solutions to a chronic labor shortage.

The most obvious reason for Singapore's labor shortage is its small population of only 3 million people. Moreover, as Singaporeans have become better educated and accustomed to higher living standards, they have opted to have fewer children. Basically, there are too few Singaporeans to keep Singapore running at top speed, even were they to continue to work at their accustomed pace (and the powers that be are already concerned that the new, softer generation is slacking up to a greater or lesser extent).

The shortage has been felt the most in the low-skill employment sector, although the movement offshore of more labor-intensive, low-wage production has served to ameliorate this problem to some extent. There are getting to be relatively fewer of these jobs to be filled, making the shortage slightly less critical. However, as more of Singapore's educated class travel and see the world, more of its highly skilled workers are emigrating to foreign lands to try their luck in different climes. This exacerbates the shortages in the high-skill, high-wage sectors that Singapore is counting on to drive its next growth stage and in which there is already a labor shortage due to the rapid growth it has already experienced.

Importing Labor

Singapore first began to face labor shortages in certain areas of its economy as early as the 1970s. The government responded by relaxing immigration laws and work permit issuance. By 1972 the immigrant workforce had risen to 12 percent of the total workforce. By 1993 the rate had grown to almost 20 percent. The shortage not only persists but is increasing, albeit at a somewhat slower rate. Nevertheless, the government is reluctant to simply bring in more foreign workers whom, they believe, also bring in more problems. In fact, the government established laws in 1992 to more strictly regulate and lower the population of foreign workers. Foreigners may still comprise up to 67 percent of a firm's total workforce, but use of subcontracted unskilled foreign laborers is being strictly limited by degree and duration as well as discouraged through high permit fees, which can amount to a 30 percent payroll surcharge.

Plan B . . . and C, and D

The government has now shifted its efforts to building up the domestic labor pool. In particular, women are being encouraged to enter the workforce in greater numbers and to go on for additional education to facilitate their entry at higher levels. However, the government is simultaneously urging educated women to procreate more. In fact, educated women are given distinct incentives to have children, while at the same time less-educated woman are being discouraged from having them. The ambivalence evidenced by this contradiction between the family values policy and the women in the workplace policy underscores another problem that especially affects women in the Singaporean business world. As a traditional male-dominated society, Singapore has been leery of letting women into the workforce as equal participants. Although less discriminatory than some Asian and other societies, Singapore continues to effectively restrict women to certain categories of jobs and sets the unofficial glass ceiling at relatively low levels.

Another marginal, but somewhat more promising near-term scheme is one encouraging people to work longer before retirement. There is even talk of raising the service requirements to force people to do so. Retirement in Singapore is a function of years of service and retirement benefit accumulation rather than of age per se. Retired people are also being encouraged to continue to participate in the workforce at least on a part-time basis. This older generation is more likely to respond to appeals to keep working because they are used to working, have a strong work ethic, and still remember what it was like in the bad old days.

Another method, strongly encouraged by Prime Minister Goh would develop technology further and faster in order to streamline jobs and compensate for the limited labor force by raising productivity. Technology, Goh believes, is the answer to making the economy more efficient and the people more productive.

And if you can't beat them, join them. Expand your perimeter. Singapore has no desire to make territorial acquisitions, but it is turning outside its borders to expand its effective workforce, especially in the lower level areas where it is most deficient. The government has been working at promoting the "Growth Triangle," a joint venture production zone which includes Johore in Malaysia and Batam in Indonesia. Singapore provides the bulk of the capital, infrastructure—such as communication and transportation facilities—and managerial and technical skill, while Malaysia and Indonesia provide land and cheap labor for these offshore operations. Unfortunately, this program has become entangled in a web of problems and has yet to get off the ground as a going concern.

Regardless of what methods are used, as Singapore's economy continues to develop and grow at an ever more rapid rate, and as it becomes a business center for more multinational corporations, the shortage of laborers is likely to be a perennial thorn in Singapore's side.

Opportunities

OPPORTUNITIES FOR IMPORTING FROM SINGAPORE

Owing to its prime location and well-developed infrastructure, Singapore has emerged as a regional trade center for petroleum products, spices, and timber, as well as numerous manufactured goods and types of machinery. Many international companies have established manufacturing and technical service operations alongside domestic producers in Singapore.

Oil and refined oil products represent the largest portion of Singapore's commodity trade. Principal non-oil exports include jewelry, electronic components, computer peripherals, household appliances, and aerospace equipment.

In line with Singapore's importance as an international and regional trade center, reexports account for a significant portion of its total export volume. Given this emphasis on regional processing and distribution services, reexports are likely to continue to be a substantial factor driving total export growth. The following section describes Singapore's most important industries and the opportunities they offer to foreign importers.

OPPORTUNITIES TABLE OF CONTENTS

Opportunities for Importing from Singapore	23
Opportunities for Exporting to Singapore	27
Opportunities for Growth	32
Public Procurement Opportunities	34
Public Procurement Process	36
Special Trade Zones	38

OIL AND REFINED PETROLEUM PRODUCTS

Oil refining and related activities are increasing rapidly throughout Asia but particularly in Singapore. Although Singapore has no known reserves of oil or natural gas, its strategic location, open economy, and industrial facilities have contributed to its status as an important refining, oil trading, and petrochemical production and distribution center. As the world's third-largest refining center, Singapore supplies almost 40 percent of Asia's imports of refined petroleum products. Singapore is also one of the world's three main oil trading centers. Output of refined petroleum products continues to record moderate increases as of mid-1993, although local refineries are already operating at capacity. Product exports include gasoline, kerosene, petroleum lubricants, automotive and marine fuels, and petrochemical by-products.

Some of the HOT items:
- asphalt and bitumen
- automotive and marine fuels
- gasoline
- kerosene
- lubricants, greases, and brake fluids
- petrochemicals
- petroleum

ELECTRONICS

The electronics industry accounts for 38 percent of Singapore's total manufacturing output and more than 40 percent of domestic exports. In recent years Singapore has also become a major electronics reexport center for the Asia Pacific region. The principal sectors of the electronics industry include consumer electronics, electronic components, telecommunications equipment, and computers and peripherals (covered separately).

Consumer Electronics

Singapore's consumer electronics sector is undergoing readjustment as a result of declining exports. Growth is expected to rebound as local producers diversify into higher value-added products, such as high-definition television. Major consumer exports include color televisions, car audio systems, tape recorders, and cassette components and parts. Singapore also has a lucrative reexport market for such items as blank videotapes.

Electronic Parts and Components

Restructuring is also taking place within Singapore's electronic components sector. Production is currently shifting from low-end sub-assemblies to more complex and competitive products. Electronic components include semiconductors, diodes, tubes, and transistors as well as capacitors, connectors, coils, resistors, switches, transformers, and printed circuit boards (PCBs).

Telecommunications Equipment

Singapore features some of the world's most modern telecommunications systems and is both a major producer and consumer of telecommunications products. The country is also a significant reexporter of telecommunications equipment. Reexports include communications equipment, facsimile machines, audiovisual equipment, and telephone systems and accessories.

Some of the HOT items:
- active components
- audiovisual equipment
- calculators
- car stereos
- color picture tubes
- electrical machinery
- electrical wiring and cables
- electronic valves
- fax machines
- high-fidelity equipment
- household appliances
- magnetic tape products
- passive components
- printed circuit boards
- radios, cassette, and CD players
- radio telephonic and telegraphic receivers
- tape recorders
- telephone systems and accessories
- telephones
- television sets
- video cameras and cassette recorders

COMPUTERS AND PERIPHERALS

Singapore's computer industry has enjoyed strong growth over the past decade. Although foreign companies have an established presence, domestic companies also manufacture sophisticated high-end computer systems and peripheral products, including printed circuit boards, network cards, printers, modems, fax cards, and monochrome and color monitors. Consistent with global trends, many local computer manufacturers are currently switching from the production of mainframes to smaller, less expensive computer hardware.

In addition, Singapore produces about half the world's supply of computer disk drives. This sector alone accounts for at least 15 percent of Singapore's non-oil exports. However, in an effort to remain competitive, many manufacturers are relocating local operations to Malaysia and Indonesia, where labor and other production costs are lower.

Some of the HOT items:
- computer central processing units (CPUs)
- computer parts
- disk drives
- integrated circuits
- microcomputers
- printers
- recorded computer tapes and diskettes

CHEMICALS

Exports of chemical products from Singapore continue to register moderate growth. Industrial and organic chemicals are used widely in the petrochemical and pharmaceutical industries. Inorganic (heavy) chemicals include chloralkalides, industrial gases, inorganic compounds, and rare earth metal salts. Despite an oversupply of such petrochemicals as synthetic resins, output in the industrial gas and chemical sectors increased by 2.5 percent between mid-1992 and mid-1993. Other chemical products include industrial and household cleaners, chemical solvents (aromatics, chlorinated hydrocarbons, alcohols, ethers, ketones, amines, esters, and glycols), paints and coatings (pigments, thinners), adhesives and sealants, water treatment chemicals, and plastic raw materials.

Some of the HOT items:
- adhesives and sealants
- detergents, soaps, and fats
- flavors and fragrances
- heavy chemicals
- industrial and household cleaning products
- industrial and specialty gases
- paints and coatings
- petrochemicals

- plastic raw materials
- solvents
- specialty plastics

MACHINERY AND EQUIPMENT

Due to a rising demand for machine tools and industrial machinery in Asia, Singapore's industrial machinery exports should enjoy continued growth. At present, Singapore is capable of producing a full range of machinery and equipment, including computerized numerical control (CNC) machinery centers, measuring and testing instruments, devices and equipment, material handling equipment, and oil rig fabrication machinery. Singapore is also very competitive in the production of such industrial machinery as high-precision surface grinding machines, CNC milling machines, and die-casting and precision machine parts. Singapore's industrial machinery producers are capable of undertaking equipment manufacturing and turnkey projects on a contractual or subcontractual basis.

Some of the HOT items:
- CNC vertical machinery centers
- die-casting parts
- grinding machines
- hydraulic cylinders
- milling machines
- oil rig fabrication machinery
- positioners
- precision machine parts
- press brakes
- surface finishing equipment

FABRICATED METAL PRODUCTS

Strong demand in both the domestic and regional construction industries is contributing to steady growth in Singapore's fabricated metals industry. The industry supports a broad product base, including pipes and tubes manufacturing, die-casting, aluminum extrusion, heaters, boiler manufacturing, and the fabrication of plant and building structures. Additional products available for export include welded steel wire mesh products, aluminum grilles, anodized aluminum extruded products, wall claddings, riveting tools, building hardware, and metal containers.

Some of the HOT items:
- aluminum grilles, grids, and gratings
- anodized aluminum extruded products
- electroplating
- hardwares
- lead frames
- metal containers
- pipes and fittings
- riveting tools
- roofing and walling materials
- stainless steel sinks
- structural deckings
- welded steel wire mesh products

TEXTILES AND APPAREL

Although no textiles are produced in Singapore, there are a substantial number of garment manufacturers, all of whom use imported fabrics. Many export ready-made clothing to Europe and the United States. Exports include infant's and children's clothing, batik garments, bridal gowns, designer fashionwear, casual wear, and garment accessories, such as zippers, buttons, sewing thread, and elastic bands.

Flat sales and rising labor costs have caused Singaporean firms to begin subcontracting such assembly work to manufacturers in neighboring countries. As Singapore continues to diversify beyond the manufacturing of apparel and textiles, its role as a trading and service center is likely to become increasingly important.

Some of the HOT items:
- batik garments and materials
- bridal gowns
- canvas products
- casual clothing and sportswear
- children's clothing
- designer fashionwear
- garment accessories
- infants' clothing
- leather footwear

PRINTED MATERIALS AND PAPER PRODUCTS

There are more than 350 domestic and foreign printers in Singapore, and the printing and publishing industry is one of the nation's top ten in terms of value-added products and output. Information and production technology upgrades have resulted in significant increases in demand for high-quality printed products. Among the major exports are trade, general, and souvenir publications as well as textbooks, educational books, and children's books. Although the domestic market is small, Singapore is becoming the Asian reprinting and distribution center for international book, magazine, and newspaper publications. Assorted paper products include tissues, disposable waste products, facsimile paper, paper tapes and labels, and computer forms. The Singaporean government exercises control over editorial content with respect to materials to be disseminated in Singapore, which can potentially cause problems.

Some of the HOT items:
- children's books
- computer forms
- educational reference materials
- general printed books and materials
- regional research papers and topical books
- textbooks and academic publications
- trade journals and books

ORCHIDS

Orchid-growing in Singapore is now a multi-million-dollar industry, with exports expected to reach US$66 million by 1995. Recognizing the immense potential for exports of cut flowers, public and private research and promotion are intensifying. As a result of research in orchidology and biotechnology, new orchid hybrids and species are continually being developed. Popular traditional varieties under cultivation include Dendrobium, Vanda, Oncidium, Mokara, and Aranda. According to government sources, 20 of the approximately 85 licensed orchid growers and exporters account for about 86 percent of Singapore's total domestic exports of cut orchids. Among them are Toh Orchids, Multico Orchids, and Asia Orchids.

Some of the HOT items:
- Aranda
- Dendrobium
- Mokara
- Oncidium
- Vanda

JEWELRY

Singapore's total trade in jewelry has almost doubled in the last decade, with more than half of its jewelry imports being reexported. Noted for quality craftsmanship, Singaporean jewelers design and manufacture a wide range of precious and semi-precious stones and both fine and costume jewelry. Many jewelers rely on both experienced in-house as well as foreign designers. In addition to custom-made jewelry for such renowned names as Franklin Mint, Heiwado, Mikimoto, and Fortuna, some Singaporean jewelers are marketing their own collections. A large range of items are available, including necklaces, rings, earrings, brooches, pendants, sliders, bangles, bracelets, watches, and crafted ornaments.

Some of the HOT items:
- cast jewelry
- gem-set jewelry, especially diamonds
- gold and silver ornaments
- gold, platinum, and silver jewelry
- pearl and small-stones jewelry

ORNAMENTAL FISH

Singapore ranks among the world's top live fish exporters, accounting for some 20 percent of the total world export of freshwater ornamental fish. Singapore is also an important marketing and distribution center. In addition to general fish traders, there are some 36 fish breeders in Singapore, specializing in only a few kinds of ornamental fish. Overall, exports comprise 300 different varieties and species, including armored catfish, barbs, discus, fighting fish, gouramies, platys, swordtails, and tetras. Singapore is also the world's leading producer of guppies.

Some of the HOT items:
- armored catfish
- barbs
- discus
- fighting fish
- gouramies
- guppies
- platys
- swordtails
- tetras

TEN EXTRA PROSPECTS FOR IMPORTING FROM SINGAPORE

- foods
- furniture
- lighting systems and equipment
- medical equipment
- office equipment and supplies
- security systems
- sewing machines
- stationery products
- timber and wood products
- transport equipment

OPPORTUNITIES FOR EXPORTING TO SINGAPORE

Singapore relies heavily on foreign trade. In addition to importing many raw materials and resources for domestic use, the country is a regional commercial hub that imports and then reexports a multitude of goods and materials. Among its reexports are petroleum, telecommunications equipment, electronic parts, and electrical machinery. Attracted by Singapore's excellent infrastructure and port facilities, many foreign manufacturers are contributing to Singapore's development as a regional transportation and service base.

Singapore's international manufacturing base is now shifting toward more capital-intensive and high-technology production. The government strongly encourages imports of aircraft and avionics, computers, electronic components, robotics, industrial automation, chemicals, and plastics items to support such production. Development of the health care, pharmaceutical, and biotechnology industries is a goal of government policy. The following section describes the industries in Singapore that offer the best prospects for foreign exporters.

AIRCRAFT AND AVIONICS EQUIPMENT

The internationalization of Singapore as a transportation hub and the increased economic vitality of the Southeast Asian region have created strong demands for more aircraft and related equipment in Singapore. Because domestic production of avionics equipment is minimal, the market is dominated by imports. The largest export opportunities in terms of dollar value are for passenger and cargo aircraft and aircraft parts. Numerous other opportunities are available to foreign exporters of airborne and ground-based avionics, such as flight and navigation equipment, as well as support and maintenance equipment.

Some of the HOT items:
airborne avionics
- automatic dependence systems
- cabin management information systems
- ground proximity systems
- satellite communication systems
- traffic alert and collision avoidance systems (TCAS)

ground-based avionics
- air traffic control systems
- electrooptical camera systems
- flight information displays (Terminal I of Changi Airport)
- integrated landing systems
- surface movement radar systems
- VOR/DME systems
- weather information systems

ground support and maintenance equipment
- advanced materials and components testing and inspection equipment
- avionics testing, checkout, repair, and calibration apparatus
- composite materials repair equipment
- converter units
- electronics hardware and software for automated maintenance and repair operations
- hydraulics checkout and repair equipment
- power plant and electrical system monitoring and repair equipment

TELECOMMUNICATIONS EQUIPMENT

Singapore is both the telecommunications center of Southeast Asia and an important world communications hub, with telephone, telex, and telefax services to more than 200 countries. As such, Singapore is a major consumer of telephone, telex/telegraph, facsimile, and switching equipment. The country also reexports a considerable amount of telecommunications equipment.

In an effort to maintain its position as a telecommunications leader, Singapore is in the process of upgrading and expanding its national and international telecommunications systems. Current plans include complete digitalization of the national network, installation of fiber-optic cable throughout the island, and commercial operation of a digital mobile network. These plans have increased the demand for an Integrated Services Digital Network (ISDN) system, a satellite telex system, closed circuit surveillance systems, traffic sensors, pulse code modulation (PCM) systems, and trunk telephone cable. Cellular phones and public pay telephones offer additional opportunities for foreign exporters.

Some of the HOT items:
- cellular phones and pay telephones
- computer data processing equipment
- data communications equipment
- facsimile communications equipment
- mobile radio (conventional) equipment
- pulse code modulation systems
- telecommunications test and measurement equipment
- telephone, telex, telegraph, switching, and switchboard equipment
- transmission equipment
- video and radio broadcasting equipment

ELECTRONICS INDUSTRY PRODUCTION AND TEST EQUIPMENT

The Singaporean government has targeted the electronics industry as a priority manufacturing sector and is actively promoting its growth and development. Programs focusing on precision engineering, factory automation and advanced manufacturing technology, mechatronics, and industrial electronics are under way. Because automation remains the key to quality in high-technology manufacturing, the market for test equipment suitable to applications in electronics production and R&D should continue to expand.

Some of the HOT items:
- automated assembly and test center for semiconductors and opto-electronics
- ceramic chip capacitor production equipment
- gallium arsenide LED wafer fabrication equipment
- integrated circuit (IC) production
- IC wafer fabrication equipment
- low-end PCs
- microwave IC technology
- sub-micron-level wafer fabrication equipment
- advanced microwave products
- automotive electronics
- industrial electronics (keyboards, printers)
- laptop computers
- large-screen liquid crystal displays for PCs
- new series PCs that facilitate 386 & 486 upgrades
- phones, answering machines, and security systems
- radio and pager products
- switching power supplies and high-resolution monitors

ELECTRONIC COMPONENTS

Despite its small size, Singapore has developed into a major producer and consumer of electronic components. Most of the world's major electronic companies—including Apple, Hewlett Packard, Seagate, Texas Instruments, Philips, Siemens, Thomson, Hitachi, Mitsubishi, and NEC—have operations in Singapore.

The presence of multinational electronic component manufacturers in Singapore complements the production of electronic equipment. Due to the strong demand in all end-user markets and the continuous encouragement of industry upgrades, Singapore's importation of electronic components is likely to remain high. The potential for foreign suppliers is excellent, particularly for computer and computer-related product components, including multilayer printed circuit boards, connectors, resistors, and capacitors.

Some of the HOT items:
- cellular components
- high-powered resistors
- high-voltage disk capacitors
- high-end semiconductors
- semi-programmable and programmable devices

COMPUTER SOFTWARE

The Singaporean government has made significant investments of money and resources to bring the country into the information age. Singapore's National Information Technology (IT) Plan, announced in 1986, and the more recent IT 2000 Plan establish a framework by which the country's public and private sectors can make greater use of information technology. The objective is to create a network-based system throughout Singapore's private and public sectors.

Given this strong emphasis on computerizing and automating office operations, the market for computer software is expected to experience rapid growth in the coming years. In particular the demand for network software and for applications software for microcomputers and business applications is expected to increase at a brisk rate. UNIX-based software and document image processing (DIP) systems are two relatively new areas that should also experience particularly strong growth in the next decade.

Some of the HOT items:
- applications software, especially for spreadsheets, word processing, communications, general business, and education
- customized software
- database management systems
- document image processing (DIP)
- integrated applications software
- operating systems software
- UNIX-based software
- utilities software

ROBOTICS AND AUTOMATION

Companies in Singapore are realizing the importance of automation as a means of staying competitive in the world market. At the same time, the manufacturing sector is moving toward higher value-added products, which require the application of newer, more advanced technologies. These trends suggest that there will be a continual need to replace and upgrade existing technologies as well as to pioneer new ones. Because the Singaporean government strongly encourages automation, and many manufacturers have already embraced the concept, the country's automated services market is likely to continue growing for some time. Demand will be particularly strong for process controls, measuring, material handling, and systems integration equipment.

Some of the HOT items:
- automatic testing, inspection, and measuring equipment
- computers used for manufacturing systems integration
- custom-made automatic machines
- electro-mechanical robots
- material handling, parts feeders, and conveying equipment
- programmable process control equipment
- surface-mount technology assemblies

LABORATORY SCIENTIFIC INSTRUMENTS

Singapore's strong emphasis on improving product quality should continue to boost, the sales of laboratory scientific instruments. To promote R&D and the upgrading of technology in industry, the government is offering companies various tax and financial incentives. Because Singapore is dependent on imported instrumentation, the government's measures should be favorable to foreign sales of chromatographs, spectrophotometers, centrifuges, oscilloscopes, and other precision measuring instruments. These types of products are used extensively in the oil refining, food and beverage preparation and the packaging, biotechnology, and electronics industries. Good export potential also exists for markets in neighboring countries.

Some of the HOT items:
- balances, viscometers, refractometers, and thermometers
- electrochemical monitors
- electronic analyzers
- electronic recorders
- gas and liquid analyzers
- gas and liquid chromatographs
- laboratory centrifuges
- microscopes
- oscilloscopes
- oxygen analyzers
- precision laboratory measuring instruments
- spectrofluorimeters
- surface analysis systems
- test and inspection instruments
- ultraviolet-visible and atomic absorption spectrophotometers

MEDICAL EQUIPMENT AND PRODUCTS

Medical facilities in Singapore are excellent, and the country aims to become the medical treatment and distribution center for Southeast Asia. To meet this goal, Singapore's hospitals and clinics are constantly expanding, improving their facilities, and acquiring new medical equipment. Imports for a wide range of equipment are expected to grow by an average of 10 percent a year through 1995. Among the best prospects are those for foreign suppliers of high-tech products, such as surgical equipment, x-ray apparatus, and electro-medical equipment. Disposable medical products (gloves, syringes, and gauze) represent another area of opportunity.

Some of the HOT items:
- advanced cardiac output computers
- advanced patient-monitoring equipment
- artificial respirators
- blood pressure equipment
- blood transfusion equipment
- cardiology equipment
- electro-medical apparatus and parts
- hearing aids and parts
- opthalmological equipment
- sterilizers, autoclaves, and parts
- surgical and medical gloves
- surgical and medical instruments and apparatus
- surgical, orthopedic, and prosthetic appliances and supplies
- syringes and parts
- waddings and gauze
- x-ray apparatus and parts
- x-ray films

DENTAL CARE AND HYGIENE SUPPLIES

Singapore currently has approximately one dentist per 5,000 inhabitants, compared to an estimated one dentist per 1,000 inhabitants in the United States. The rising affluence of the citizenry is expected to bring about an increasing demand for quality dental care, including orthodontics, implants, and cosmetic surgery.

Singapore has developed a reputation for having the highest-quality dental care in Southeast Asia. Expatriates of many Western countries who live in neighboring Southeast Asian countries make regular visits to Singapore for dental treatment. Singapore also serves as the region's leading supplier of quality dental supplies and training services, and it places relatively few restrictions on the importation of dental and dental hygiene equipment. Foreign-brand home dental care products (toothbrushes, toothpaste, dental floss, and mouth rinses) are becoming increasingly popular.

Some of the HOT items:
- advanced dental aesthetic procedures
- dental imaging systems
- dental implants
- dental insurance plans
- home dental care products
- infection control products
- laser-based dental procedures
- waste disposal systems

FOOD PROCESSING AND PACKAGING EQUIPMENT

Trends toward smaller families, more women in the labor force, and supermarket shopping have contributed to the demand for processed food in Singapore. Consumers are also becoming more interested in health foods. As a result of all these factors, the market for food processing equipment is expected to grow. Opportunities exist for foreign exporters of preparation, processing, and packaging equipment, especially equipment used in the beverage, dairy, and baked-goods industries.

Although Singapore is a relatively small market, it is well-suited to act as a distribution center for the growing economies of the Asian Pacific region.

Some of the HOT items:
beverages
- blending equipment
- heat treatment equipment
- pasteurization equipment

dairy products
- heat exchangers for margarine
- ice cream manufacturing equipment
- refrigeration equipment

bakery products
- boilers
- chillers
- compressors
- heat exchangers
- pressers

automated processing equipment
- aseptic processing equipment
- canning equipment
- extruders
- freeze dryers
- high speed filling machines

packaging equipment
- pallet stretch wrap machines
- pillowpackers (horizontal)
- stretch wrap machines (manual)

REFRIGERATION EQUIPMENT

Singapore's market for household and institutional refrigeration equipment is expected to show modest but steady growth over the next several years. The expansion of the institutional market will reflect the ongoing construction and renovation of hotels and restaurants. In addition to a mini-refrigerator in each guest room, Singapore's top hotels feature from two to as many as eight dining establishments. The main factors bearing on market growth in the household sector include continued projected increases in per capita income and the trend toward larger dwellings and larger kitchens. Due to the local preference for fresh food, domestic freezers are not as prevalent as refrigerators. However, with more households making weekly (as opposed to daily) shopping trips, larger refrigerators are in demand.

Some of the HOT items:
- commercial refrigerating and freezing equipment
- consumer refrigerators (over 3.2 cubic meters)

PRINTING AND GRAPHIC ARTS EQUIPMENT

Singapore has emerged as an international printing and graphic arts center. Singaporean printers are among the best in the world. Many are equipped with up-to-date sheet-fed presses and computerized prepress equipment, and the industry has invested more than US$120 million in advanced equipment to meet the increasingly sophisticated needs of clients.

A number of companies are already using computerized typesetting machines, computer-aided design and measuring (CAD/CAM) systems, laser printers, and multicolor and web offset presses. Modern printing techniques include the use of electronics in the fields of color separation, phototypography, and press control. Because printing and graphic arts companies are mindful of keeping their operations up-to-date, they are committed to ongoing purchases of new, more advanced equipment.

Some of the HOT items:
- computerized book binders
- computerized typesetting machines
- electronic scanning equipment
- flexographic press plates
- graphic arts film
- laser typesetting, phototypesetting, and composing machines
- micro-computer-based information systems
- multi-color offset presses
- offset-type duplicating machines
- printing blocks and plates

- printing ink
- process cameras and photographic equipment
- roll-fed and web offset presses
- scanners and integrated systems

SECURITY AND SAFETY EQUIPMENT

As Singapore has become more urbanized, its infrastructure more sophisticated, and its people and businesses more affluent, security and safety systems have become common among the mechanical and electrical services in many buildings. Consonant with Singapore's crime-free reputation, there is great emphasis on security for individuals, private dwellings, and industrial and commercial buildings. Singapore's labor shortage has strengthened the demand for advanced equipment and systems that can replace conventional guard services. The market offers good potential for all types of electronic security and safety equipment including access control, intruder detection, central monitoring and management systems, and surveillance equipment.

Some of the HOT items:
- advanced monitoring systems
- alarms
- anti-intruder detection systems
- card-access systems for single door or multi-door applications
- centrally controlled security systems (telemetry components)
- fire detection equipment
- wireless systems

SKIN CARE PRODUCTS

The market for skin care products in Singapore should experience moderate growth over the next several years as consumer spending continues to increase, especially among women. The strong tourist market (Singapore hosted more than five million tourists last year) has also contributed to the demand.

Skin care products are supplied primarily through imports. European, US, and Japanese companies dominate the high-value brand-name market, while smaller companies from Taiwan, Hong Kong, and Malaysia supply less expensive products. Although long-term growth will probably be slow, the skin care market in Singapore still presents significant opportunities to foreign firms.

Some of the HOT items:
- basic skin cleansers
- eye cream and wrinkle cream
- facial masks
- toners and moisturizers

FOURTEEN EXTRA PROSPECTS FOR EXPORTING TO SINGAPORE

- audiovisual equipment
- biotechnology
- building products
- chemical production machinery
- computers and peripherals
- construction equipment
- drugs and pharmaceuticals
- electrical power systems
- hotel and restaurant equipment
- household consumer goods
- industrial chemicals
- oil and gas field machinery and services
- pollution control equipment
- sporting and recreation goods

OPPORTUNITIES FOR GROWTH

SERVICES

The outlook for the service sector is positive. Singapore's banking and financial services are currently among the most open in the world. Fiscal and tax incentives are also available for advertising, public relations, and exhibition services; agrotechnology; transportation and distribution; computer services; education and training; laboratory and testing services; leisure, entertainment, and cultural services; management consulting; medical services; and information services.

Furthermore, Singapore's service sector is closely tied to the rapid growth of East Asia, which is expected to be the world's most dynamic area for the next 20 to 30 years. A strong commitment by governments and businesses toward the development of Asia's infrastructure will also generate high demand for engineering and technical services.

ACCOUNTING AND TAXATION

A tax on goods and services is scheduled to be implemented in Singapore in April 1994. In anticipation, companies have begun reviewing and amending their accounting systems, which has created a demand for accounting and tax consulting services. Accountants, in particular, predict heightened demand for their services immediately after the new tax takes effect.

FINANCIAL MARKETS

The Singaporean government has placed a strong emphasis on establishing the country as a leading Asian securities center. In a recent poll by securities regulators, the Singapore stock exchange ranked highest overall in East Asia. Among the criteria were the amount of investor protection, and controls on insider trading and share manipulation. The Singapore stock exchange also received the highest rating for its regulation of participants and even-handedness. In a significant development, the Singapore government recently opened up membership on the stock exchange to foreign firms. Foreign participation, however, is still subject to various restrictions.

Shearson Lehman Brothers rated Singapore as the most attractive equity market in the world in 1993. According to the US-based stock brokerage firm, the Singapore market had just entered an extended period of rising prices in which companies' shares for the most part closely matched their true market value. Finally, the Singapore International Monetary Exchange (SIMEX) is the tenth fastest-growing futures exchange in the world. By offering a more competitive and lower-cost market, SIMEX has begun to compete against the Nikkei and Euroyen markets in East Asia.

LIFE INSURANCE

Since the early 1980s, the life insurance business in Singapore has been growing by more than 20 percent annually, and prospects for the next few years remain bright. Compared to the advanced industrialized countries, Singapore is underinsured, but interest in insurance is growing rapidly. Single-premium policies combined with high-profile marketing are expected to provide additional buying incentives, as are such traditional growth factors as rising affluence and a greater awareness of health and wellness.

SHIP CONVERSION

Ship repair accounts for more than 60 percent of the revenue of Singapore's marine industry. The medium- to long-term prospects continue to be bright, as many ship owners will require repairs to keep their vessels in operation and to meet new regulations. An upsurge in demand for repairs is also predicted because the cost of building new ships is expected to climb sharply. Singapore's Trade Development Board anticipates strong growth in the ship conversion market in response to new environmental regulations, which require new tankers to be outfitted with double hulls and double bottoms by 1995. Perhaps more importantly, Singapore is well-positioned to tap the growing global market in ship conversion because of its high dry dock capacity and strategic location.

PHARMACEUTICALS

Singapore has become a key offshore manufacturing and support base for multinational pharmaceutical companies. Many pharmaceutical products, including semi-synthetic penicillin, antibiotic intermediates, and anti-ulcer medicines are manufactured in Singapore, largely for export. In contrast, local consumption of pharmaceuticals is met mainly through imports. The domestic market is expected to grow by 10 percent between 1993 and 1995, due to the rising standard of medical care. Singapore's role as a regional medical and distribution center makes it an excellent springboard to many Asian markets.

ENVIRONMENTAL SERVICES

According to the US-ASEAN Council for Business and Technology, the market for environmental control and services in the ASEAN region—which includes Thailand, the Philippines, Malaysia, Indonesia, Brunei, and Singapore—is estimated to be worth more than US$1.5 billion. Singapore's establishment of a central authority to advise, monitor, and enforce national environmental standards has been critical to the country's developing role as a gateway to Asian markets. In the coming years, Singapore is likely to become the regional headquarters for many environmental products and services. There is currently a high demand for industrial waste management and pollution control.

FRANCHISING

Franchising is a highly accepted way to do business in Singapore. Singaporeans' increasing disposable income and their tendency to adopt features of a Western lifestyle have prompted more business ventures to develop in the area of franchising. Furthermore, franchising is encouraged through various incentives offered by the Singapore government to local businesses, which stand to gain enormously from foreign expertise in this field. Food franchises are common in Singapore, including such American chains as McDonald's, Kentucky Fried Chicken, and Pizza Hut. Franchises are also becoming popular in retail apparel, laundry services, health and recreation services, and shoe repair.

DESIGN SERVICES

The Singaporean government has been promoting the concept of improving the design of domestic products as one way of remaining competitive in the international market. The Design Venture Program, administered by the Singapore Trade Development Board (STDB), was drawn up in the late 1980s to assist local companies in upgrading the design features of their products, product packaging, promotional literature, and corporate brochures. In 1991 the TDB opened the National Design Center to showcase innovative and attractive products, all made in Singapore. The fierce competition for eye-catching new designs for all types of consumer products provides diverse opportunities for foreign design services, particularly those that focus on product and packaging design.

SPORTS AND LEISURE

The Singaporean government has begun to actively promote physical fitness among schoolchildren and in the workplace as part of efforts to create a "rugged society." Existing sports facilities are being expanded, and plans for new housing estates incorporate sports facilities. Among the sports and leisure activities gaining in popularity are golf, tennis, fishing, and boating. Opportunities exist for foreign franchisers of fitness centers, sporting goods, and physical fitness equipment. Singapore also serves as an important distribution point for sports equipment destined for Southeast Asia.

CINEMATOGRAPHY

Singapore's video and film industry will benefit from the establishment of a diploma course in film, sound, and video, which is being sponsored jointly by government and industry. The Ngee Ann Polytechnic will spend US$4.4 million over the next five years to build and equip studios for training directors and cinematographers. Singapore's Economic Development Board will support the diploma program through scholarships and by helping to link foreign expertise with the new program. Down the road, prospects are excellent for the development of a billion-dollar film industry in the Asian region.

PUBLIC PROCUREMENT OPPORTUNITIES

A key factor in Singapore's economic success has been a sound investment strategy within the public sector. A long-standing goal of Singapore's government has been to become a complete business center, offering an international manufacturing base, a well-developed financial infrastructure, and excellent communications. In addition to public investment in the domestic infrastructure, the government is actively promoting Singapore as a high-tech manufacturing and service hub for the Southeast Asian region. Concurrent with the government's domestic and international goals, its current fiscal budgets call for increased spending on education, training, and research and development. The government also offers numerous incentives to encourage foreign participation in Singapore's infrastructure projects.

CONSTRUCTION

Singapore's construction industry has expanded considerably in the past few years, and the Construction Industry Development Board (CIDB) is predicting double-digit growth through 1994. Opportunities for construction-related services and building materials are abundant.

While construction of private sector office and residential buildings is expected to level off, the outlook for public housing construction remains positive. Between 1992 and 1997 the government is planning for the construction of 90,000 new units accompanied by renovations of existing housing. The CIDB has urged contractors to pay special attention to the retrofitting market, which is poised to expand as the building stock ages. It is estimated that retrofitting contracts will total almost US$1.25 billion over the next several years. Other major infrastructure projects to be undertaken include the construction of hospitals and schools as well as an expansion of the road system.

TELECOMMUNICATIONS

Singapore has one of the world's most modern telecommunications systems. The Telecommunications Authority of Singapore (TAS) is in the process of being privatized, which should eventually increase the market share of foreign equipment and services. Despite privatization, the Singapore government is expected to remain a strong promoter of telecommunications infrastructure and industry. Government plans call for US$1.7 billion in investment between 1992 to 1995 to upgrade business and consumer telecommunications services. In addition, the international arm of TAS, Singapore Telecom International, continues to pursue joint ventures throughout southern Asia. It is believed that the tremendous telecommunications needs in neighboring countries can be best served through a combination of equipment upgrades and foreign expertise.

UTILITIES

Singapore's Public Utilities Board (PUB), a quasi-governmental organization, is chiefly responsible for providing service to the country's largely urban population and industrial users. Moderate procurement levels are expected through 1996. New transmission lines and electricity power systems account for a major portion of allocated spending by the PUB. With a forecasted average annual increase in electricity use of 8 percent over the next decade, the PUB is considering building a new power station on one of Singapore's outlying islands. Planning has already begun for a power station in the mainland district of Tuas. The new plant's capacity will far exceed that of existing facilities, which are scheduled to be phased out gradually over the next decade.

The Linggiu reservoir project in Johor, also under the direction of the PUB, was recently completed. The Johor river waterworks extension scheme (stage three) will be completed by late 1994. Finally, a new transmission pipeline system is planned, and a new gasworks will be built in Senoko to replace the existing gasworks in Kallang.

INDUSTRIAL LAND DEVELOPMENT

Singapore's largest land holder, Jurong Town Corporation (JTC), is engaged in industrial and infrastructure development both locally and throughout Asia. In Singapore, JTC is planning to acquire land on Pulau Seraya, Changi South, Tanglin Halt, Boon Lay Way, and Bishan. Regionally, JTC's subsidiary, Jurong Environmental Engineering (JEE), is seeking the cooperation of the government and private firms in the development of estate projects. JEE has participated in projects in Indonesia and Thailand, and now plans to work on development projects in Taiwan and China.

ROAD TRANSPORTATION

As part of a major transportation project emphasizing all infrastructure areas, the Singapore government will spend US$1.2 billion between 1992 through 1996 to expand the capacity of the existing road network. All 20 existing roadways will be widened, and there are plans to construct 14 new roads and expressways and 35 new interchanges. The capacity of the underground road network, which totals 15 km (9 miles), will be increased by at least 40 percent, to keep some 80 percent of the city's major roads free of congestion. Work in downtown Singapore is expected to take place over the next decade. Finally, construction of the Mass Rapid Transport System (MRT) is nearly complete, and future ongoing improvements are expected.

PORTS

The Port of Singapore Authority and the Civil Aviation Authority of Singapore are responsible for managing the country's seaports and airport operations. Although Singapore's seaports and airports are capable of handling some of the highest volumes of traffic in Asia, there is an urgent need to expand cargo handling and container services. Several major projects scheduled to begin over the next few years offer foreign procurement opportunities. Singapore plans to expand its port facilities to increase cargo capacity. A major portion of the development includes a new container terminal, to be constructed on an offshore island by 1998. Work will also be undertaken at Singapore's Changi Airport to expand such services as cargo and baggage handling. On a related note, the Singapore Postal Service plans to spend US$50 million to construct a new mail-handling facility.

AEROSPACE

The aerospace industry in Singapore has gained momentum over the past two decades. Increased offshore oil exploration and expansion of the domestic airline were responsible for the industry's initial growth. More recently, greater numbers of multinational aerospace companies have established operations in Singapore and have contributed considerable expertise in the areas of manufacturing, overhauling, and refurbishing. The Singapore government has targeted the country's aerospace industry for even further development and is therefore encouraging foreign firms to introduce advanced aviation equipment as well as new airframe and engine designs.

ENVIRONMENT

Singapore has some of the highest standards for environmental protection among Asian nations. To supplement private sector efforts, the Singaporean government is actively seeking foreign partners who can meet the region's growing environmental needs. The Ministry of Environment recently formed an Environmental Policy and Management Division to develop environmental protection strategies for the next century. This Ministry has also formed a quasi-government corporation aimed at attracting Western technology for application in Singapore and throughout Southeast Asia. Possible areas for foreign procurement include infrastructure works, handling and treatment of waste in tropical conditions, and development of waste water treatment plants. Present government plans also call for construction of a fourth incinerator plant by the year 2000 to handle refuse disposal. However, this project could be delayed due to the success of strong efforts to reduce resource waste and to promote recycling.

BIOTECHNOLOGY

Programs sponsored by the National Science and Technology Board and the Economic Development Board are designed to encourage domestic and foreign firms to pursue R&D activities in the area of biotechnology. Furthermore, the National Biotechnology Master Plan describes five strategies to enhance the level of biotechnology in Singapore: (1) the use of more advanced technologies, (2) the training of more skilled manpower, (3) industry development, (4) infrastructure development, and (5) the promotion of greater public awareness. In support of these efforts, the Singaporean government has earmarked US$12 million to encourage biotechnology efforts related to health care, agriculture, marine life, and food processing. Numerous opportunities exist for joint ventures, the licensing of foreign technology, and the development of tropical products.

PETROCHEMICALS

The Petrochemical Corporation of Singapore (PCS) is planning a multi-billion-dollar expansion of its complex at Pulau Ayer Merbau, an island south of Singapore. The feasibility study was completed in 1992, and the PCS is now in the process of defining the design, environmental, engineering, and construction requirements for the expansion. Construction is scheduled to last through 1995. Opportunities also exist for foreign equipment manufacturers within the refining sector.

TOURISM

In an effort to capture a portion of the dynamic Southeast Asian tourism market, the government of Singapore is set to acquire and develop land for an adventure park on Pulau Ubin, a sparsely populated island located between Singapore and Malaysia. The Ministry of National Development has not yet announced the specific details of the project but plans to acquire 125.43 hectares for the park.

PUBLIC PROCUREMENT PROCESS

REGULATORY AUTHORITIES

Three agencies are responsible for procurement on behalf of the Singapore government: the Central Procurement Office of the Ministry of Finance, the Construction Industry Development Board, and the Pharmaceutical Department of the Ministry of Health. Firms wishing to supply goods or services to the Singapore government must register with one of these three agencies. In order to qualify for registration, foreign firms must have a registered office in Singapore or must appoint a local agent. Quasi-government agencies (such as the Telecommunication Authority of Singapore or the Civil Aviation Authority of Singapore) may also use formal procurement procedures, especially for non-recurrent items.

CENTRAL PROCUREMENT OFFICE (CPO)

The CPO is responsible for tenders involving products and services such as stationery and printing, furniture, office systems equipment, communication and navigation equipment, electronic components, food and beverages, safety equipment, engineering and building materials, household materials, textiles and apparel, electrical items, laboratory testing and survey equipment, military spare parts, shipbuilding and repair, photographic and optical supplies, and weapons and ammunition.

General Procedures

The following are the minimum requirements for firms wishing to register with the CPO:

- The firm must be solvent, and financially and technically competent.
- The firm is expected to have a record of providing the types of materials, equipment, or supplies for which it intends to register.
- The firm must furnish summaries of its trading transactions that demonstrate supplies or trading capacity of the required magnitude. The firm must also provide the names and addresses of two reputable clients who can testify to the extent and quality of the applicant's supplies and services.
- A firm seeking to register for a financial category above S$500,000 (about US$311,500) is required to be incorporated under the laws of Singapore or under the laws of a foreign jurisdiction that are recognized as having a similar legal effect.

Detailed information and applications for registration may be obtained from:

Central Procurement Office
Depot Road
Singapore 0410
Tel: [65] 2721655 Fax: [65] 2790524.

CONSTRUCTION INDUSTRY DEVELOPMENT BOARD (CIDB)

In 1984 the CIDB established a contractor's registry for firms wishing to provide construction and related goods and services to the public sector. Contractors intending to register with the CIDB must have relevant experience in construction projects and must prove that they have the financial, technical, and management capability to execute construction projects. As of January 1, 1993 two sets of tendering limits went into effect. One set is applicable to construction work and the other to such construction-related activities as mechanical, electrical, and maintenance work, and supply.

General Procedures

The following are the minimum requirements for registration with the CIDB:

- A firm must be registered with the Singapore Registry of Companies and Businesses before applying to be registered as a CIDB contractor. The former requires that foreign firms maintain a registered office in Singapore.
- The firm must submit its latest audited financial report or accounts in order to prove that it meets the requirements.
- A firm applying under the category of construction work or such construction-related activities as mechanical and electrical work must employ sufficient numbers of full-time, qualified technical personnel in the relevant disciplines.
- The firm must have at least a three-year record of having completed projects within the relevant category. Relevant letters of award for previous such contracts must be submitted with the application.
- Additional requirements for certain categories of work may apply. For example, electrical engineering firms must possess a valid electrical contractor license issued by the Public Utilities Board (PUB) and must employ a full-time employee who holds a valid electrician's license or its equivalent, issued by the PUB.

Detailed information and applications for registration may be obtained from:

Contractors Registry
Construction Industry Development Board
National Development Building
Annex A, 3rd Story, 9 Maxwell Road
Singapore 0106
Tel: [65] 2256711 Fax: [65] 2257301

PHARMACEUTICAL DEPARTMENT

The Pharmaceutical Department of the Ministry of Health procures medical and surgical sundries, instruments, and other equipment for government hospitals and institutions.

General Procedures

The basic requirements for registration are:

- A foreign firm that does not have offices in Singapore must appoint a local agent, distributor, or dealership to handle its products. This local agent must register with the Pharmaceutical Department.
- There are 16 categories of supplies. For some categories, including dental, laboratory, and medical equipment, the applicant must employ appropriate technical personnel and have relevant facilities to service and maintain the medical equipment.
- A firm is limited to participation in a tender based on the financial category in which it is registered.

Detailed information and applications for registration may be obtained from:

Procurement and Distribution Section
Pharmaceutical Department
2 Jalan Bukit Merah
Singapore 0316
Tel: [65] 2297607 Fax: [65] 2226797

INVITATIONS TO BID

The Singapore government attempts to influence the pattern of investment and industrial activity through the use of incentives, but it does not engage in detailed central planning. Almost all government purchasing is done by tenders. Invitations to bid on government projects are published in the *Singapore Government Gazette*. Some invitations are restricted to contractors registered with the department initiating the request; others are open to general bidding. Complete instructions for obtaining tender documents and specifications are issued at the time invitations are published.

The *Singapore Government Gazette* is available from:

Singapore National Printers Ltd.
8 Shenton Way #B1-07
Treasury Building
Singapore 0106
Tel: [65] 2230834

SPECIAL TRADE ZONES

OVERVIEW

Singapore is generally classed as a free port and has an open economy. There are no restrictive or discriminatory trade or investment policies of any consequence, except with regard to a selected number of products and some service industries. There is no capital gains tax, turnover tax, value-added tax, development tax, or surtax on imports. Nor are there requirements that foreign investors have local joint venture partners in either the private or public sectors, or relinquish control of management to local interests. Foreign firms have ready access to credit as well as modern banking and financial services.

FREE TRADE ZONES

Singapore's free trade zones, which have been in operation since 1969, provide a full range of facilities and services for the storage and reexport of dutiable goods. There are five free trade zones for seaborne cargoes (PSA gateways and Jurong Wharves) and one for air cargoes (Singapore Changi Airport).

The free trade zones for seaborne cargoes facilitate entrepôt trade and promote the efficient handling of transshipments. They offer 72 hour storage for conventional import and export cargoes and containerized cargoes, and 28 days' free storage for transshipment cargoes intended for reexport.

SINGAPORE SCIENCE PARK

Since its inception in 1983 the Singapore Science Park has become the hub of the republic's research efforts. It is the focal point of high-tech R&D and information services, such as software development and consulting.

The demand for R&D space in Singapore is expected to grow by some 45,000 square meters annually over the next 10 years. This surging demand has prompted the Singapore government to embark on the S$291 million (about US$181 million) phase two of the Singapore Science Park even before the first phase is fully completed.

Phase I

A 30 hectare park offers ready-built research units, land leases, and office units designed exclusively for activities relating to software development. The park has consistently had occupancy rates exceeding 90 percent.

Located in southwestern Singapore along the Ayer Rajah Expressway, the Singapore Science Park is close to the central business district, institutions of higher learning, and many high-technology companies.

Phase II

Covering about 20 hectares along Bona Vista and Pasir Panjang Roads, the second phase will add 120,000 square meters of R&D space by the year 2001.

Phase III

A 13 hectare plot of land has been cleared for the third phase, and another 49 hectare plot is earmarked for future development, as needed.

R&D companies in the fields of biotechnology, computer and information technology, microelectronics, marine technology, and chemical and petrochemical engineering make up the majority of tenants in the Singapore Science Park.

Application Process

Prospective tenants wishing to lease land or facilities in the Singapore Science Park must submit an application to the Technology Parks Pte Ltd. and the National Science and Technology Board. If the application is approved, a letter of offer will be sent out within seven days of receipt of the application.

Prospective tenants have two weeks to accept the offer. Upon acceptance, they must make the following payments:

(a) three months' rent and service charge as deposit
(b) one month's rent and service charge in advance
(c) cost to prepare the tenancy agreement
(d) stamp duty

Once the offer has been accepted and all payments made, Technology Parks Pte Ltd. prepares the tenancy agreement. The agreement is forwarded to the tenant for signature and must be returned within two weeks to Technology Parks Pte Ltd. for stamping.

After a tenant takes possession of the premises, the tenant must submit any renovation plans to Technology Parks Pte Ltd. for approval. Technology Parks Pte Ltd. usually obtains approval for the plans within three weeks of submission. The tenant must also apply to the Singapore Telecom and the Public Utilities Board for utility supplies.

For further information, contact:

Technology Parks Pte Ltd.
Marketing Department
The Pasteur, 16 Science Park Drive, #02-01
Singapore 0511
Tel: [65] 7741033 Fax: [65] 7784761

Industrial Zones

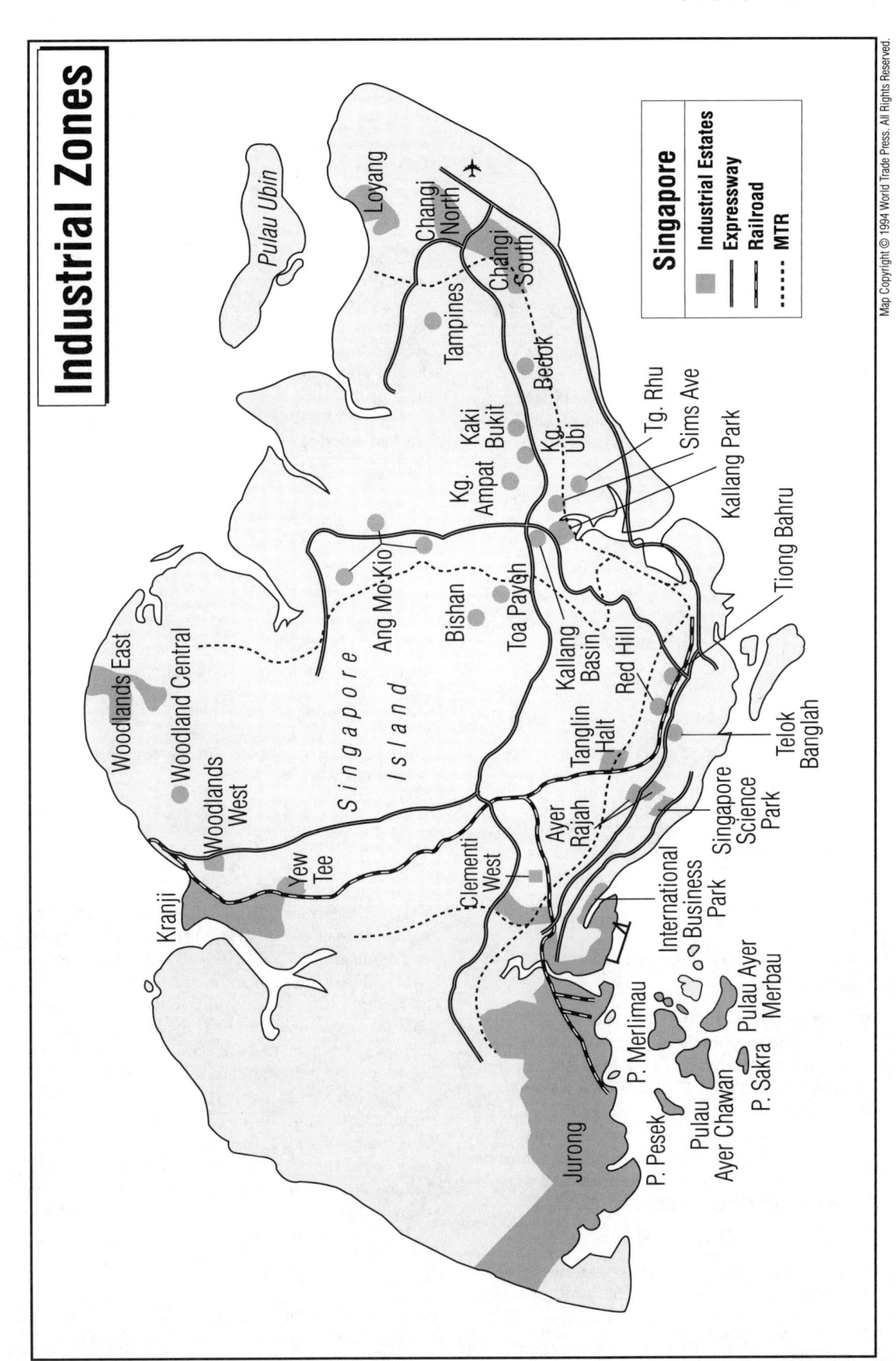

Singapore

- Industrial Estates
- Expressway
- Railroad
- MTR

Loyang, Changi North, Changi South, Pulau Ubin, Tampines, Kg. Ampat, Kaki Bukit, Kg. Ubi, Bedok, Tg. Rhu, Sims Ave, Kallang Park, Ang Mo Kio, Bishan, Toa Payoh, Kallang Basin, Red Hill, Tiong Bahru, Tanglin Halt, Ayer Rajah, Telok Blangah, Singapore Science Park, Woodlands East, Woodland Central, Woodlands West, Yew Tee, Clementi West, International Business Park, Kranji, P. Merlimau, Pulau Ayer Chawan, P. Sakra, Pulau Ayer Merbau, P. Pesek, Jurong

Singapore Island

INDUSTRIAL ESTATES

Currently 30 industrial estates house more than 4,500 international companies in Singapore. These industrial estates are managed by the Jurong Town Corporation (JTC). Established in 1968, JTC is Singapore's principal developer and manager of industrial estates and related facilities.

Manufacturers who wish to set up their operations in Singapore can choose from a wide range of industrial facilities:

- Industrial land is for companies that prefer to build their own factories.
- Standard factories are pre-built, single-story facilities.
- Flatted factories are multi-story facilities for light, non-polluting industries, and are generally located in centers of high population density.

The three largest industrial estates in terms of area are Jurong (5,436 hectares), Sungei Kadut (366 hectares), and Sungei East (248 hectares). The three largest industrial estates in terms of number of companies are Jurong (2,256), Kallang Basin (754), and Sungei Kadut (204).

For further information, contact:

Mr. David Lim, Chief Executive Officer
Jurong Town Hall
301 Jurong Town Hall Road
Singapore 2260
Tel: [65] 5600056 Fax: [65] 5653906

INTERNATIONAL BUSINESS PARKS

A business park is an area set aside for non-polluting industries and businesses engaged in high-technology, R&D, or other high value-added, knowledge-intensive activities.

A business park differs from an industrial estate in the following ways:

- The range of permitted uses is generally limited to non-production activities, which are characteristic of high-technology and research industries.
- The emphases on landscaping, quality building design, and amenities are intended to reflect the added importance placed by companies on the image of the business park and the welfare of the companies' employees.

Qualifications

A company that engages in any of the following activities is eligible for a site within a business park:

- advanced manufacturing of high-technology products
- laboratory testing
- research and development
- product design and development
- data processing and other computing activities
- software development
- industrial training
- central distribution

Development Plans

Business parks are developed on sites that have good support facilities and that capitalize on linkages with complementary activity centers. Business park sites must also have excellent accessibility to public transport, good quality recreational and residential areas, and proximity to major educational institutions.

Some of the sites that development agencies are now considering for business park development include:

- Ayer Rajah
- Jalan Bahar
- Jurong East
- Outram Park
- Seletaar
- Tampines
- Yishun North

For further information, contact:

Mr. Han Chiaw Juan
Senior Executive Lands Officer
Jurong Town Corporation, Jurong Town Hall
301 Jurong Town Hall Road
Singapore 2260
Tel: [65] 5600056 Fax: [65] 5681992

TELETECH PARK

The Singapore Telecommunications Authority has announced plans to establish a complex for telecommunications R&D by the end of 1995. Teletech Park will be located in Jurong East and will consist of a single low-rise building with 24,000 square meters of rental space for R&D and small-scale manufacturing. The facilities will feature satellite communications and video conference capabilities.

Teletech Park is part of the government's plan to establish Singapore as a global and regional hub for telecommunications. The new complex was first announced at the opening of the Asia Telecom 93 Exhibition, which attracted 360 exhibitors from more than 60 countries.

For further information, contact:

Mr. Lim Choon Sai, Director
Telecommunication Authority of Singapore
31 Exeter Road, #05-00 Comcentre
Singapore 0923
Tel: [65] 7387788 Fax: [65] 7330073

Foreign Investment

INVESTMENT CLIMATE AND TRENDS

Singapore excels at attracting foreign investment. Not only does the city-state give investors numerous tax and nontax incentives, but it also allows extensive freedom of commercial activity, has a highly developed infrastructure, has an educated and productive work force, and offers political and economic stability. In 1992 for the fourth consecutive year, the *World Competitiveness Report* rated Singapore the most competitive of 14 newly industrializing economies. The *Report* is produced annually by two Swiss-based organizations, the Institute for International Management Development and the World Economic Forum.

The Singaporean government encourages business and investment by keeping regulations to a minimum. There are no exchange controls, and foreign investors may either import capital or raise funds locally. Capital and profits are allowed to move freely within the country or abroad, and there is no capital gains tax, turnover tax, value-added tax, development tax, or surtax on imports entering Singapore.

For years Singapore has pursued an aggressive industrialization program, and it hopes to reach the status of advanced industrialized nation by the turn of the century. To achieve this goal, the country will continue to promote foreign investment in skill-intensive and technology-intensive industries and enterprises. Moreover, its investment promotion strategies are increasingly focused on the development of such service industries as banking, accounting, the transport of air and sea cargo, and on making Singapore a regional operating center for major international corporations.

The high-technology manufacturing industries in which Singapore has established itself and continues to promote include automation and artificial intelligence, biotechnology, communications technology, information technology, lasers and optics, and microelectronics.

The service sector has become increasingly important in Singapore, and it now contributes roughly twice as much to GNP as does manufacturing. The service industries in which Singapore already has a reputation and that it continues to promote are advertising, public relations, and exhibition services; agrotechnology services; air and sea transportation and distribution services; banking and financial services; education and training services; laboratory and testing services; legal and accounting services; leisure, entertainment, and cultural services; management consulting services; medical services; technical and engineering consulting services; and telecommunications and information services.

Specialized financial services are particularly important to Singapore's development strategy. The plan is to make the country the primary financial center for Southeast Asia and one of the leading financial capitals of the world. Specialized financial services that are considered to have the greatest potential for growth are capital markets, financial and commodity futures, financing of third-country trading, funds management, reinsurance and captive insurance, and risk management.

Singapore will undoubtedly remain an attractive location in which foreigners can invest. Its proximity to China, India, Indonesia, and other rapidly growing Asian markets makes it an excellent springboard for operations in that region. In fact, the Singaporean government recently began to promote outward investment to such countries as China. In summer 1993 a group of Singaporean companies signed a joint venture agreement with provincial leaders in Suzhou, near Shanghai in China. The plan is to develop a self-contained industrial, commercial, and residential estate modeled on Singapore itself.

LEADING INVESTORS

While foreign direct investment in Singapore has generally been both consistent and substantial during the last couple of decades, investment has increased dramatically since the mid-1980s. In 1985 total foreign direct investment amounted to approximately US$1 billion. Between 1988 and 1992, annual investment fluctuated between US$3 billion and US$4 billion. Of the roughly US$4 billion invested from abroad in 1992 the United States accounted for 38 percent, and Japan and the European Community accounted for 22 percent and 16 percent respectively. Commitments by overseas investors play a critical role in the expansion of Singapore's economy. In 1992 foreign investments represented 81 percent of all investment in the country, with only 19 percent coming from local investors.

INVESTMENT POLICY

The Singaporean government actively promotes and facilitates foreign investment. Investment in manufacturing and service industries that utilize high technologies is particularly encouraged. Foreign ownership and participation is restricted for national security reasons in such areas as air transport, public utilities, weapons and munitions manufacture, newspaper publishing, and banking (a 40 percent limit has been placed on the foreign ownership of locally incorporated banks). Specific firms can and do restrict the amount of foreign participation through their bylaws. However in general, foreign capital is treated equally with local capital, and foreign nationals may acquire partial or total ownership of local enterprises.

INVESTMENT INCENTIVES

Singapore offers many incentives aimed at attracting foreign investment into the country. These include tax concessions, research and development incentives, export incentives, and special loan and grant programs. Incentives are used both to promote new investment in industries and services and to encourage existing companies to upgrade through automation or the introduction of new products and services.

Tax Incentives

The Economic Development Board (EDB) has primary responsibility for the planning and promotion of industrial and commercial development. This section reviews 10 incentive programs administered by the EDB.

Incentive Pioneer Status.
Eligibility Requirement Enterprises engaged in manufacturing or service activities that introduce advanced technology, know-how, or skills to the industry.
Tax Concession Exemption for a period between five and ten years from the 31 percent tax on profits arising from the pioneer activity.

Incentive Postpioneer Status.
Eligibility Requirement Enterprises that enjoyed pioneer status as of April 1, 1986 or later and in which additional investments are made.
Tax Concession Corporate tax rate of 15 percent for up to 10 years after expiration of pioneer status.

Incentive Investment Allowance.
Eligibility Requirement Qualifying period of up to five years within which specified investments (research and development, new construction, or a project to reduce the use of potable water) must be made.
Tax Concession Exemption of taxable income by an amount equal to as much as 50 percent of new fixed investment.

Incentive Expansion.
Eligibility Requirement Minimum investment of S$10 million (about US$6.2 million) in productive (manufacturing) equipment and machinery.
Tax Concession Exemption for as many as five years from the 31 percent tax on profits in excess of the those made prior to expansion.

Incentive Export of Services.
Eligibility Requirement Enterprises providing services from a Singaporean base to an overseas project; at least 20 percent of total revenues must result from exports.
Tax Concession Ninety percent of the qualifying export income exempted from tax for five years, with the possibility of extension.

Incentive Operational Headquarters.
Eligibility Requirement Enterprises having their operational headquarters in Singapore, managing related companies outside Singapore, and providing approved services to the overseas companies from Singapore.
Tax Concession Taxation of income arising from the provision of approved services at a rate of 10 percent for up to 10 years; other income arising from overseas subsidiaries and affiliated companies may also be eligible for tax relief.

Incentive Approved Foreign Loan.
Eligibility Requirement Receipt of a minimum loan of S$200,000 (about US$125,000) from a foreign lender for the purchase of manufacturing equipment; tax relief should not result in an increase in tax liability in the foreign country.
Tax Concession Exemption of withholding tax on interest.

Incentive Approved Royalties.

Eligibility Requirement A foreign individual or firm that derives royalties as a result of manufacturing or service activities in Singapore; tax relief should not result in an increase in tax liability in the foreign country.

Tax Concession Full or partial exemption of withholding tax on royalties.

Incentive Venture Capital.

Eligibility Requirement Enterprises must be at least 50 percent owned by citizens of Singapore and incorporated in Singapore for tax purposes.

Tax Concession Losses incurred from the sale of shares, up to 100 percent of invested equity, can offset the investor's other taxable income.

Incentive International Direct Investment.

Eligibility Requirement Enterprises must be at least 50 percent owned by citizens of Singapore and incorporated in Singapore for tax purposes.

Tax Concession Losses incurred from the sale of shares or liquidation of the overseas company, up to 100 percent of invested equity, can offset the investor's other taxable income.

Incentive Schemes

Singapore also offers a variety of incentive schemes to encourage investment in local industries and enterprises. Investors interested in being considered for the incentive programs reviewed in the paragraphs that follow should apply using the appropriate forms to the EDB, unless otherwise indicated.

Product Development Assistance Scheme The EDB established the Product Development Assistance Scheme to encourage local product design and development capability and enhance Singapore's technological base. The scheme provides cash grants to Singapore-based enterprises that develop new products (including computer hardware and software) or that substantially improve existing products or processes related to their manufacturing activities.

Companies 30 percent or more owned by Singaporean citizens with a proven track record and sufficient resources to develop and market their products or services are eligible to apply. Individuals are also eligible to apply, but they must be prepared to form a business entity (proprietorship, partnership, or company).

The proposed project must meet four basic criteria. First, the proposed product must be of high technical standards and conform to sound design practice. Second, the development team undertaking the project must demonstrate an in-depth knowledge of the technical and commercial aspects of the product. Third, the development work must be carried out mainly in Singapore by local engineers or designers, although foreign experts may lead or supervise the project if appropriate. Fourth, the product must be a marketable item suitable for more than a single customer.

The Product Development Assistance Scheme provides a grant equal to 50 percent of the approved, direct development costs. Most projects require grants less than S$200,000 (about US$125,000), that is, development costs of less than S$400,000 (about US$250,000).

Research and Development Assistance Scheme The Research and Development Assistance Scheme is a grant program designed to encourage medium-term research and development projects. Administered by the National Science and Technology Board, the program provides financial grants for enterprises that conduct research and development of technological and commercial value.

National Science and Technology Board
The Pasteur, 16 Science Park Drive, #01-03
Singapore 0511
Tel: [65] 7797066 Fax: [65] 7771711

Enterprises registered in Singapore can apply on their own or in collaboration with the public sector. The level of funding depends on the project's overall contribution toward Singapore's technological capability, its technological content, and its commercial potential. Products arising from the project should be manufactured in Singapore, and processes developed should be utilized there as well.

The scheme provides funding of up to 70 percent of the total direct project cost, including the cost of manpower, equipment, and miscellaneous R&D expenses. If the project results in a marketable product or process, the enterprise pays a royalty fee ranging from 0.5 percent to 3 percent of the annual revenue derived from the sales of the product or process developed and a proportion of the licensing fees. The royalty payment period is 15 years from the date on which the item is brought to market. The maximum sum that can be collected is twice the amount of the grant funding.

Accelerated Depreciation Allowances An annual allowance of 33.3 percent can be claimed over three years for a plant and all associated machinery. An accelerated depreciation allowance of 100 percent (a one-year write-off) can be claimed for computers, prescribed automation equipment, and robots. Industrial building allowances can be depreciated over 25 years.

Local Enterprise Finance Scheme The Local Enterprise Finance Scheme is a fixed-rate, low-cost financing program administered by the Enterprise Development Division of the EDB. Designed to assist and encourage small- and medium-sized local enterprises to upgrade and expand their operations, it was formerly known as the Small Industry Finance Scheme. The program was renamed and expanded to increase accessibility and raise the loan limit.

To make the program available to as many qualifying enterprises as possible, the EDB operates it jointly with 33 participating banks and other financial institutions: Bank of Singapore, Chung Khiaw Bank, DBS Bank, DBS Factors, DBS Finance, ECICS Holdings, Far Eastern Bank, FOCAL Finance, Four Seas Bank, Great Pacific Finance, Hong Leong Finance, Indian Bank, Industrial & Commercial Bank, International Bank of Singapore, International Factors Marine, International Factors, Keppel Bank, Keppel Factors, Keppel Finance, Lee Wah Bank, OCBC Finance, OUB Factors, Overseas-Chinese Banking Corporation, Overseas Union Bank, Overseas Union Trust, SAL Industrial Leasing, Sing Investments & Finance, Singapura Building Society, Standard Chartered Bank, Standard Chartered Finance, Tat Lee Bank, United Overseas Bank, and United Overseas Finance.

To be eligible for a loan under the program, a company must have at least 30 percent active local ownership and fixed productive assets (defined here as factory building, machinery, and equipment) worth not more than S$12 million (about US$7.5 million). Service enterprises may not employ more than 50 workers. The loans must be used for one of the following purposes: to establish a viable new business, to modernize and automate plant and machinery, to expand existing manufacturing capacity, to diversify into other product lines, or to augment working capital.

Interest rates range from 6.5 percent to 7 percent. However, projects that involve the purchase of equipment or machinery that can achieve one or more of the following goals may enjoy a preferential rate of 3.5 percent: substantial savings in labor usage, significant increase in output per worker, or introduction of more sophisticated and skilled operations.

An enterprise with fixed productive assets not exceeding S$3 million (about US$1.9 million) can apply for a maximum loan of S$8 million (about US$5 million). An enterprise with fixed productive assets of more than S$3 million but below S$12 million (about US$7.5 million) can apply for a maximum loan of up to S$6 million (about US$3.75 million).

Local Enterprise Technical Assistance Scheme The Local Enterprise Technical Assistance Scheme, formerly known as the Small Industry Technical Assistance Scheme, assists small- and medium-sized local enterprises in seeking external expertise, including the expertise of a foreign national, for the purpose of modernizing and upgrading their operations. It is administered by the EDB. To be eligible, a company must have at least 30 percent active local ownership and fixed productive assets not exceeding S$12 million (about US$7.5 million). Service enterprises may not employ more than 50 workers.

Under this program, the EDB reimburses up to 70 percent of the cost of employing an external expert for an approved limited-term assignment. The purpose of the assignment must be to upgrade business operations or impart technology or skills. The size of the grant awarded depends on the scope, depth, and effectiveness of the short-term assignment and on its relevance to Singapore's economic development objectives. Generally, grants are based on 30, 50, or 70 percent of the fees paid to the external expert(s) and the cost of their expenses (return airfares and accommodations) in Singapore.

Business Development Scheme The Business Development Scheme was introduced to help local firms offset the cost of pursuing opportunities that fall outside the scope of other schemes. Its primary focus is in identifying international markets. Assistance takes the form of grants covering the costs of studies or overseas visits to explore new technologies or markets, establish new business contracts, pursue joint venture arrangements, or participate in approved business development seminars.

To be eligible, a company must have at least 30 percent active local ownership and fixed productive assets not exceeding S$12 million (about US$7.5 million). Service enterprises may not employ more than 50 workers. Under this program, the EDB covers up to 50 percent of the cost of return airfare and a cost-of-living allowance for overseas trips if applicable (in practice, only one overseas trip per project is supported, and each company is limited to one overseas trip per year); up to 50 percent of the cost of acquiring reports, data, and intelligence on markets, new technologies, and joint venture partners; and up to 70 percent of the cost of participating in approved business development seminars or workshops.

Double Tax Deduction Scheme The Singapore Trade Development Board (STDB) administers the Double Tax Deduction Scheme under sections 14B and 14C of the Income Tax Act. It aims to assist manufacturers and traders for the purpose of promoting the export of products made in Singapore. Companies that are involved in the following activities can apply for a double tax deduction: overseas trade fairs and trade missions, approved local trade fairs, overseas trade offices, development of export markets, advertisement in approved Singapore export promotional publications, or publication of promotional brochures.

Tax Concessions for Nonresidents

There are no restrictions on the remittance of interest earned by nonresidents on accounts with Singaporean banks, and such interest is not liable to taxation in Singapore. Nonresident deposits with Asian Currency Units (ACUs) and holdings of approved Asian dollar bonds and Singaporean government tax-free bonds are exempt from Singaporean estate duty.

LOANS AND CREDIT AVAILABILITY

Many sources of finance are available to foreign investors in Singapore. The options include equity and loan financing through local and foreign financial institutions; such capital market instruments as note issuance facilities (NIFs), revolving underwriting facilities (RUFs), floating rate notes (FRNs), and negotiable certificates of deposit (NCDs); and venture capital and development financing.

The Monetary Authority of Singapore (MAS), which serves as the country's central bank, requires all foreign and domestic banks to observe its policy of discouraging the internationalization of the Singapore dollar. Banks must consult the MAS before granting nonresidents credit in Singapore dollars in excess of S$5 million (about US$3.1 million) when they plan to use Singapore dollars outside the country. Prior consultation is not required for credit in the local currency, regardless of the amount, when the funds are used for direct exports from or imports into Singapore or as guarantee of payment arising from construction or other economic activities in Singapore. Enterprises incorporated in Singapore that are majority owned, jointly owned, or otherwise controlled by foreign nationals are classified as nonresident.

Monetary Authority of Singapore (MAS)
10 Shenton Way, MAS Building
Singapore 0207
Tel: [65] 2255577 Fax: [65] 2299491
Tlx: 28174 ORCHID

With this exception, foreign investors generally have equal access to local finance, may invest in government securities, and may purchase shares in companies listed on the stock market. However, foreigners are limited to a cumulative stake of 40 percent in Singaporean incorporated banks and many listed Singaporean firms limit foreign participation through their bylaws.

COMMERCIAL AND INDUSTRIAL SPACE

Singapore's large export trade and the rapid increase in investment in manufacturing and services have created strong demand for commercial and industrial space. Most commercial and industrial property is owned by the government and managed by the Jurong Town Corporation (JTC), a government development agency.

Jurong Town Corporation (JTC)
301 Jurong Town Hall Road
Jurong Town Hall
Singapore 2260
Tel: [65] 5600056 Fax: [65] 5655301 Tlx: 35733 JTC

Factories and Industrial Estates The JTC runs 30 industrial estates, in which more than 4,500 companies are located. Manufacturers who wish to set up industrial facilities have three options: They can lease industrial land and build their own factories; small- and medium-sized companies can lease standard single-story factories of various sizes and designs; or they can lease multistory facilities, which are suitable for light, nonpolluting industries and which are normally located in areas of high population density.

Factory buildings in the industrial estates range in floor area from 913 square meters to 4,055 square meters, while the factory sites themselves range in size from 1,420 square meters to 8,000 square meters. Entire factory buildings sell for anywhere from about S$911,000 (about US$568,000) for the smallest buildings to S$4,016,300 (about US$2.5 million) for the largest. Monthly rentals for land and buildings in the industrial estates (inclusive of property tax) ranges from S$12,630 (about US$7,900) to S$58,480 (about US$36,400).

Companies that rent a factory building commonly have one of two types of leases: a 30-year lease or a 30-plus-30-year lease. For a 30-year lease, the company must meet the following within five years after the date on which the lease begins: it must invest a minimum of S$625 (about US$390) per square meter of the gross building floor area in the building and a minimum of S$200 (about US$125) per square meter of the allocated land in machinery and equipment. A company with a 30-plus-30-year lease must invest a minimum of S$700 (about US$436) per square meter of the gross building floor area in the building and a minimum of S$400 (about US$250) per square meter of the allocated land in machinery and equipment.

If the tenant qualifies for the next 30-year option, the rental rate on the land is revised to the market rate prevailing on the date on which the option term began. Thereafter, the annual increase is limited to 7.6 percent of the preceding year's rent.

International Business Park The JTC is developing Singapore's first business park at a 40 hectare site in Jurong East. The International Business Park is currently under construction, and the first business building will be available for rent in early 1995. The intent of the park is to integrate a variety of business functions in a single location. Once the park is complete, firms will be able to conduct research and development, make use of advanced testing laboratories, and conduct industrial training. Prepared land sites are also available for lease to companies that wish to build their own facilities.

Office Rental Rates In Singapore's central business district, office space rents from a low of S$80 (about US$50) per square meter per month to a high of S$120 (about US$75) per square meter. In areas outside the central business district, monthly rates range from S$62 (about US$39) to S$71 (about US$44) per square meter.

The purchase prices, rental rates, and space availability cited in this section were correct as of the time of writing. Changes are to be expected.

INVESTMENT ASSISTANCE

Two official government agencies have primary responsibility for promoting trade and assisting foreign investors: the Economic Development Board (EDB) and the Singapore Trade Development Board (STDB).

Economic Development Board The Economic Development Board specializes in aiding small- and medium-sized businesses in the manufacturing and service sectors. Large companies will also find the EDB a useful source of information. Aside from evaluating applications for tax and other incentive programs, the EDB assists investors in obtaining land, factories, and office space; financing; and skilled workers; as well as in locating suppliers, subcontractors, and joint venture partners. The EDB has offices in major cities throughout the world. (Refer to "Important Addresses" chapter for a listing of EDB offices worldwide).

Economic Development Board
250 North Bridge Road, #24-00
Raffles City Tower
Singapore 0617
Tel: [65] 3362288 Fax: [65] 3396077
Tlx: 26233 SINEDB

Singapore Trade Development Board The Singapore Trade Development Board aims to promote Singapore as an international trade center. It helps both local and foreign businesses and other investors to establish a base of operations in Singapore. The STDB disseminates timely trade and market information through its on-line information services, newsletters, news releases, talks, and seminars. Like the EDB, the STDB maintains a network of offices around the world. (Refer to "Important Addresses" chapter for a listing of STDB offices worldwide.)

Singapore Trade Development Board
World Trade Centre
1 Maritime Square, #10-40
Telok Blangah Road
Singapore 0409
Tel: [65] 2719388 Fax: [65] 2740770
Tlx: 28614 TRADEV

Singapore International Chamber of Commerce The Singapore International Chamber of Commerce (SICC), established in 1837, is the oldest private representative organization in Singapore. Membership is open to companies and individuals of all countries. Currently, 44 nationalities are represented, including US (15 percent), British (12 percent), and Japanese (10 percent). The SICC enjoys an unusually high status in Singapore for an organization of its kind. Regularly consulted by government officials on matters that have a direct bearing on the interests of the private sector, including draft legislation or new government rules and regulations, its members include representatives of government advisory committees and statutory boards.

The SICC provides four publications that are of use to investors:

- *The Economic Bulletin* is a monthly publication providing news, including the latest available trade statistics, about Singapore and other surrounding regions of interest to businesspeople and industrialists.
- *The Investor's Guide* to Singapore is an annual publication that explains in detail the investment requirements, incentives, and opportunities in Singapore.
- *Expatriate Living Costs in Singapore* is updated regularly and provides information on the costs of living of expatriates in Singapore.
- Finally, the SICC's *Annual Report* reviews the performance of Singapore's major economic sectors each year.

The SICC also offers numerous services. For example, it provides certificates of origin and other shipping documents, maintains a trade inquiry register that can match importers and exporters with local agents and outlets, and, in conjunction with the International Chamber of Commerce Court Arbitration in Paris, arbitrates trade disputes.

Singapore International Chamber of Commerce
6 Raffles Quay #05-00
Denmark House
Singapore 0104
Tel: [65] 2241255 Fax: [65] 2242785
Tlx: 25235 INTCHAM

Foreign Trade

Singapore has depended on trade since its founding by the British in 1819. With a limited area, consisting mostly of a low, swampy main island, and no natural resources, Singapore finds its only real advantages in its first-class harbor and its strategic location at the opening of the Strait of Malacca. This waterway connects the Indian Ocean and the South China Sea, making it a natural gateway to Southeast Asia. First the British and now modern-day traders have been quick to capitalize on these natural assets to turn an otherwise unprepossessing piece of real estate into a thriving trading port and international business and finance center.

Like Hong Kong, Singapore had no real economy prior to its founding as a European treaty port. It still had no self-sustaining economy when it became independent in 1965. In 1969 94 percent of Singapore's trade consisted of reexports—transshipments of items originating elsewhere and destined for delivery to another location without any significant intermediate processing in Singapore. Essentially no local economy existed to produce export goods or support the import of either consumer goods or inputs for production. By 1991 Singapore had such a booming domestic and international economy that reexports accounted for only 35 percent of its trade, far less than the almost 75 percent of trade accounted for by reexports in Hong Kong's economy. But this reduction represents not only the growth of a healthy domestic economy; it also reflects increased competition and a loss of revenues and influence, as exporters develop independent relations with their customers and bypass Singapore.

Despite its degree of development, Singapore's economy remains dependent on trade and therefore at the mercy of external factors. The city-state has had to get itself noticed and offer a unique selling proposition to sustain itself. One means of doing this has been Singapore's maintenance of a free port. Few restrictions exist on foreign exchange or imports, 96 percent of which enter duty-free. The only substantial export controls relate to restrictions imposed by other nations that require Singapore to limit its own exports of certain products, such as textiles.

Singapore also offers a range of incentives to attract overseas operations. It is a politically stable—some would say authoritarian—well-equipped business and financial center, with up-to-date infrastructure, communications, production, port, and related facilities.

During the late 1960s Singapore put its resources into developing light and medium industry to take advantage of its low-cost workforce. It has since made the development of that workforce a priority, and it is now highly educated and skilled but no longer as competitively priced. During the 1970s and 1980s Singapore's primary goal was to further develop its manufacturing economy, while adding another, secondary, dimension by becoming a service center as well. Although Singapore's manufacturing economy continued to grow rapidly in both the size and sophistication of its output, by the end of the 1980s it was becoming clear that lower-cost labor was draining basic production away from the city-state.

In the 1990s Singapore has shifted its attentions to high-technology manufacture and high-touch service, particularly financial and business services. It has also renewed its efforts to strengthen its position as the trade hub of Southeast Asia and serves as a base for more than 100 international trading firms. Singapore has offered incentives for approved oil traders, international commodities traders, and shippers that offer large companies hefty tax breaks based on the volume of business they conduct through Singapore. Similar incentives are available for firms engaging in Singapore-based countertrade activities. Singapore has also established a system of free trade zones, which include facilities for storage and handling of goods destined for reexport, and TradeNet, a computerized paperless system, to standardize and speed up the processing of documents in locally conducted trade transactions.

Although Singapore directed its resources to developing its domestic manufacturing capacity in the

Top 10 Imports by % Increase	
Commodity	**% Increase**
Animal & vegetable oils	36%
Non-ferrous metals	29
Perfume	20
Chemical products	17
Road vehicles	17
Transport equipment	16
Office & data machines	12
Fish	12
Rubber manufactures	12
Dyes & colors	12

Source: Foreign Trade Magazine

Top 10 Exports by % Increase	
Commodity	**% Increase**
Transport equipment	42%
Perfume	27
Dyes & colors	25
Office & data machines	20
Tobacco & manufactures	20
Road vehicles	18
Chemical products	13
Animal vegetable oils	13
Crude rubber	11
Misc. manufactured articles	10

Source: Foreign Trade Magazine

1960s through most of the 1980s, it never forgot its origins in or dependence on trade. In 1968 Singapore established a national trading company, INTRACO to facilitate and upgrade its presence in and service of trade. Between 1960 and 1991 total trade grew by an annual average rate of 11.5 percent.

EXPORTS AND IMPORTS

In 1991 Singapore, despite its small size, was the world's seventeenth-largest exporter and fifteenth-largest importer. In 1992 it had the highest ratio in the world of total trade to total economy, with trade representing a multiple rather than a fraction of its gross national product (GNP): 274 percent. By comparison, Hong Kong's total trade represented 246 percent of its GNP, while trade represented 76.9 percent of GNP in Taiwan and 53.8 percent in South Korea, both Asian countries that ostensibly focus their economies on international trade. By contrast total trade represents about 15.2 percent of GNP in Japan and 17.5 percent of that of the United States.

Nearly two-thirds of Singapore's imports represent capital goods destined either to upgrade its domestic economy, or as reexports, to develop the economies of its downstream trading partners. Roughly one-quarter represent raw material inputs, and only about 7 percent represent consumer goods destined for local consumption (even the majority of foodstuffs imported are processed for export). Electrical machinery, petroleum, and office and data processing machines account for somewhat more than one-third of imports, with the top ten categories representing about two-thirds of the total.

Because of its intermediate processing role, many of Singapore's exports represent different value-added forms of its imports. Singapore's top three exports—office and data processing machines, electrical machinery, and petroleum products, the same three categories that represent its top three im-

Top 10 Imports by Commodity (in S$ millions)			
Commodity	**1992**	**1991**	**% Change**
Electrical machinery	$17,396	$16,067	8%
Petroleum products	14,970	16,040	-7
Office & data machines	9,116	8,134	12
Telecommunications equipment	8,804	9,272	-5
General industrial machinery	5,618	5,449	3
Transport equipment	4,525	3,896	16
Misc. manufactured items	3,979	3,584	11
Power-generating machinery	3,798	3,762	1
Industrial machinery	3,458	3,653	-5
Textile manufactures	3,255	3,285	-1

Source: Foreign Trade Magazine

ports—account for nearly half of its exports, with office and data processing machines being the biggest value-added item. Exports within this category are more than 130 percent the value of imports. Together, the top ten categories of imports represent about three-quarters of the total.

Singapore specializes in the assembly of high-end consumer and business electronics and is the premier regional oil refiner and supplier. As such, it is highly dependent on external conditions and demand. For example, it experienced a surge in volume and profitability from petroleum sales during the Persian Gulf crisis in 1990 and 1991, although this business contracted after international oil markets returned to normal. Singapore has seen reduced demand for many of its end-user electronics products due to the global recession, changing tastes, and increased competition, although it has also seen a partially compensating rise in downstream demand for intermediate electronic components, such as semiconductors and printed circuit boards, from both specialized domestic manufacturers and overseas assembly competitors.

BALANCE OF TRADE

Dependent as it is on imported industrial inputs and consumer goods, Singapore has racked up a series of merchandise trade deficits that continue to grow in absolute terms if not in percent of total trade represented. Since the mid-1980s imports have outpaced exports by an average of 6.2 percent of total trade per year. Recently, the deficit has begun to accelerate, with the single largest problem being the strength of the yen. Japan is Singapore's largest supplier, and, although Japan buys a considerable amount from Singapore, trade between the two leaves Singapore in a substantial chronic deficit po-

Leading Exporters to Singapore* (in S$ millions)
*January-August 1992

Japan	
Machinery & transport equipment	$10,801
Transport equipment	1,352
Chemicals	985
Iron & steel	794
Scientific instruments	742
United States	
Machinery & transport equipment	$6,959
Chemicals	1,858
Transport equipment	1,254
Scientific instruments	727
Food, drink & tobacco	536
Malaysia	
Machinery & transport equipment	$5,268
Petroleum products	1,288
Food, drink & tobacco	801
Clothing & footwear	735
Animal & vegetable oils & fats	475
Taiwan	
Machinery & transport equipment	$1,556
Textile yarn, fabrics & manufactures	444
Chemicals	201
Food, drink & tobacco	160
Scientific instruments	126

Note: Oil imports from Saudi Arabia in this period totaled S$3.7 billion.
Source: Foreign Trade Magazine

Singapore's Top 10 Exports by Commodity (in S$ millions)

Commodity	1992	1991	% change
Office & data machines	$21,113	$17,652	20%
Electrical machinery	13,656	12,745	7
Petroleum & products	13,360	17,191	-22
Telecommunications equip.	11,897	11,797	1
General industrial machinery	3,462	3,289	5
Misc. manufactured articles	3,045	2,757	10
Clothing	2,948	3,008	-2
Organic chemicals	2,347	2,276	3
Industrial machinery	1,833	1,963	-7
Power-generating machinery	1,821	1,696	7

Source: Foreign Trade Magazine

sition. Fortunately for Singapore, world oil prices have remained relatively low and oil can be purchased with relatively cheap US dollars.

Despite Singapore's growing structural deficit, its overall accounts remain favorable and its international reserves are not only strong but growing, having more than doubled since 1985. Singapore had foreign currency reserves of US$41.4 billion in March 1993, almost as large of those held by China. These reserves were almost equal to its GDP and had grown by nearly 20 percent from the previous year. All in all, Singapore's international reserves have increased at an average rate of 10.9 percent since 1985.

Singapore has been able to sustain substantial merchandise trade deficits while increasing its foreign reserves primarily because it has negligible foreign debt to act as a drain on its holdings. With its high internal savings rate, much of it due to mandatory contributions to the Central Provident Fund, Singapore has not had to borrow on international markets for years. It is also becoming less dependent on revenues from merchandise sales because of burgeoning intangibles revenues, such as business and financial services. The service sector now accounts for more than 60 percent of the economy, and it is growing at a rate faster than that of the economy as a whole. Much of that growth comes from servicing overseas traders and other businesses, providing added revenues to offset the merchandise deficit.

TRADE PARTNERS

The United States, Malaysia, the European Community (EC), and Japan have been Singapore's main trading partners for the last several years. However, in 1992 Hong Kong edged past Japan in its consumption of Singaporean exports. In 1992 the top three export markets collectively took 43 percent of the total, down from 57 percent in 1991, an indication of greater diversification in Singapore's export trade. Singapore's top three sellers of imports provided 54.4 percent of the total in 1992, down from 64.4 percent in 1991. Among exporters, the United States provided 21.9 percent, the largest percentage. Neighboring Malaysia was next (12.9 percent), followed by Hong Kong (8.2 percent), Japan (7.9 percent), Thailand (6.5 percent), Germany (4.4 percent), Taiwan (4.3 percent), the Netherlands (3.3 percent), the United Kingdom (2.9 percent), and Australia (2.5 percent) in 1992. Together, these 10 countries account for 74.8 percent of all export purchases. Remaining buyers of Singaporean goods each account for less than 2.5 percent of total exports.

In 1992 major import sources included Japan (21.9 percent), the United States (17.2 percent), Malaysia (15.3 percent), Saudi Arabia (5.3 percent), Taiwan (4.2 percent), Thailand (3.9 percent), South Korea (3.5 percent), Germany (3.5 percent), China (3.3 percent), and Hong Kong (3.2 percent). The top ten individual importing countries together account for

Singapore's Leading Trade Partners

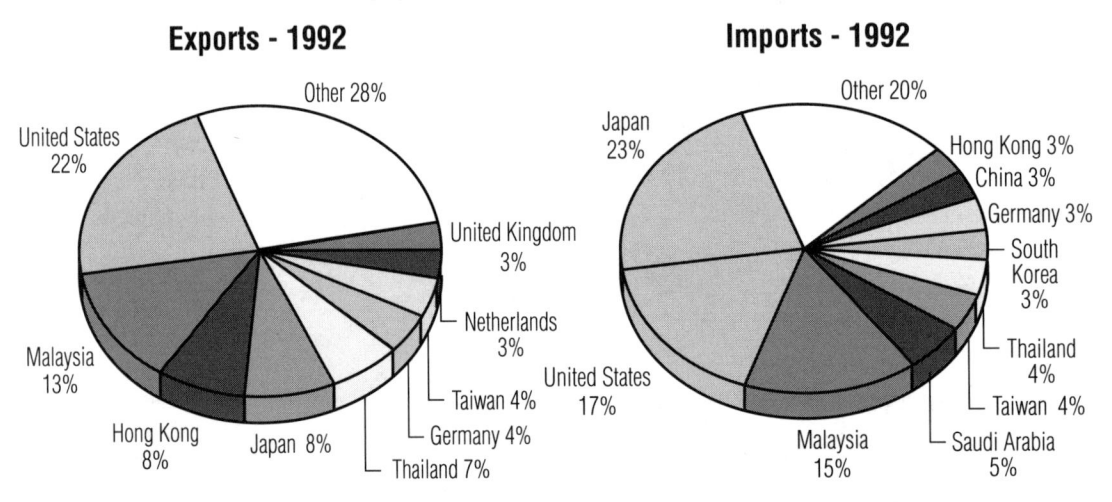

Exports - 1992

- Other 28%
- United States 22%
- Malaysia 13%
- Hong Kong 8%
- Japan 8%
- Thailand 7%
- Germany 4%
- Taiwan 4%
- Netherlands 3%
- United Kingdom 3%

Total 1992 Exports: US$ 61.1 Billion

Imports - 1992

- Other 20%
- Japan 23%
- United States 17%
- Malaysia 15%
- Saudi Arabia 5%
- Taiwan 4%
- Thailand 4%
- South Korea 3%
- Germany 3%
- China 3%
- Hong Kong 3%

Total 1992 Imports: US$ 69.3 Billion

Source: Foreign Trade Magazine
note: Shares rounded to the nearest whole percent

Singapore's Imports by Country (in S$ millions)			
Country	1992	1991	% Change
Japan	$24,753	$24,370	2%
United States	19,339	18,030	7
Malaysia	17,287	17,382	-1
Saudi Arabia	6,018	5,864	3
Taiwan	4,721	4,681	1
Thailand	4,365	3,629	20
South Korea	3,869	3,241	19
Germany	3,839	3,650	5
China	3,668	3,839	-4
Hong Kong	3,587	3,434	4
United Kingdom	3,281	3,286	0
France	2,926	2,932	0
Italy	2,088	1,575	33
Australia	1,996	2,149	-7
Switzerland	1,203	1,052	14

Source: Foreign Trade Magazine

Singapore's Exports by Country (in S$ millions)			
Country	1992	1991	% Change
United States	$21,779	$20,103	8%
Malaysia	12,925	15,236	-15
Hong Kong	8,081	7,346	10
Japan	7,857	8,836	-11
Thailand	6,442	6,401	1
Germany	4,389	4,263	3
Taiwan	4,188	3,621	16
Netherlands	3,217	2,620	23
United Kingdom	3,003	3,082	-3
Australia	2,457	2,516	-2
China	1,811	1,485	22
France	1,582	1,190	33
India	1,524	1,727	-12
Brunei	1,079	956	13
Italy	1,065	963	11

Source: Foreign Trade Magazine

81.3 percent of Singapore's total imports. Remaining suppliers each account for less than 3 percent of total imports.

One of Singapore's stated goals is to diversify its trade to reduce its current dependence on major trade partners. As part of this effort, it has been making overtures to a variety of nations, most of them underserved, emerging economies. Singapore has sent trade delegations to countries in Africa, Latin America, and the Middle East. It is also trying to penetrate markets in Eastern Europe and is eyeing developments in the former Soviet republics. But its primary goal is to exploit relationships with neighbors in Southeast Asia and other Asian markets to which it has cultural links, such as China and India. Singapore had remained aloof from Laos, Cambodia, and Vietnam largely for ideological reasons, but it has recently opened negotiations to trade and manage investment projects in these countries and in Myanmar.

GOVERNMENT STRATEGY AND DEVELOPMENT PROJECTS

Singapore hopes to become the Switzerland of the Far East and to equal the economic level of the United States by no later than 2030. To accomplish these goals, it plans to upgrade its manufacturing capabilities, increase its attractiveness as a regional operating center for international businesses, and boost its role in international trade. Singapore's stated position is that it relies solely on the operation of free markets to accomplish its developmental goals. In reality, it has been somewhat more directive in its management of the economy. The government provides a large number of incentives, and it exerts a considerable degree of control in the process. Its current focus is on the development of a high technology, high-value-added manufacturing industry, coupled with a state-of-the-art services sector.

Singapore does appear to be priming the pump through massive government spending on construction and infrastructure projects to provide a floor under economic activity while it adjusts to market fluctuations and new international market realities. Although Singapore has extensive internal managerial skill in such development projects, it will require considerable inputs of materials, many of them fairly sophisticated items, to construct the built-in smart information and communications capabilities that it envisions for its new projects. Although the government wants to provide a very high level of such amenities for its people, its primary aim is to make Singapore a user-friendly business environment to encourage international finance, business, and trade operations to use it as a regional operating base.

INTERNATIONAL TRADE ORGANIZATION MEMBERSHIPS

Singapore's primary affiliation is with the Association of Southeast Asian Nations (ASEAN), a developing regional trade block. Although ASEAN's effectiveness as a trade bloc is still in its infancy, it has set up a system of preferential duties and is actively trying to promote trade and development of markets among its member nations. Singapore is also a contracting member of the General Agreement on Tariffs and Trade (GATT). It has diplomatic relations with more than 60 nations and is represented at the United Nations, participating in most of that body's major agencies. The Singapore Trade Development Board (STDB), an official agency charged with promoting trade with Singapore, and the Economic Development Board (EDB), the agency responsible for facilitating and regulating foreign investment, maintain offices worldwide. The Singapore Federation of Chambers of Commerce and Industry (SFCCI) is a private organization with quasi-official standing that also promotes Singaporean businesses.

Import Policy and Procedures

INTRODUCTION

Singapore currently has no import quotas, and almost all products—except those controlled for health, safety, or security reasons—can be imported without restriction. Singapore continues to seek ways to improve its warehousing and distribution facilities, simplify customs operations and import documentation procedures, and attract respected international trading houses. One example of such efforts is TradeNet, an electronic trade documentation system that allows businesses to receive and submit documents via modem.

REGULATORY AUTHORITY

The Singapore Trade Development Board (STDB) is part of the Ministry of Trade and Industry (MTI) and has jurisdiction over all aspects of foreign trade.

Singapore Trade Development Board
1 Maritime Square
#10-40 (Lobby D), World Trade Centre
Telok Blangah Rd.
Singapore 0409
Tel: [65] 2719388 Fax: [65] 2740770
Tlx: 28617 TRADEV

Singapore Trade Development Board
(Changi Airport Office)
115 Airport Cargo Road
#04-18 Cargo Agents Building C
Singapore 1781
Tel: [65] 5427179 Fax: [65] 5425385

The Customs and Excise Department is responsible for collecting customs duties and excise taxes. It publishes updated lists of dutiable goods.

Customs and Excise Department
1 Maritime Square #03-01 & #10-01
World Trade Centre
Singapore 0409
Tel: [65] 2728222 Fax: [65] 2779090

IMPORT POLICY

Tariff Structure

In recent years Singapore has moved to an emphasis on export-oriented industries, and no longer considers high import duties an important means of protecting developing industries. Trade classification and custom duties are based on the Harmonized System, adopted in January 1989. Only a few categories of imports are subject to tariffs. The following items are subject to minimal import duties:

- accumulator plates, other than nickel cadmium, and fluorescent tubes
- alcoholic beverages
- birds' eggs and egg yolk
- bread, biscuits, pastry, and cakes
- chairs, other furniture of cane or wood, filing cabinets, mattress supports, and other articles of bedding
- clothing of all materials
- headgear and hats
- household refrigerators
- hydrocarbons, dulcin, saccharine, and cyclamates
- imitation jewelry
- leather handbags, purses, wallets, and pochettes
- motor vehicles
- petroleum products
- sugar, confectionery, and chocolate
- tobacco

Approximately 50 percent of the duties are levied on a specific basis (per unit of measure); 40 percent are levied on an ad valorem basis (percentage of the total invoice value); and 10 percent of the duties are either specific or ad valorem, whichever is higher, or a combination of the two. Ad valorem rates range from 5 to 45 percent.

A current copy of the *Singapore Trade Classification and Customs Duties* is available from:

Singapore National Printers Ltd.
8 Shenton Way #B1-07, Treasury Building
Singapore 0106
Tel: [65] 2230834

Basis of Duty Assessment

For direct importation of goods from suppliers or manufacturers, ad valorem duties are based on the Singapore customs open-market value, that is, at the cost, insurance, and freight (CIF) value, plus 1 percent of the CIF (to cover handling and other incidental expenses). For importation through an agent, ad valorem duties are based on CIF value plus cost and agent commission. Duties are payable in Singapore dollars at the time dutiable goods are cleared through customs. For information on dutiable imports in Singapore, contact the Customs and Excise Department.

ASEAN Tariff Amendments

Under an agreement on preferential trading arrangements by the Association of Southeast Asian Nations (ASEAN), Singapore grants preferential rates of duty only on certain goods originating in Singapore or the other ASEAN countries—Indonesia, Malaysia, the Philippines, and Thailand. These goods include sarongs, batik items, shirts, under and outer garments, cotton handkerchiefs, and goods made of leather, composition leather, or textiles.

Food, Health, and Safety Regulations

Imports of animals, birds, meat, meat products, living plants or parts thereof, fodder of animal origin, colored skimmed milk, animal fertilizers, and veterinary medicaments require prior permission from the Primary Production Department.

Prior permission from the Ministry of Health is required for importation of skimmed milk, chemicals, and pharmaceuticals. Processed foods and pharmaceuticals must be inspected and approved by the Ministry of Health. Flunitrazepam cannot be imported by any unauthorized person. Wholesalers who import the drug may do so only on special import authorizations issued by the Ministry of Health. Importation of the drugs amidopyrine and noramidopyrine is also prohibited, as well as any medicines containing these agents.

Special regulations from the Ministry of the Environment govern the use of the food stabilizer additive calcium disodium ethylent diamine tetra acetate (EDTA). EDTA is usually permitted only in canned fish and only in a limited amount.

Arms and explosives can be imported only by licensed dealers with a permit from the police; poisons require a permit from the Ministry of Health.

The following regulations control the import of rendered edible and inedible fat of animal origin, as well as products containing such fat:

- Special permits must be obtained from the Director, Primary Production, prior to importation of the goods.
- The pertinent items must be accompanied by certificates, signed by the government veterinary authority in the country of origin, stating that the various rules of processing, production, and preservation have been complied with.
- The Director may require the testing of these products and may require the disposal of any products that do not meet standards. Typically, the importer/owner will not be compensated for disposed items.
- Edible products must bear labels showing the name of the manufacturer and the country of origin, the trade designation of the product, and the nature of the fats contained. Labels of nonedible products must state, in capital letters, NOT FOR HUMAN CONSUMPTION.

The importation (except for transshipment) of the following dutiable goods is prohibited:

- Brandy or whisky that is not accompanied by a certificate—issued by the proper customs or other authority of the country of origin and acceptable to a senior customs officer—that the brandy or whisky has been stored in wood for a period of not less than three years.
- Intoxicating liquors containing arsenic, copper, lead, or any compound of these in excess of specified amounts.
- Intoxicating liquors whose bottles or labels have been marked *Singapore Duty Not Paid* on the direction of the Comptroller of Customs.
- Locally made cigarettes with the prefix letter "E" before the code numbers or any other designation embossed on the packet on the direction of the Comptroller of Customs.
- Shag tobacco.

Standards

Singapore uses the metric system. Industrial standards applied in the engineering and construction fields are basically those used in the advanced industrialized countries, but the Singapore Institute of Standards and Industrial Research (SISIR) has developed standards for certain electrical, sanitary, and building products. Foreign suppliers of these products planning to expand sales into Singapore should check with SISIR before exporting.

Singapore Institute of Standards and
Industrial Research (SISIR)
1 Science Park Drive
Singapore 0511
Tel: [65] 7787777 Fax: [65] 7780086
Tlx: 28499 SISIR

Electric Current

The electric current used in Singapore is AC, 50 cycles, 230/400 volts, 1,3 phases, 2, 4 wires.

Environmental Protection

Most industrial and commercial enterprises are covered by antipollution legislation. Regulatory measures for motor vehicles include air pollution standards.

Countertrade

No regulations or requirements currently govern countertrade transactions in Singapore. The country has recently begun promoting itself as a countertrade center—a reversal of earlier policies that discouraged such business because it was considered to violate free trade.

Preshipment Inspection

There are no government requirements for preshipment inspections, but an importer may request an inspection.

Samples and Advertising Matter

Samples having no commercial value are admitted duty-free. Dutiable samples (such as liquor and tobacco items) may be brought in by commercial travelers under bond or under deposit of duty. The bond is canceled or the deposit refunded if the samples are exported within six months or within such time as the authorities may grant.

Commercial travelers' samples and samples imported by parcel post may enter duty-free. No restrictions are imposed on the reexportation of duty-free goods. No refund of duty is granted on liquor and tobacco samples. Only advertising materials relating to a limited number of items, including tobacco and liquor, is subject to customs duties.

ATA Carnets

Singapore is a signatory to the Customs Convention on the ATA Carnet for Temporary Importation of Goods. (The ATA carnet is a standardized international customs document that provides for duty-free, temporary admission of certain goods into signatory nations.) The ATA carnet may be used for transporting commercial samples, tools of the trade, advertising material, audiovisual, medical, scientific, or other professional equipment. ATA carnets can be obtained from:

Singapore International Chamber of Commerce
6 Raffles Quay #05-00
Denmark House
Singapore 0104
Tel: [65] 2241255 Fax: [65] 2242785
Tlx: 25234 INTCHAM

Advanced Ruling on Classification

When an importer is unsure of the classification of goods, a sample and description of the goods may be sent to appropriate customs officers. Any such ruling is purely advisory and non-binding. A customs officer may elect to send a sample of the goods to the Comptroller of Customs for a binding ruling on the classification of the article.

Fines and Penalties

With the exception of customs officers who are performing their duties, no persons may have in their possession or control in a customs or licensed warehouse any dutiable goods or denatured spirits that are imported contrary to the provisions of the customs regulations. All goods deemed in contravention of these regulations are liable to seizure, and any person found guilty of possessing such dutiable goods is liable to a fine.

Any person who makes false or fraudulent alterations on any document required by the Import and Export Control or customs authorities is subject to punishment with imprisonment or fine.

Penalties are levied for the importation of prohibited goods, removal of goods before examination by a customs officer, the illegal removal of goods from a warehouse or other place of security, the deliberate concealment of prohibited or undeclared goods or goods that have been removed illegally, and the fraudulent evasion or attempt at evasion of customs duty. The Singapore Customs (Dumping and Subsidies) Ordinance of 1962 provides for the imposition of antidumping and countervailing duties—in addition to normal import duties—on dumped and subsidized goods if their import is likely to endanger an established industry or retard the establishment of an industry.

Internal Taxes

Excise taxes equal to import duties are levied on locally manufactured petroleum products. An excise tax is also imposed on certain locally manufactured alcoholic beverages. A censorship fee is levied on films, and a tax is levied on admission tickets.

On first registration of a new motor vehicle, a fee of 175 percent (plus a 45 percent duty for imports) of the import market value (determined by the Registrar) of all passenger cars is levied. Annual vehicle registration fees (road taxes) are based on the cubic capacity of the engine; fees range from S$0.60 to S$1.50 (from about US$0.37 to US$0.94) per cubic centimeter. An additional tax, calculated at six times the road tax levied on gasoline-powered vehicles, is imposed on motor vehicles fitted with engines using heavy or diesel oil.

IMPORT PROCEDURES

Application for a Central Registration Number

Anyone who imports or exports products or services in Singapore must register with the Singapore Trade Development Board and obtain a central registration number (CR No.). The CR No. can be applied for in person, by fax, or by mail.

Urgent requests for a CR No. are best handled in person by submitting the application form to either of the STDB's two offices in Singapore. The application must be signed by a manager, sole proprietor, partner, director, or chairman of the company or establishment. An applicant must also bring a photocopy of the company's incorporation, registration certificate, and company rubber stamp with address. Firms that are registered with the STDB or the Monetary Authority of Singapore (MAS) should bring the approval letter from either of these agencies to the STDB's Imports and Exports Office at the World Trade Centre. Submission deadlines are 11:45 am and 4:15 pm on weekdays and 12:15 pm on Saturdays. The office is closed for lunch from 12:30 pm to 1:30 pm on weekdays. For additional information, the Imports and Exports Help Desk telephone number is [65] 2790350.

Firms wishing to apply for a CR No. by fax should transmit a note to the Imports and Exports Office, attention Mrs. Elsie Ong, requesting an application form; the fax number is [65] 272-4720. The form will arrive within 24 hours, and the completed form together with the company's incorporation and registration certificate should be faxed to the Imports and Exports Office. Firms that are registered with the STDB or the MAS should also fax the approval letter from either of these agencies.

Firms wishing to apply for a CR No. by mail should send a completed application form together with the company's incorporation and registration certificates to:

Mrs. Elsie Ong, Licensing Section
Trade Development Board
Imports and Exports Office
1 Maritime Square #10-40 (Lobby D)
World Trade Centre, Telok Blangah Rd.
Singapore 0409

A business must reapply for a CR No. if the number is lost, if the company's name has been changed but the registration number with the Registrar of Companies and Businesses remains the same, or if the company's name and registration number have been changed. This last condition applies especially to sole proprietorships and partnerships that change to private limited or limited companies.

Inward Declaration

Importers must complete the Singapore Inward Declaration form and make four copies. The following documents must be attached:

- invoice
- certificate of origin (if required)
- import license (if required)
- catalogue/sample (for new products)
- air waybills (for air consignments).

The items declared must correspond with such supporting documents as the invoice and import license.

An importer must then affix a S$10 (about US$6.25) STDB revenue stamp or a postal franking machine impression on the original copy of the declaration. The declaration is then submitted to the Imports and Exports Office for endorsement; for sea consignments, the declaration should be submitted a few days before the goods arrive.

Endorsement of the declaration by the Imports and Exports Office is based on the following conditions:

- The approved permit is valid for one month from the date of endorsement.
- All unused permits must be returned by the date of expiration to the Imports and Export Office for cancellation.
- All valid used permits must be returned to the shipping agents within 10 days of importation.
- Permits are approved subject to the condition that the products declared do not require Import Certificate and Delivery Verification (ICDV).

Failure to comply with these conditions is a violation of Singapore law.

To take delivery of imported goods, the importer either hands the original copy of the endorsed declaration to the shipping agent in exchange for the order (for sea consignments), shows the original copy to the airline or cargo agent (for air consignments), or shows the original copy to customs at the checkpoint or gate when the goods are taken out of the controlled area or a free trade zone.

Electronic Document Processing—TradeNet

In January 1989 Singapore introduced the world's first nationwide electronic trade documentation system, known as TradeNet. Subscribers to TradeNet can exchange business documents electronically with government agencies as well as with local and overseas trading partners. Importers and exporters can use TradeNet to submit their declarations electronically to the STDB and other government agencies. Declaration approvals and other permits are returned electronically. The system is administered by Singapore Network Services Pte Ltd. and provides the following services for TradeNet users:

- central mailbox system for the exchange of mail
- information services, including cargo information system inquiry, database services, and company billing inquiry
- password management
- information exchange archive to retrieve lost permits
- acknowledgment from the Exports and Imports Office.

To obtain more information or to register with TradeNet, contact Singapore Network Services, Tel: [65] 778-5611.

Transit, Warehousing, Transshipment, and Reexport

Singapore's traditional role as a transit zone and transshipment point for Southeast Asian trade is enhanced by the country's minimal trade controls. Most of its trade moves without customs duties or other restrictions. Government and private bonded warehouses are available both inside and outside port areas for the storage of dutiable goods.

Much of Singapore's transit trade is handled under the jurisdiction of the Port of Singapore Authority. Within its controls, goods may be loaded, unloaded, stored, sorted, repackaged, and transshipped with minimum customs involvement. Goods do not pass through Singapore customs unless they are removed from this area.

Sampling of cargoes held on Port Authority premises may be approved upon written application. While other forms of processing are generally not permitted, in some cases (again subject to written application and approval), assembly may be undertaken. Manufacturing is not permitted in port authority public warehouses but may be performed (with a license from the Customs and Excise Department) in the free trade zones and in other buildings specifically leased for such purposes within the port area or in other areas approved by the government.

Port of Singapore Authority
460 Alexandra Road, PSA Building
Singapore 0511
Tel: [65] 274-7111 Fax: [65] 2744677
Tlx: 21507 PORT

The entry, departure, or transfer of goods to or from transit sheds and storage warehouses within the general port area by shippers, agents, or consignees is subject to documentary requirements. Prior to the exportation, reexportation, or transshipment of goods, an export permit from the Imports and Exports Office is required. Shipping agents are also allowed to ship cargo overland on transshipment declaration permits.

Dutiable goods on which duties have not been paid may be reexported without payment of duty only if they are moved under customs control and checked out at the port of boundary station. All or part of the customs duty paid on the goods may be refunded, as prescribed by the Minister of Finance, if imported and reexported within a prescribed period. This policy applies also to goods imported, processed, and then reexported.

The transshipment of non-controlled goods does not require a declaration, as long as goods are shipped on a through bill of lading or a through air waybill. The goods may be unloaded in the free trade zone and reloaded in the same zone. However, it is necessary to submit a transshipment declaration (pink form) for:

- controlled goods;
- overland cargo to be rerouted on the shipper's instruction;
- goods transshipped via the Customs Territory other than for interzone movements undertaken by approved agents;
- movements of transshipment goods under the Singapore Port Authority's Service Charge Scheme; and
- movements for which transshipment status is proved.

The procedure for declaring goods that are transshipped is:

- Complete the transshipment form and make five copies.
- Affix a S$10 (about US$6.25) STDB revenue stamp or postal franking machine impression on the first copy (not the original) of the declaration.
- Submit the declaration to the Imports and Exports Office at the World Trade Centre for approval. The Imports and Exports Office retains the original copy of the declaration for nondutiable goods. For dutiable goods, the office retains the last copy and customs retains the original.
- Present the first copy of the approved declaration to the outward carrier's agent and the third copy to the inward carrier's agent.

Consignments that do not have a through bill of lading or a through air waybill require an inward and an outward declaration when the goods are being imported and reexported. This requirement applies even if the consignment remains inside the free trade zone.

Import Licensing

The STDB under the MTI and Industry administers import licenses for the few products that require one. Products controlled for health, safety, or security reasons include rice, films and videos, and telecommunications equipment. The government has banned the importation, sale, and manufacture of chewing gum (effective January 3, 1992) and of motor vehicles more than three years old (effective September 1, 1992). Import licenses are also required for all imports originating in, or consigned from, Albania, the Laos People's Democratic Republic, Mongolia, and Vietnam. Licenses for imports from these countries are subject to a surcharge of 0.5 percent of the CIF value. Licenses must be obtained prior to the opening of the respective letter of credit. Licenses are usually valid for six months, but extensions may be granted. All imports from Iraq are prohibited.

Additional Documentation

Documents required by Import and Export Control and Customs authorities for incoming air and surface shipments include the commercial invoice, the bill of lading for surface shipments, the air waybill for air freight, import declaration, packing lists, and insurance documents. Special documents are also required for the importation of certain plant materials, birds, and animals.

Bill of Lading

The bill of lading should show the names of the shipper, consignee, and ship; the exporter's mark; the number of packages; and a description of the goods. *To Order* bills are usually acceptable and should be issued in a minimum of two copies. Additional copies may be requested by the importer.

Certificate of Origin

A certificate of origin is required only for banking purposes when a dollar exchange is supplied by the local control authorities and for goods for which a preferential tariff rate is to be claimed. There is no special form for the certificate, which may be certified by a chamber of commerce, a recognized bank, or by a firm of international repute.

Import Declaration

Importers are required to present an import declaration for all imports.

Insurance Certificates

Normal commercial practices apply. Foreign exporters are advised to follow the instructions of the local importer or insurance company.

Special Import Documents

Imports of animals and birds require import permits, health certificates (issued within seven days of export), and CITES certificates (for endangered species). Meat and meat products require veterinary certificates, meat inspection certificates, and method of processing certificates, and other certificates as appropriate; living plants and parts thereof require import permits and phytosanitary certificates.

Slaughter dates, establishment codes, and the address of the approved processing plants of origin must be shown on the export certificates and shipping cartons of frozen or chilled meat and poultry products. For shipments containing products from animals or birds slaughtered on different days, the first and last date of slaughter must be shown.

Marking and Labeling Requirements

Labels are required on imports of food, drug, liquors, paints, and solvents, and must specify the country of origin. Prepackaged foods must be labeled in English, with the appropriate designation of the food content printed in capital letters at least 1/16 inch high. Labels must also state whether foods are compounded, mixed, or blended; the minimum quantity, in metric net weight or measure (intoxicating liquors, soft drinks, and condensed or dried milk are exempt from this provision if prepackaged in a container for retail sale); the name and address of the manufacturer or seller; and the country of origin. Sugar confectionery, chocolate, and chocolate confectionery are exempted from all general labeling requirements for prepackaged foods.

A description (in English) of the contents of the package may appear on the face of the label, provided that the description is not contrary to, or a modification of, any statement on the label. Pictorial illustrations must not be misleading as to the true nature or origin of the food. Foods having defined standards must be labeled to conform to these standards and must be free of added foreign substances. Packages of food labeled *enriched*, *fortified*, *vitaminized*, or with any other words that imply that the article contains added vitamins or minerals, must show the quantity of vitamins or minerals added per metric unit.

Special labels are required for certain foods, medicines, and goods, including edible and nonedible animal fats as well as paints and solvents. Processed foods and pharmaceuticals must be inspected and approved by the Ministry of Health. Electrical goods must be checked by the Public Utilities Board's engineers before they can be installed in government establishments. Paints and solvents are the responsibility of the Chief Inspector of Factories, Ministry of Labor.

Anti-Dumping Duties, Subsidies, and Countervailing Duties

Singapore is an adherent to the General Agreement on Tariffs and Trade (GATT) Convention covering Anti-Dumping Duties (negotiated under the Multilateral Trade Negotiations of the Tokyo Round). In addition, the Singapore Customs Ordinance (Dumping and Subsidies) of 1962 provides for the imposition of countervailing duties—in addition to normal import duties—on subsidized goods, if their importation is likely to endanger an established industry or retard the establishment of an industry.

Abandoned and Reexported Goods

Dutiable goods on which duties have not been paid may be reexported without payment of duty only if they are moved under customs control and checked out at the port of boundary station. All or part of the customs duty paid may be refunded on goods, as prescribed by the Minister of Finance, if imported and reexported within a prescribed period. This policy also applies to goods imported, processed, and then reexported.

Export Policy & Procedures

INTRODUCTION

Singapore once had a reputation as an assembly and processing center. At that time some 90 percent of the country's exports were reexports of products sent to Singapore for finishing. Currently, reexports constitute about one third of total exports, and Singapore is an important designer, manufacturer, and exporter of advanced electronic, telecommunication, and computer products as well as medical equipment and industrial machinery in its own right. Its telecommunications infrastructure is one of the most technologically advanced in the world. For example, all required trade documentation can be acquired and declared electronically by modem through the TradeNet system.

Although the structure of the domestic economy has changed, the country's policy of free and open trade is the same. There are few export controls, and formal export procedures are kept to a minimum.

REGULATORY AUTHORITY

The Singapore Trade Development Board (STDB) is part of the Ministry of Trade (MIT) and Industry has jurisdiction over all aspects of foreign trade.

Singapore Trade Development Board
1 Maritime Square, #10-40 (Lobby D)
World Trade Centre, Telok Blangah Rd.
Singapore 0409
Tel: [65] 2719388 Fax: [65] 2724720
Tlx: 28617 TRADEV

Singapore Trade Development Board
(Changi Airport Office)
115 Airport Cargo Road
#04-18 Cargo Agents Building C
Singapore 1781
Tel: [65] 542-7179 Fax: [65] 542-5385

EXPORT POLICY

Singapore has very few export restrictions. The country adopted the Harmonized System for tariff classification and coding of goods on January 1, 1989. The Harmonized System is designed to classify goods in international trade for customs purposes and for developing trade statistics. It is arranged into 99 chapters. The sections are grouped in categories such as agriculture, chemicals, chief material of the product, or type of manufacturing industry.

In order to export, a business must register with the STDB and obtain a central registration number (CR No.). All exported goods must also be declared, which is a relatively simple procedure. The nationwide electronic trade documentation system—TradeNet—allows subscribers to complete all registration and trade documentation electronically.

Export Controls and Licensing

The exportation of certain textiles and garments to Canada, the European Community, Norway, and the United States is subject to quota restriction. Other items—including rubber, timber, and chlorofluorocarbons—are also subject to export controls, and granite and sand are subject to export licensing requirements.

Countertrade

No regulations or requirements currently govern countertrade transactions in Singapore. The country has recently begun promoting itself as a countertrade center—a reversal of earlier policies that discouraged such business because it was though to violate free trade.

EXPORT PROCEDURES

Application for a Central Registration Number

Anyone who imports or exports products or services in Singapore must register with the Singapore

Trade Development Board and obtain a central registration number (CR No.). The CR No. can be applied for in person, by fax, or by mail.

Urgent requests for a CR No. are best handled in person by submitting the application form to either of the Trade Development Board's two offices in Singapore. The application must be signed by a manager, sole proprietor, partner, director, or chairman of the company or establishment. An applicant must also bring a photocopy of the company's incorporation, registration certificate, and company rubber stamp with address. Firms that are registered with the STDB or the Monetary Authority of Singapore (MAS) should bring the approval letter from either of these agencies to the Trade Development Board's Imports and Exports Office at the World Trade Centre. Submission deadlines are 11:45 am and 4:15 pm on weekdays and 12:15 pm on Saturdays. The office is closed for lunch from 12:30 pm to 1:30 pm on weekdays. For additional information, the Imports and Exports Help Desk telephone number is [65] 2790350.

Firms wishing to apply for a CR No. by fax should transmit a note to the Imports and Exports Office, attention Mrs. Elsie Ong, requesting an application form; the fax number is [65] 272-4720. The form will arrive within 24 hours, and the completed form together with the company's incorporation and registration certificate should be faxed to the Imports and Exports Office. Firms that are registered with the STDB or the MAS should also fax the approval letter from either of these agencies.

Firms wishing to apply for a CR No. by mail should send a completed application form together with the company's incorporation and registration certificates to:

Mrs. Elsie Ong, Licensing Section
Trade Development Board
Imports and Exports Office
1 Maritime Square #10-40 (Lobby D)
World Trade Centre, Telok Blangah Rd.
Singapore 0409

A business must reapply for a CR No. if the number is lost, if the company's name has been changed but the registration number with the Registrar of Companies and Businesses remains the same, or if the company's name and registration number have been changed. This last condition applies especially to sole proprietorships and partnerships that change to private limited or limited companies.

Outward Declaration

All goods for export or reexport must be declared. The procedures are:

- Complete the Singapore Outward Declaration form and make four copies. Separate declaration forms are needed for non-controlled goods and controlled goods; an ocean bill of lading or master air waybill is needed for direct consignments or a house bill of lading or one house air waybill for consolidated shipments.
- Affix a S$10 (about US$6.25) STDB revenue stamp or a postal franking machine impression on the original copy of the declaration. Separate processing fees are payable for each declaration.
- Submit the declaration form promptly to the Trade Development Board's Imports and Exports Office for endorsement.
- Present the original copy of the endorsed declaration (with the STDB revenue stamp or postal franking machine impression) to the carrier's agent or the cargo agent.

Endorsement of the declaration is based on the following conditions:

- The approved permit is valid for one month from the date of endorsement.
- All unused permits must be returned, by the date of expiration, to the Imports and Export Office for cancellation.
- All valid used permits must be returned to the shipping/airline agents within seven days of exportation.
- Permits are approved subject to the condition that the products declared do not require Import Certificate and Delivery Verification (ICDV).

Failure to comply with these conditions is a violation of Singapore law.

Transit, Transshipment, and Reexport

Singapore's traditional role as a transit zone and transshipment point for Southeast Asian trade is enhanced by the country's minimal trade controls. Most of its trade moves without customs duties or other restrictions. Government and private bonded warehouses are available both inside and outside port areas for the storage of dutiable goods.

Much of Singapore's transit trade is handled under the jurisdiction of the Port of Singapore Authority. Within its controls, goods may be loaded, unloaded, stored, sorted, repackaged, and transshipped with minimum customs involvement. Goods do not pass through Singaporean customs unless they are removed from this area.

Sampling of cargoes held on Port Authority premises may be approved upon written application While other forms of processing are generally not permitted, in some cases (again subject to written application and approval), assembly may be undertaken. Manufacturing is not permitted in port authority public warehouses but may be performed (with a license from the Customs and Excise Department) in the free trade zones and in other buildings specifically leased for such purposes within the port area or in other areas approved by the government.

Port of Singapore Authority
460 Alexandra Road, PSA Building
Singapore 0511
Tel: [65] 2747111 Fax: [65] 2744677
Tlx: 21507 PORT

The entry, departure, or transfer of goods to or from transit sheds and storage warehouses within the general port area by shippers, agents, or consignees is subject to documentary requirements. Prior to the exportation, reexportation, or transshipment of goods, an export permit from the Imports and Exports Office is required. Shipping agents are also allowed to ship cargo overland on transshipment declaration permits.

Dutiable goods on which duties have not been paid may be reexported without payment of duty only if they are moved under customs control and checked out at the port of boundary station. All or part of the customs duty paid on the goods may be refunded, as prescribed by the Minister of Finance (MOF), if imported and reexported within a prescribed period. This policy applies also to goods imported, processed, and then reexported.

The transshipment of non-controlled goods does not require a declaration, as long as goods are shipped on a through bill of lading or a through air waybill. The goods may be unloaded in the free trade zone and reloaded in the same zone. However, it is necessary to submit a transshipment declaration (pink form) for:

- controlled goods;
- overland cargo to be rerouted on the shipper's instruction;
- goods transshipped via the Customs Territory other than for interzone movements undertaken by approved agents;
- movements of transshipment goods under the Singapore Port Authority's Service Charge Scheme; and
- movements for which transshipment status is proved.

The procedure for declaring goods that are transshipped is:

- Complete the transshipment form and make five copies.
- Affix a S$10 (about US$6.25) STDB revenue stamp or postal franking machine impression on the first copy (not the original) of the declaration.
- Submit the declaration to the Imports and Exports Office at the World Trade Centre for approval. The Imports and Exports Office retains the original copy of the declaration for nondutiable goods. For dutiable goods, the office retains the last copy and customs retains the original.
- Present the first copy of the approved declaration to the outward carrier's agent and the third copy to the inward carrier's agent.

Consignments that do not have a through bill of lading or a through air waybill require an inward and an outward declaration when the goods are being imported and reexported. This requirement applies even if the consignment remains inside the free trade zone.

Electronic Document Processing—TradeNet

In January 1989, Singapore introduced the world's first nationwide electronic trade documentation system—TradeNet. Subscribers to TradeNet can exchange business documents electronically with government agencies as well as with local and overseas trading partners. Importers and exporters can use TradeNet to submit their declarations electronically to the Trade Development Board and other government agencies. Declaration approvals and other permits are returned electronically. The system is administered by Singapore Network Services Pte Ltd. and provides the following services for TradeNet users:

- central mailbox system for the exchange of mail;
- information services, including cargo information system inquiry, database services, and company billing inquiry;
- password management;
- information exchange archive to retrieve lost permits; and
- acknowledgment from the Exports and Imports Office.

To obtain more information or to register with TradeNet, contact Singapore Network Services, Tel: [65] 7785611.

Additional Documentation

Documents required by customs authorities for air and surface shipments out of Singapore include the commercial invoice, the bill of lading for surface shipments, the air waybill for air freight, export declaration (if necessary), packing lists, and insurance documents. A certificate of origin may be necessary when exporting certain products to certain countries.

Commercial Invoice

There is no special form for the commercial invoice. A minimum of two to three copies of the invoice should be issued on the shipper's letterhead and signed by the exporter or an approved representative of the exporter. The invoice should either accompany or precede the shipment; otherwise, clearance of the shipment is likely to be delayed.

The invoice must contain at least the following information:

- marks, numbers, name, tariff classification, and an accurate description of the goods;
- quantities, gross and net weights, unit value of the goods, and total value;
- itemized expenses (including freight, insurance, and shipping charges);
- place and date of preparation of the invoice; destination and consignee; and
- conditions of the contract that relate to the determination of value of the goods.

Additional information may be required by the importer or the importer's bank, or to meet the requirements for particular commodities

Bill of Lading

The bill of lading is the most important document in a transaction involving transport by sea or by air (in which case it is called an air waybill). As the document of title to the merchandise, it must conform strictly with the terms in the letter of credit. For shipments under cost, insurance, and freight (CIF), cost, freight/carriage, and insurance paid (CIP), or other related methods, the supplier contracts and pays for the freight. However, many buyers prefer to arrange for the shipment themselves in cooperation with a local freight forwarder, consolidator, or shipping line. In this case, payment is made under such terms as free alongside ship (FAS), free on board (FOB), free on board airport (FOA), or other related methods. Buyers should make certain that the shipping agent is aware of the correct terms and how the freight charges will be paid. This information will help the carrier prepare the bill of lading in accordance with the conditions of the letter of credit, purchase contract, and other documents.

The bill of lading lists the port of departure, port of discharge, name of the carrying vessel, and date of issue. The date of issue is very important because it indicates whether goods have been shipped within the time period required in the letter of credit. Suppliers must submit all required documents on time to receive payment under the terms of the credit.

Bills of lading can be either negotiable or nonnegotiable. A negotiable bill of lading is made to *the order* of the shipper, who makes a blank endorsement on the back, or it is endorsed to the order of the bank that issues the letter of credit. A nonnegotiable bill of lading is consigned to a specific party (to the buyer or buyer's representative) and endorsement by the shipper is not required. In this case, the consignee must produce the original bill of lading in order to take delivery.

Packing List

Two copies of the packing list are recommended and should be forwarded with the commercial invoice and other shipping documents to assist in customs clearance. The following information should be included in the packing list: an exact description of all items in the shipment, the gross and net weight of each package, the exterior measurements of each package, the total number of shipping containers, and the gross weight and measurement.

Certificate of Origin

A certificate of origin is sometimes required by certain countries for specific products. The certificate must be notarized by a local chamber of commerce or by a consular or other diplomatic official. In most cases, a copy of the certificate of origin should be made and kept with the original. Suppliers and shippers should follow the advice of their principals, as the foreign importer or representative is often in the best position to determine whether a certificate of origin is necessary.

Generally, there is no standard form for a certificate of origin. Shippers can use the general form available from commercial stationers or submit a document with the following information: place of origin, marks or numbers of the commodities, commodity description, number of packages, quantities of the merchandise, total value, port of shipment, and destination. The certificate should also include a statement that the commodities listed were produced or manufactured in the place of origin shown on the document. The document must be signed and dated by the applicant and the certifying officer.

Abandoned and Reexported Goods

Dutiable goods on which duties have not been paid may be reexported without payment of duty only if they are moved under customs control and checked out at the port of boundary station. All or part of the customs duty paid may be refunded on goods, as prescribed by the MOF, if imported and reexported within a prescribed period. This policy also applies to goods imported, processed, and then reexported. Singapore's special trade zones provide specific services and facilities for export processing and other reexport trade. Cargo that is transported by sea may be stored free of charge for 28 days. There are no requirements for the local content of reexported goods.

Industry Reviews

This chapter describes the status of and trends in major Singapore industries. It also lists key contacts for finding sources of supply, developing sales leads and conducting economic research. We have grouped industries into 11 categories, which are listed below. Some smaller sectors of commerce are not detailed here, while others may overlap into more than one area. If your business even remotely fits into a category don't hesitate to contact several of the organizations listed; they should be able to assist you further in gathering the information you need. We have included industry-specific contacts only. General trade organizations, which may also be very helpful, particularly if your business is in an industry not covered here, are listed in the "Important Addresses" chapter at the end of this book.

Each section has two segments: an industry summary and a list of useful contacts. The summary gives an overview of the range of products available in a certain industry and that industry's ability to compete in worldwide markets. The contacts listed are government departments, trade associations, publications, and trade fairs which can provide information specific to the industry. An entire volume could likely be devoted to each area, but such in-depth coverage is beyond the scope of this book. Our intent is to give you a basis for your own research.

All addresses and telephone numbers given are located in Singapore, the of Singapore, unless otherwise noted. The telephone country code for Singapore is [65]; other telephone country codes are shown in square brackets where appropriate. Telephone city codes, if needed, appear in parentheses.

We highly recommend that you peruse the chapters on Trade Fairs and Important Addresses, where you will find additional resources including a variety of trade promotion organizations, chambers of commerce, business services, and media.

INDUSTRY REVIEWS TABLE OF CONTENTS

Aircraft Parts 65
Computers and Other
    Information Products 66
Electronic Consumer and
    Component Products 68
Food and Beverage Products 69
Industrial Chemicals and Materials 70
Industrial Machinery 72
Medical and Pharmaceutical Products 73
Ships and Marine Equipment 74
Telecommunications Equipment 75
Textiles and Apparel 76
Tools and Instruments 77

AIRCRAFT PARTS

As a key transportation hub for passengers and freight in the Asia-Pacific area, Singapore is extensively improving its aerospace and aviation industry, which now ranks as one of the fastest-growing sectors in its economy. Singapore's aviation industry is world-renowned for quality, and many local firms hold US Federal Aviation Administration certification.

Approximately 50 companies, many of which are foreign-invested, are involved in aircraft or aerospace maintenance and component manufacturing in Singapore. These aerospace and aviation firms have concentrated on aircraft repair, making Singapore one of the region's most comprehensive aircraft overhaul centers. About 15 percent of the aerospace and aviation industry is devoted to local manufacturing of avionic parts and supplies, primarily in support of domestic repair and maintenance facilities. Exports are minimal and primarily consist of reexported aircraft repair and maintenance equipment, which is shipped to neighboring Asian countries.

Products Aircraft parts available in Singapore include components for helicopters, military fighters, and commercial airplanes. Support centers for domestic aircraft facilities supply flying controls, servosystems, electrical generation systems, fuel system components, and other parts. However, aircraft repair and maintenance equipment is not made in Singapore and therefore must be imported.

Competitive Situation

All aircraft repair facilities in Singapore continue to expand their capabilities and factory space. To provide faster, more comprehensive, and more sophisticated repair and overhaul services, many aerospace and aircraft companies have established their own product support facilities. To address the shortage of skilled workers, a training academy has been established in Singapore.

Government Agencies

Civil Aviation Authority of Singapore
Singapore Changi Airport
Singapore 1781
Tel: 5421122 Fax: 5421231 Tlx: 21231
Air cargo Tel: 5412179 Fax: 5425390
Air transport Tel: 5412390 Fax: 5456515

Trade Associations

International Air Transport Association
331 North Bridge Rd. #20-00
Singapore
Tel: 3399978 Fax: 3390855

Directories/Publications

Asian Aviation
(Monthly)
Asian Aviation Publications
2 Leng Kee Rd. #04-01, Thye Hong Centre
Singapore 0315
Tel: 4747088 Fax: 4796668

Singapore Shipping & Air Transportation Industries Directory
(Bimonthly)
Victor Kamkin Inc.
4956 Boiling Brook Parkway
Rockville MD 20852, USA
Tel: [1] (301) 881-5973

Trade Fairs

Refer to the "Trade Fairs" chapter for complete listings, including contact information, dates, and venues. Trade fairs with particular relevance to this industry include the following, which are listed in that chapter under the headings given below:

Aerospace & Oceanic
- Asian Aerospace incorporating Asian Airport Equipment & Technology and Asia Defence Technology Exhibition

Health & Safety
- International Airport Emergency Exhibition

Machines & Instruments
- InstrumentAsia
- Transfluid
- TurboMachinery Asia

For other trade fairs that may be of interest, we recommend that you also consult the headings Factory Automation, Industrial Materials & Chemicals, and Metal & Metal Finishing.

COMPUTERS AND OTHER INFORMATION PRODUCTS

Singapore is the world's largest manufacturer of disk drives, as well as a major suppler of computer equipment. Much of the local production is exported, and nearly 40 percent of all imports are reexported. Major consumers of Singapore's computer products are England, Germany, the United States, the Netherlands, and Japan. Reexports are primarily marketed in Malaysia, Australia, Hong Kong, Germany, and the United States.

Domestic production of computers is dominated by foreign-invested companies, although about seven small local manufacturers also operate in Singapore. About 150 computer companies in Singapore provide packaged software products.

Hardware Singapore's computer hardware exports include disk drives, microcomputers, minicomputers, network systems, and peripheral units, such as printers, keyboards, and high-resolution monitors. Disk drives account for more than 70 percent of domestic computer hardware production; most of these drives are exported. Singapore's computer companies also offer low-cost unbranded computer products, which are locally assembled or imported from Taiwan. A large number of central processing units are produced in Singapore, most of which are exported to Australia, Japan, and the United States. Singapore's computer outlets reexport diskettes and computer tapes, local production of which is minimal.

Software Software suppliers in Singapore continue to provide value-added localization, customization, and installation services. In addition, these companies are developing and exporting their own packaged software products, many of which are specially adapted for domestic consumers. Software packages include payroll systems, applications for garment and textile industries, and general business applications.

Competitive Situation

Computer firms in Singapore face a shortage of skilled labor, high software development costs, and declining worldwide hardware and software prices. The government is actively promoting the domestic computer and information product industry, and it has instituted programs to encourage the use of computers in the private and public sectors. The government has also tried to foster development in the computer software industry through public acquisition, grants, research centers, and tax incentives. Copyright laws have been enacted to protect domestic and foreign software makers in Singapore. New software products under development emphasize applications for international general business markets. Singapore retains a competitive edge in world markets by offering brand name, low-cost computer products without import or export duties.

Government Agencies

Information Communication Institute of Singapore
1 Hillcrest Rd. #08-00
Singapore
Tel: 4676000 Fax: 4676601

Ministry of Communications
460 Alexandra Rd., PSA Bldg. #39-00
Singapore 0511
Tel: 2707988 Fax: 2799734 Tlx: 25500
Corporate Services & Sea Transport Division
Tel: 2799716

National Computer Board
71 Science Park Dr., NCB Bldg.
Singapore 0511
Tel: 7782211 Fax: 7789641 Tlx: 38610 NCB

National Science & Technology Board
The Pasteur, 16 Science Park Dr. #01-03
Singapore 0511
Tel: 7797066 Fax: 7771711

Trade Associations

Microcomputer Trade Association
211 Henderson Rd. #01-01
Singapore
Tel: 2782855

Singapore Computer Society
71 Science Park Dr.
Singapore
Tel: 7783901 Fax: 7788221

Singapore UNIX Association (SINEX)
190 Clemenceau Ave #05-33/34
Singapore
Tel: 3343512

Directories/Publications

Asian Computer Directory
(Monthly)
Washington Plaza
1/F., 230 Wanchai Rd.
Wanchai, Hong Kong
Tel: [852] 8327123 Fax: [852] 8329208

Asia Computer Weekly
(Bimonthly)
Asian Business Press Pte., Ltd.
100 Beach Rd. #26-00 Shaw Towers
Singapore 0718
Tel: 2943366 Fax: 2985534

Asian Computer Monthly
(Monthly)
Computer Publications Ltd.
Washington Plaza, 1st Fl.
230 Wanchai Road
Wanchai, Hong Kong
Tel: [852] 9327123 Fax: [852] 8329208

Asian Sources: Computer Products
(Monthly)
Asian Sources Media Group
22nd Fl., Vita Tower
29 Wong Chuk Hang Road
Wong Chuk Hang, Hong Kong
Tel: [852] 5554777 Fax: [852] 8730488

Journal—Singapore Computer Society
(Three times a year)
Singapore Computer Society
71 Science Park Dr.
Singapore
Tel: 7783901 Fax: 7788221

Times Guide to Computers
(Annual)
Times Periodicals Pte. Ltd.
1 New Industrial Rd.
Times Centre
Singapore 1953
Tel: 2848844 Fax: 2874720 Tlx: 25713 TIMESS

What's New in Computing
(Monthly)
Asian Business Press Pte. Ltd.
100 Beach Rd. #26-00, Shaw Towers
Singapore 0718
Tel: 2943366 Fax: 2985534 Tlx: 25280 ABPSIN

Trade Fairs

Refer to the "Trade Fairs" chapter for complete listings, including contact information, dates, and venues. Trade fairs with particular relevance to this industry include the following, which are listed in that chapter under the headings given below:

Communications & Networks
- CITY-IQ (Advanced Technology for Infrastructure Development)
- CommunicAsia
- MobilCommAsia
- NetworkAsia
- NOW
- OFFICE-IQ (Office Management Technology and Systems)
- Singapore Informatics

Computer & Information Industries
- APCIC (Automatic Processors, Computers & Instrumentation in Construction)
- AUTO-ID Asia
- Comex
- International Symposium on IC Technology Systems & Applications Exhibition (ISIC)
- Manusoft
- Pre Press Asia / Multi-Media Asia Exhibition

- Singapore Information Technology Exposition (SITEX)
- The PC Show Singapore

For other trade fairs that may be of interest, we recommend that you also consult the headings Electronic & Electric Equipment; Factory Automation, Machines & Instruments; and Multimedia & Audiovisual Equipment.

ELECTRONIC CONSUMER AND COMPONENT PRODUCTS

Electronics is Singapore's largest manufacturing industry, accounting for almost 40 percent of all industrial output and over 35 percent of all manufacturing output. Singapore's electronics industry also accounts for 42 percent of total exports. Overall the sector grew by at least 25 percent per quarter in 1993. Moreover, about one-third of Singapore's electronic product imports are reexported to Malaysia, Thailand, China, and other neighboring countries. Computer products account for approximately 47 percent of total electronics industry output.

More than 200 enterprises in Singapore produce consumer electronic goods and electronic components. Most large international electronics firms have manufacturing operations in Singapore, and foreign investment in Singapore's electronics industry is high. Major markets for Singapore-made electronic goods include Malaysia, Hong Kong, and the United States.

Consumer Electronic Goods Consumer goods represent approximately 16 percent of Singapore's total output of electronic products. Finished consumer products include office equipment, microphones, loudspeakers, amplifiers, audio and video equipment, radio cassette recorders, tape drives, and televisions.

Electronic Components Electronic components constitute more 30 percent of the total output of Singapore's electronics industry. Domestic supplies are insufficient to meet demands from Singapore's industrial and electronic equipment manufacturers, and so the market for imports of electronic components remains strong.

Active and Passive Electronic Components Singapore ranks as one of the world's top five semiconductor producers. More than 10 semiconductor centers in Singapore both design and manufacture high-tech integrated circuits (ICs) and other semiconductors. Two plants are fabricating wafers and are designing advanced wafer fabrication processes. Color television picture tubes are also key exports.

In the passive electronic component sector, significant exports include industrial grade connectors, resistors, inductors, and ceramic chip capacitors.

Printed Circuit Boards More than 20 printed circuit board (PCB) firms and 60 printed circuit board assembly (PCBA) firms support Singapore's electronics industry. Most of these firms are small- to mid-size locally owned companies, although a few are multinational corporations. Most major electronics firms in Singapore also have their own inhouse PCBA lines. Singapore's PCB firms are following market trends toward denser, multi-layer, fine-line, and smaller-hole boards. Through government financial incentives, Singapore's electronics firms are upgrading from making one-sided boards to four- and six-layer boards. Most equipment required to produce high-tech PCBs is imported from Germany, Japan, and the United States.

Competitive Situation

Singapore's strategic location, good infrastructure, and proximity to regional electronic production centers have made it a regional trading center for electronic equipment and components. Despite a worldwide slowdown in the electronics industry, capital investment in Singapore-based electronics firms is high, and a large number of foreign firms continue to invest substantially in Singapore's electronics manufacturing sector. Encouraged by government financial incentives to upgrade and automate, electronics firms are focusing on surface mount technology. The government is also encouraging local firms to increase manufacturing of equipment needed to produce electronic products and to seek transfer technology agreements with overseas companies in this area. Faced with labor shortages and rising costs, electronics firms have relocated their labor-intensive production facilities, such as factories for manufacturing lower-end disk drives and semiconductors, to Malaysia.

Government Agencies

National Science & Technology Board
The Pasteur, 16 Science Park Dr. #01-03
Singapore 0511
Tel: 7797066 Fax: 7771711

Trade Associations

Air-conditioning & Refrigeration Association
58 Kensington Park Rd.
Singapore 1955
Tel: 2885491

Singapore Electrical Contractors Association
315 Outram Rd. #10-09A, Tan Boon Liat Bldg.
Singapore 0316
Tel: 2263216 Fax: 2237568

Singapore Electrical Traders Association
35A/B Truro Rd.
Singapore 0821
Tel: 2990355 Fax: 2995495

Association of Electronic Industries in Singapore
470 North Bridge Rd. #03-09
Singapore
Tel: 3374643 Fax: 3399341

Radio & Electrical Traders Association of
Singapore
68 Lor 16 Geylang #04-01
Singapore
Tel: 7470971, 7457568

Directories/Publications

Asian Electricity
(11 per year)
Reed Business Publishing Ltd.
5001 Beach Rd. #06-12, Golden Mile Complex
Singapore 0719
Tel: 2913188 Fax: 2913180

Asian Electronics Engineer
(Monthly)
Trade Media Ltd.
29 Wong Chuck Hang Rd.
Hong Kong
Tel: [852] 5554777 Fax: [852] 8700816

Asian Sources: Electronic Components
Asian Sources: Gifts & Home Products
(Monthly)
Asian Sources Media Group
22nd Fl., Vita Tower
29 Wong Chuk Hang Road
Wong Chuk Hang, Hong Kong
Tel: [852] 5554777 Fax: [852] 8730488

International Journal of High Speed Electronics
(Quarterly)
World Scientific Publ Company
Farrer Rd., PO Box 128
Singapore 9128
Tel: 3825663 Fax: 3825919

Trade Fairs

Refer to the "Trade Fairs" chapter for complete listings, including contact information, dates, and venues. Trade fairs with particular relevance to this industry include the following, which are listed in that chapter under the heading given below:

Electronic & Electric Equipment
- Asia Electronics
- BroadcastAsia
- DISKCON Singapore
- ELECTRIC ASIA
- Electronics Subcontracting Asia
- Electrotest Asia
- Globatronics
- International Procurement
- International Symposium on the Physical & Failure Analysis of Integrated Circuits Exhibition (IPFA)
- Nepcon Asia-Pacific
- Rigid Disk Drive Components Technology Review Exhibition
- SEMICON/Singapore
- Semitech Asia

For other trade fairs that may be of interest, we recommend that you also consult the headings Computer & Information Industries; Construction & Housing; Environmental & Energy Industries; Factory Automation; Furniture & Housewares; and Multimedia & Audiovisual Equipment.

FOOD AND BEVERAGE PRODUCTS

Food and beverage production is among Singapore's top ten industry sectors. These products have steadily gained acceptance in international markets. About 70 percent of Singapore's food and beverage output is exported to Japan, the United States, Australia, Europe, and the Middle East.

Nearly 300 companies produce food and beverage products in Singapore. A few of these companies are large multinational firms, but most are small enterprises.

Products Traditional food products made in Singapore include sauces, fish and beef balls, fish cakes, and noodles, but the industry also markets edible oils, frozen seafood, frozen prepared foods, chocolate and cocoa products, snack foods, and food flavorings. Foods are packaged in a wide assortment of materials, including metal, glass, paper, and plastic. Locally produced beverages include soft drinks, mixes for alcoholic drinks, and beer.

Competitive Situation

Singapore has no large-scale agricultural activities or natural resources to support its food processing and beverage industry. Thus, the industry depends heavily on imports. In addition, the industry faces labor shortages, increasing labor costs, and a limited domestic market. Although Singapore's foreign-invested food producers have invested in automation and expansion of production capacity, the smaller firms lack capital to purchase new equipment and have remained fairly labor-intensive. As a consequence, investment in Singapore's food and beverage industry began to decline in 1989, and the industry's exports and production fell.

To increase profitability, the industry is focusing on responding to on market trends. For example, it has diversified into new product lines to attract more affluent, health-conscious consumers and to accommodate growing customer demands for fast, frozen, and processed foods. Such diversification has increased Singapore's competitiveness in the international food and beverage markets and the industry is now experiencing gradual renewed growth. To secure a larger share of world markets, Singapore's food and beverage industry is also employing such strategies as brand stretching, product repositioning, package embellishment, and flavor enhancments. In addition, the government's strict enforcement of health standards is generating Singapore's worldwide reputation for high-quality foods. The industry is concentrating R&D on new techniques and better technology for food processing and improved product yield. Increasing emphasis is placed on food preservation, longer shelf-life, lower fat content, better taste, and other performance-related consumer considerations. Stability of stored food is particularly important because of Singapore's high ambient temperatures and distance from many of its markets.

Government Agencies

Ministry of Health
16 College Rd., College of Medicine Bldg.
Singapore 0316
Tel: 2237777 Fax: 2241677 Tlx: 34360

Ministry of the Environment
40 Scotts Rd., Environment Bldg.
Singapore 0922
Tel: 7327733 Fax: 7319456

Trade Associations

Malayan Pineapple Industry Board
10 Collyer Quay, #19-06, Ocean Bldg.
Singapore 0104
Tel: 5338827

Restaurant Association of Singapore
11 Dhoby Ghaut #04-03, Cathay Bldg.
Singapore 0922
Tel: 3383774 Fax: 3390903

Singapore Grocer's Association
33A Lor 15 Geylang
Singapore
Tel: 7451821

Trade Fairs

Refer to the "Trade Fairs" chapter for complete listings, including contact information, dates, and venues. Trade fairs with particular relevance to this industry include the following, which are listed in that chapter under the headings given below:

Agriculture, Forestry & Aquaculture
- Agrotechnology in the Commonwealth: Focus in the 21st Century Exhibition
- Aquarama
- Fish Asia

Food, Beverages & Food Processing
- Brew Drink Tech Asia Exhibition
- Chinese Beverage Exhibition
- Food & Hotel Asia
- Food Ingredients Asia (FIA)
- IMFEX
- ProPakAsia
- Salon Culinaire

INDUSTRIAL CHEMICALS AND MATERIALS

The industrial chemical industry ranks among Singapore's top five manufacturing sectors in terms of output. Much of this sector's production is from Singapore's petrochemical plants, which are among the world's most competitive on a cost basis. Singapore ranks third in the world in refining petrochemicals and related products. A lack of raw materials has kept Singapore's other industrial material industries small in size and output, although metal fabrication is growing slowly.

Petrochemicals Singapore's petrochemical plants produce such products as polyacetal, specialty fibers, polyvinyl chloride, polystyrene, and polycarbonate. Many of these petrochemical companies are foreign-invested. Most firms are investing heavily in expansion and in upgrading refining capacity with new aromatic plants. To meet anticipated new environmental requirements, these companies are increasing their desulphurizing capacities.

Fabricated Metal Products Growth in Singapore's metal fabrication industry depends largely on public and private construction projects. Metal fabricators in Singapore primarily support domestic manufacturers, including pipe and tube makers, die-casting and aluminum-extrusion plants, heater and boiler companies, and plant and building structure fabricators. Singapore's metal fabricators also export some products to construction industries in neighboring countries.

Government Agencies

Ministry of the Environment
40 Scotts Rd., Environment Bldg.
Singapore 0922
Tel: 7327733 Fax: 7319456

National Science & Technology Board
The Pasteur, 16 Science Park Dr. #01-03
Singapore 0511
Tel: 7797066 Fax: 7771711

Trade Associations

Granite Quarry Owners & Employers Association
141 Cecil St. #05-00, Tung Ann Association Bldg.
Singapore 0106
Tel: 2211560

Ready-Mixed Concrete Association of Singapore
Blk. 1 Thomson Rd. #03-332E
Singapore
Tel: 2568359

Rubber Association of Singapore
79 Robinson Rd. #14-01, CPF Bldg.
Singapore 0106
Tel: 2219022 Fax: 2241641, 2215316 Tlx: 20554

Singapore Building Materials Suppliers'
Association
426 Race Course Rd.
Singapore 0821
Tel: 2984660

Singapore Rattan Industry Association
12 Lor 24A Geylang
Singapore
Tel: 7441853

Directories/Publications

Asian Oil & Gas
(Monthly)
Intercontinental Marketing Corp.
PO Box 5056
Tokyo 100-31, Japan
Fax: [81] (3) 3667-9646

Asian Plastic News
(Quarterly)
Reed Asian Publishing Pte., Ltd.
5001 Beach Rd. #06-12, Golden Mile Complex
Singapore 0719
Tel: 2913188 Fax: 2913180

Building & Construction News
(Weekly)
Al Hilal Publishing (FE) Ltd.
50 Jalan Sultan #20-06, Jalan Sultan Centre
Singapore 0719
Tel: 2939233 Fax: 2970862

Industrial News And Research
Singapore Institute of Standards & Industrial Research (SISIR)
1 Science Park Dr.
Singapore 0511
Tel: 7787777 Fax: 7780086 Tlx: 28499 SISIR

Oil & Gas News
(Weekly)
Al Hilal Publishing (FE) Ltd.
50 Jalan Sultan #20-06, Jalan Sultan Centre
Singapore 0719
Tel: 2939233 Fax: 2970862

Petrochemicals & Refining
(Monthly)
Petroleum News Publishing Pte. Ltd.
41 Middle Rd. #01-00
Singapore 0718
Tel: 3361728 Fax: 3367919

Petroleum News
Petroleum News Publishing Pte. Ltd.
41 Middle Rd. #01-00
0718 Singapore
Tel: 3361728 Fax: 3367919

Polymers & Rubber Asia
(Bimonthly)
Upper West St.
Reigate Surrey RH2 9HX, UK
Tel: [44] (737) 242599 Fax: [44] (737) 223235
Tlx: 932699

Singapore Builders Directory
(Annual)
Far East Media Representatives Bldg.
320 Serangoon Rd.
Singapore

Trade Fairs

Refer to the "Trade Fairs" chapter for complete listings, including contact information, dates, and venues. Trade fairs with particular relevance to this industry include the following, which are listed in that chapter under the headings given below:

Environmental & Energy Industries
- ChemAsia—Asian International Chemical Process Engineering & Contracting Exhibition
- ENEX —ASIA
- POWER GENERATION ASIA
- POWER—GEN ASIA
- Pumps & Systems Asia Exhibition
- Refining, LNG & PetrochemAsia

Industrial Materials & Chemicals
- Asian Industrial Coatings
- Asiaplas
- International Conference on Geotextiles, Geomembranes & Related Products Exhibition
- International Rubber & Plastics Exhibition
- Metals, Manufacturing & Engineering Asia Exhibition
- Plastec Asia
- World of Concrete Asia

Metal & Metal Finishing
- Asia Weldex/Asia Fastener & Hardware
- Metalasia
- Metals, Manufacturing & Engineering Asia Exhibition
- Mould & Die Asia
- Stamptec Asia
- The International Advanced Materials & Manufacturing Technology Exhibition (AMT)
- Weldtechasia

Petroleum & Gas Industries
- Asia Pacific Oil and Gas Exhibition
- International Exhibition on Safety, Health and Loss Prevention in the Oil, Chemical and Process Industries
- Offshore South East Asia Exhibition & Conference
- Petro-Safe Asia

For other trade fairs that may be of interest, we recommend that you also consult the headings Agriculture, Forestry & Aquaculture; Construction & Housing; and Packaging, Printing & Paper.

INDUSTRIAL MACHINERY

A variety of industrial machinery is produced in Singapore, contributing about 5 percent of total manufacturing output. Domestic production is relatively limited and low-tech, and most exports are in fact reexports.

Cranes Companies in Singapore do not manufacture heavy industrial cranes, but they do assemble, and to some extent fabricate, metal frames for integration in complete crane systems. Most Singapore-produced cranes are used domestically in construction and civil engineering projects and in shipping and port facilities. Crane exports are mainly reexports of cranes assembled from imported parts.

Electric-Power Generating Equipment Although Singapore imports most of its utility-grade power generation equipment (thermal units), several domestic manufacturers produce electric motors, generators, and other electrical apparatus for the industrial power market. Singapore also reexports a large percentage of imported electric-power generating equipment to neighboring countries. To increase domestic production, the government has slated power turbines, transformers, distribution equipment, and circuit breakers for local manufacture. Domestic projects are bid internationally, with preferential treatment given to bids that utilize ancillary equipment made in Singapore.

Food Processing Equipment More than 200 firms are engaged in food processing in Singapore, creating a fairly sizable domestic market for food processing equipment. However, domestic production of this equipment is minimal and consists of low-tech machinery. Thus, most food processing equipment is imported, but over half of all imported equipment is reexported. About 65 percent of these reexports are shipped to Malaysia, Indonesia, and Thailand.

Robotic and Automated Machinery Most of Singapore's limited production of robotic and automated machinery is destined for domestic markets. Only about one-third of the robotic and automated machines shipped from Singapore are domestically produced; the rest represent reexports. The government is aggressively promoting automation throughout the country's industrial sectors, in hopes of making Singapore an automation hub for the entire region. To this end, Singapore's manufacturers are seeking joint arrangements with foreign producers of automated industrial machinery.

Government Agencies

Ministry of the Environment
40 Scotts Rd., Environment Bldg.
Singapore 0922
Tel: 7327733 Fax: 7319456

Ministry of Trade and Industry
#33-00 Treasury Bldg.
Singapore 0106
Tel: 2259911 Fax: 3209260 Tlx: 24702
Dept. of Statistics Tel: 2259911

National Science & Technology Board
The Pasteur, 16 Science Park Dr. #01-03
Singapore 0511
Tel: 7797066 Fax: 7771711

Trade Associations

Air-conditioning & Refrigeration Association
58 Kensington Park Rd.
Singapore 1955
Tel: 2885491

Singapore Industrial Automation Association
151 Chin Swee Rd. #03-13, Manhattan House
Singapore 0316
Tel: 7346911 Fax: 2355721

Directories/Publications

Asiamac Journal: The Machine-Building and Metal Working Journal for the Asia Pacific Region (Quarterly)
Adsale Publishing Company
21st Fl., Tung Wai Commercial Building
109-111 Gloucester Road
Hong Kong
Tel: [852] 8920511 Fax: [852] 8384119, 8345014
Tlx: 63109 ADSAP HX

Trade Fairs

Refer to the "Trade Fairs" chapter for complete listings, including contact information, dates, and venues. Trade fairs with particular relevance to this industry include the following, which are listed in that chapter under the headings given below:

Construction & Housing
- APCIC (Automatic Processors, Computers & Instrumentation in Construction)
- Baucon Asia Exhibition
- BURO-IQ (Building Automation & Telecommunications for Commercial Buildings)
- China Build
- SIBEX—Southeast Asian International Building & Construction Exposition

Environmental & Energy Industries
- ENEX-ASIA
- Environmex Asia
- Enviroworld
- POWER GENERATION ASIA
- POWER-GEN ASIA

Factory Automation
- AutomAsia
- IA—The Asia Pacific Industrial Automation Exhibition
- ICCIM—International Exhibition on Computer Integrated Manufacturing
- Industrial Automation

- International Conference on Automation, Robotics & Computer Vision
- MachineAsia
- Robotics

Food Processing Equipment
- Food Ingredients Asia (FIA)
- IMFEX
- ProPakAsia

Machines & Instruments
- Asia-Pacific Forex Assembly Exhibition
- Instrumentasia
- Sign Asia
- Transfluid
- Turbomachinery Asia

For other trade fairs that may be of interest, we recommend that you also consult the headings Aerospace & Oceanic; Agriculture, Forestry, & Aquaculture; Computer & Information Industries; Electronic & Electric Equipment; Industrial Materials & Chemicals; Metal & Metal Finishing; and Packaging, Printing & Paper.

MEDICAL AND PHARMACEUTICAL PRODUCTS

Singapore's pharmaceutical industry accounts for approximately one-third of the country's manufacturing output. Singapore's health and medical products industry also produces a limited number of low-tech goods. Major export markets are in neighboring Asian countries and the United States.

Pharmaceuticals Singapore is a key offshore manufacturing and support base for multinational pharmaceutical corporations that produce medicines primarily for export. Singapore-made pharmaceuticals include semisynthetic penicillin, antibiotic intermediates, and anti-ulcer medicines. Production is concentrated on intermediate products. More than 80 percent of pharmaceuticals made in Singapore are exported, and reexports of antibiotics to markets throughout the Asia-Pacific region are growing rapidly.

Medical Equipment Production of medical equipment in Singapore is limited primarily to syringes, rubber gloves, and low-tech disposable goods. More than 80 percent of Singapore-produced disposable goods and over 90 percent of Singapore-made syringes and rubber gloves are exported, primarily to the United States. A significant portion of imported medical equipment is reexported to Indonesia and neighboring Asian countries.

Government Agencies

Ministry of Health
16 College Rd., College of Medicine Bldg.
Singapore 0316
Tel: 2237777 Fax: 2241677 Tlx: 34360

Trade Associations

Singapore Dental Association
2 College Rd.
Singapore 0316
Tel: 2202588 Fax: 2237967

The Association of Medical Practitioners of Singapore Pte. Ltd.
Alumni Medical Centre
2 College Rd.
Singapore 0316
Tel: 2230901

Singapore Society of Optometrists
Blk. 531A Up Cross St. #04-85
Singapore
Tel: 5341195 Fax: 5382027

Singapore Association of Pharmaceutical Industries
30 Shaw Rd., 5/F., Roche Bldg.
Singapore 1336
Tel: 2868277 Fax: 2802167

Directories/Publications

Asian Medical News
(Bimonthly)
MediMedia Pacific Ltd.
Unit 1216, Seaview Estate
2-8 Watson Rd.
North Point, Hong Kong
Tel: [852] 5700708 Fax: [852] 5705076

Far East Health
(10 per year)
Update-Siebert Publications
Reed Asian Publishing Pte.
5001 Beach Rd. #06-12, Golden Mile Complex
Singapore 0719
Tel: 2913188 Fax: 2913180

Medicine Digest Asia
(Monthly)
Rm. 1903, Tung Sun Commercial Centre
194-200 Lockhart Rd.
Wanchai, Hong Kong
Tel: [852] 8939303 Fax: [852] 8912591

Singapore Medical Journal
(Bimonthly)
Singapore Medical Association
4 A College Rd.
Singapore 0316 Singapore
Tel: 2231264

Trade Fairs

Refer to the "Trade Fairs" chapter for complete listings, including contact information, dates, and venues. Trade fairs with particular relevance to this industry include the following, which are listed in that chapter under the heading given below:

Medicine & Pharmaceuticals
- AIDEX
- AnaLabAsia—South East Asia International Laboratory & Analytical Exhibition

SINGAPORE Business

- ASEAN Congress of Rheumatology Exhibition
- Asia Pacific Cancer Society Exhibition
- Asia Pacific Congress of Endoscopic Surgery Exhibition
- Asian Congress of Stereotactic, Functional and Computer Assisted
- Asian Pacific Society of Respriology Exhibition
- Asian-Pacific Conference on Doppler and Echocardiography Exhibition
- Basic & Advanced Techniques in Interventional Cardiology Exhibition
- Chinese Medicine & Phamacology Show
- Congress of the Federation of Asia & Oceania Perinatal Societies Exhibition
- Congress of the International Society of University Colon & Rectal Surgeons Trade Exhibition
- Dentistry
- Expomed Singapore 2000
- Hands-on Course on Total Joint Replacement Exhibition
- International Association for Dental Research Exhibition (IADR)
- International Conference on Biomedical Engineering (ICBME) Exhibition
- International Congress of Radiology Exhibition
- International Congress on Lasers in Dentistry Exhibition
- International Congress on Transplantation in Developing Countries Exhibition
- International Exhibition of the International Society of Haematology
- International Symposium on Limb Salvage—ISOLS
- International Union of Biochemistry & Molecular Biology Exhibition
- IUVDT World STD/AIDS Exhibition
- Optics
- Pan Ophthalmologica, Singapore
- Post-Congress Meeting of the International Society of Urology (SIU) Exhibition
- Regional Conference on Dermatology Exhibition
- Renal Update
- Singapore National Eye Centre (SNEC) International Meeting
- Singapore Obstetrics & Gynaecology Exhibition
- Singapore-Malaysia Congress of Medicine Exhibition
- The Singapore International Dental Exhibition (SIDEC)
- World Meeting on Impotence Exhibition

For other trade fairs that may be of interest, we recommend that you also consult the headings Agriculture, Forestry & Aquaculture; Food, Beverages & Food Processing; and Health & Safety.

SHIPS AND MARINE EQUIPMENT

Singapore's shipyards are engaged in the construction of many specialized vessels, and they have the largest repair capacity in the world. With a combined dry and floating dock capacity of more than 3.2 million dead weight tons, Singapore's shipyards can work on very large vessels, such as crude carriers, supertankers, and naval aircraft carriers.

Ship repair accounts for more than 60 percent of revenues for Singapore's marine industry, and shipbuilding for about 30 percent.

Products Vessels constructed in Singapore's shipyards include naval craft, offshore vessels, container ships, and freighters.

Competitive Situation

After a fall off in business in the early 1990s because of the weak global economy, Singapore's shipyards are again showing profits. They generally compare favorably with those of the larger shipbuilding nations such as Japan and South Korea on cost. Orders for new ships are booming, as are repairs of aging vessels from international merchant marine fleets. Declining worldwide freight rates are also good for business. To increase capacity, one of Singapore's shipyards is planning capital investments in a third shipyard.

Government Agencies

Port of Singapore Authority
460 Alexandra Rd., PSA Bldg.
Singapore 0511
Tel: 2747111 Fax: 2744677 Tlx: 21507 PORT

Trade Associations

Association of Singapore Marine Industries
1 Maritime Sq. #09-50, World Trade Centre
Singapore 0409
Tel: 2707883 Fax: 2731867 Tlx: 37706

Singapore Association of Ship Suppliers
1 Colombo Court #07-19
Singapore 0617
Tel: 3367755 Fax: 3390329 Tlx: 22860

Singapore National Shipping Association
456 Alexandra Rd. #02-02, NOL Bldg.
Singapore 0511
Tel: 2733574, 2783464 Fax: 2745079 Tlx: 24021

Directories/Publications

Asian Shipping
(Monthly)
Asia Trade Journals Ltd.
7th Fl., Sincere Insurance Building
4 Hennessy Road
Wanchai, Hong Kong
Tel: [852] 5278532 Fax: [852] 5278753

Lloyd's Maritime Asia
(Monthly)
Lloyd's of London Press (FE)
Rm. 1101 Hollywood Centre
233 Hollywood Rd.
Hong Kong
Tel: [852] 8543222 Fax: [852] 854153

Lloyd's Singapore Port Services Index
(Annual)
LLP-Times Maritime & Business Publishing Co.
Times Centre, 1 New Industrial Rd.
Singapore 1953
Tel: 2848844 Fax: 2881186 Tlx: 25713

Shippers' Times
(Bimonthly)
Singapore National Shippers Council
47 Hill St., SCCI Bldg.
Singapore 0617
Tel: 3372441 Tlx: 24473 FRETER

Shipping & Transport News
(Monthly)
Al Hilal Publishing (FE) Ltd.
50 Jalan Sultan #20-06, Jalan Sultan Centre
Singapore 0719
Tel: 2939233 Fax: 2970862

Singapore Shipping & Air Transportation
Industries Directory
(Bimonthly)
Victor Kamkin Inc
4956 Boiling Brook Parkway
Rockville MD 20852, USA
Tel: [1] (301) 881-5973

Singapore Shipping 'n' Shipbuilder
(Monthly)
Cosmic Media
PO Box 3163
Singapore

Trade Fairs

Refer to the "Trade Fairs" chapter for complete listings, including contact information, dates, and venues. Trade fairs with particular relevance to this industry include the following, which are listed in that chapter under the heading given below:

Aerospace & Oceanic
- Boats & Aquasports
- International High Speed Surface Craft Conference & Exhibition
- SingaPort

For other trade fairs that may be of interest, we recommend that you also consult the headings Agriculture, Forestry & Aquaculture; Environmental & Energy Industries; and Sporting Goods.

TELECOMMUNICATIONS EQUIPMENT

Singapore is a major world supplier of telecommunications equipment and has become a prime telecommunications center of Southeast Asia, with telephone, telex, and telefax services to over 200 countries. Approximately 70 percent of Singapore's total production of telecommunications equipment is for export. Major destinations for exports and reexports of telecommunications equipment include Malaysia, Germany, Japan, and the United States.

Products Singapore's exports include corded and cordless telephones, modems, car and communication radios, ordinary and filed telephone sets, paging systems, coaxial and submarine cables, digital data processing equipment, high-frequency radios, and microwave transmitters.

Competitive Situation

Singapore's exports of telecommunications equipment fell in the early 1990s, but had resumed their climb by 1992. Reexports account for much of the increase, but, domestic exports have also risen slightly. Local manufacturers have neither the technology nor the skilled labor needed to make high-tech telecommunications products that are competitive in international markets. To promote R&D in the telecommunications industry and to establish Singapore as a leading international telecommunications center, the government is planning to build an industrial complex devoted to the field. In addition, Singapore's telecommunications industry is forming alliances with foreign governments and private companies to establish satellite telecommunications worldwide.

Government Agencies

Information Communication Institute of Singapore
1 Hillcrest Rd. #08-00
Singapore
Tel: 4676000 Fax: 4676601

Ministry of Communications
460 Alexandra Rd., PSA Bldg. #39-00
Singapore 0511
Tel: 2707988 Fax: 2799734 Tlx: 25500
Corporate Services & Sea Transport Division
Tel: 2799716

Singapore Broadcasting Corporation (SBC)
Caldecott Hill, Andrew Rd.
Singapore
Tel: 2560401 Fax: 2538119 (Publicity), 3551503 (Radio programs)

Telecommunication Authority of Singapore
31 Exeter Rd. #05-00 Comcentre
Singapore 0923
Tel: 7343344, 7387788 Fax: 7328428, 7330073 Tlx: 33311 TELECOM

Trade Associations

Association of Electronic Industries in Singapore
470 North Bridge Rd. #03-09
Singapore
Tel: 3374643 Fax: 3399341

Radio & Electrical Traders Association of
Singapore
68 Lor 16 Geylang #04-01
Singapore
Tel: 7470971, 7457568

Directories/Publications

Asia Pacific Broadcasting & Telecommunications
(Monthly)
Asian Business Press Pte., Ltd.
100 Beach Rd. #26-00, Shaw Towers
Singapore 0718
Tel: 2943366 Fax: 2985534

Media Asia
(Quarterly)
Asian Mass Communication Research and
Information Center
39 Newton Rd.
Singapore
Tel: 2515105 Fax: 2534535 Tlx: AMICSI 55524

Singapore Telex & Telefax Directory / Telecoms
(Annual)
Telecommunication Authority of Singapore
31 Exeter Rd. #05-00 Comcentre
Singapore 0923
Tel: 7343344, 7387788 Fax: 7328428, 7330073
Tlx: 33311 TELECOM

Trade Fairs

Refer to the "Trade Fairs" chapter for complete listings, including contact information, dates, and venues. Trade fairs with particular relevance to this industry include the following, which are listed in that chapter under the headings given below:

Communications & Networks
- Asia TELECOM Exhibition
- BroadcastAsia
- CITY-IQ (Advanced Technology for Infrastructure Development)
- CommunicAsia
- MobilCommAsia
- NetworkAsia
- NOW
- OFFICE-IQ (Office Management Technology and Systems)
- Singapore Informatics

Multimedia & Audiovisual Equipment
- Asia Pacific Imaging and Information Management Exhibition (AIM)
- Multi-Media 2000 Exhibition
- Pre Press Asia/Multi-Media Asia Exhibition
- Singapore Information Technology Exposition
- World TV, Cable & Satellite

For other trade fairs that may be of interest, we recommend that you also consult the headings Aerospace & Oceanic; Computer & Information Industries; Electronic & Electric Equipment; and Furniture & Housewares.

TEXTILES AND APPAREL

Textiles and apparel account for less than 3 percent of Singapore's total manufacturing output. Singapore's major export market for apparel is the United States, while most of its textile exports are shipped to Malaysia, Mauritius, Japan, and the United States. About 80 percent of its textile exports are reexports. About 70 textile and 370 apparel producers operate in Singapore.

Products The only textiles made in Singapore are woven fabrics with varying percentages of synthetic fibers and bleached or unbleached cotton. The single largest product is woven fabrics that are at least 85 percent bleached cotton.

Competitive Situation

Persistent labor shortages have limited the growth of Singapore's textile industry. Exports of Singapore-made apparel have declined, primarily because of weak demand in US markets and the relocation of labor-intensive cutting and sewing operations to Malaysia and other countries with low labor costs. Relocation of production facilities has, however, created a stronger demand for Singapore's textile exports in these other countries, a demand that is primarily filled by reexports of textiles imported into Singapore from the United States and Europe.

Trade Associations

Textile & Garment Manufacturers' Association of
Singapore
47 Beach Rd. #06-01/02
Singapore
Tel: 3372022 Fax: 3389179

Directories/Publications

Asia Pacific Leather Directory
(Annual)
Asia Pacific Leather Yearbook
(Annual)
Asia Pacific Directories, Ltd.
6/F. Wah Hen Commercial Centre
381 Hennessy Rd.
Hong Kong
Tel: [852] 8936377 Fax: [852] 8935752

ATA Journal: Journal for Asia on Textile & Apparel
(Bimonthly)
Adsale Publishing Company
Tung Wai Commercial Building, 21st Fl.
109-111 Gloucester Road
Wanchai, Hong Kong
Tel: [852] 8920511 Fax: [852] 8384119

Trade Fairs

Refer to the "Trade Fairs" chapter for complete listings, including contact information, dates, and venues. Trade fairs with particular relevance to this industry include the following, which are listed in that chapter under the headings given below:

Comprehensive
- Asian International Gift Fair
- Utech Asia Exhibition

Textiles & Apparel
- Fashion Asia
- International Trade Fair for Footwear, Leather Goods, Bags & Travel Goods (lbf)
- Wedding Festival

For other trade fairs that may be of interest, we recommend that you also consult the headings Furniture & Housewares; Gifts, Jewelry & Stationery; Industrial Materials & Chemicals; and Sporting Goods.

TOOLS AND INSTRUMENTS

Singapore's tool and instrument industry is a small contributor to Singapore's total exports. This sector's most significant exports are metal machine tools, which are shipped primarily to Japan, Europe, and the United States. Singapore also produces and exports low-tech chemical analytical instruments; neighboring Asian countries are the main market for these exports.

Metal Machine Tools Fewer than 30 machine tool manufacturers operate in Singapore, and fewer than 10 of these have workforces that exceed 60 employees. A majority of Singapore's metal machine tool makers are subsidiaries of international companies, which tend to export most of their production. For that reason, nearly 80 percent of all local production is exported. A large percentage of Singapore's exports of metal machine tools are reexports.

Singapore's metal machinery factories produce numerically and computer numerically controlled metal machine tools, such as lathes, tool holders, surface and internal grinders, metal stamping machines, molds, dies, blades, metal cutting, sealing, and shearing machines. End-users of these machine tools include such industries as electronics manufacturing, metal and precision engineering, shipbuilding, ship repair, oil rig construction, aviation parts manufacturing, and aircraft repair.

As a result of sluggish Japanese and Western economies, Singapore's exports have dropped in recent years, but they are again growing as worldwide economies improve. To keep exports more stable, Singapore's machine tool industry is diversifying into new export markets. Another boost in demand has come from domestic industries that are automating their production processes in order to compete with developing nations, where labor costs are lower.

Chemical Analytical Instruments Singapore's manufacturers of chemical analytical instruments produce lower-technology instruments, such as pH meters. These firms do not have sufficient technology to produce more sophisticated instruments, like infrared sensors or mass spectrometers. Most Singapore-produced pH meters are used in the pharmaceutical, petrochemical, industrial chemical, and electronics industries.

Government Agencies

Ministry of Trade and Industry
#33-00 Treasury Bldg.
Singapore 0106
Tel: 2259911 Fax: 3209260 Tlx: 24702

National Science & Technology Board
The Pasteur, 16 Science Park Dr. #01-03
Singapore 0511
Tel: 7797066 Fax: 7771711

Trade Associations

Singapore Building Materials Suppliers' Association
426 Race Course Rd.
Singapore 0821
Tel: 2984660

Singapore Contractors Association Ltd.
1 Bt. Merah Lane 2, Construction House
Singapore 0315
Tel: 2789577 Fax: 2733977 Tlx: 22406

Singapore Electrical Contractors Association
315 Outram Rd. #10-09A, Tan Boon Liat Bldg.
Singapore 0316
Tel: 2263216 Fax: 2237568

Singapore Industrial Automation Association
151 Chin Swee Rd. #03-13, Manhattan House
Singapore 0316
Tel: 7346911 Fax: 2355721

Directories/Publications

Asian Sources: Hardware
(Monthly)
Asian Sources Media Group
22nd Fl., Vita Tower
29 Wong Chuk Hang Road
Wong Chuk Hang, Hong Kong
Tel: [852] 5554777 Fax: [852] 8730488

Building & Construction News
(Weekly)
Al Hilal Publishing (FE) Ltd.
50 Jalan Sultan #20-06, Jalan Sultan Centre
Singapore 0719
Tel: 2939233 Fax: 2970862

International Construction
(Monthly)
Reed Business Publishing, Ltd.
Reed Asian Publishing Pte.
5001 Beach Rd. #06-12, Golden Mile Complex
Singapore 0719
Tel: 2913188 Fax: 2913180

Medicine Digest Asia
(Monthly)
Rm. 1903, Tung Sun Commercial Centre
194-200 Lockhart Rd.
Wanchai, Hong Kong
Tel: [852] 8939303 Fax: [852] 8912591

Singapore Medical Journal
(Bimonthly)
Singapore Medical Association
4 A College Rd.
Singapore 0316
Tel: 2231264

Southeast Asia Building Magazine
(Monthly)
Safan Publishing Pte.
510 Thomson Rd.
Block A #08-01, SLF Complex
Singapore 1129
Tel: 2586988 Fax: 2589945

Trade Fairs

Refer to the "Trade Fairs" chapter for complete listings, including contact information, dates, and venues. Trade fairs with particular relevance to this industry include the following, which are listed in that chapter under the heading given below:

Tools: Precision & Measuring
- China Exhibition For Measuring Instruments (CESMI)
- International Electro-Optics, incorporating Optics Asia
- Measuring Asia—Asia's Premier Exhibition & Conference on Measurement Technology
- Metrology Asia
- Photo Fair
- Quality Asia

For other trade fairs that may be of interest, we recommend that you also consult the headings Agriculture, Forestry & Aquaculture; Automobiles & Automotive Parts; Construction & Housing; Electronic & Electric Equipment; Factory Automation; Machines & Instruments; and Medicine & Pharmaceuticals.

Trade Fairs

Singapore hosts a wide range of trade fairs and expositions that should interest anyone who seeks to do business in this dynamic and expanding economy. Whether you want to buy Singaporean goods or exhibit your own goods and services for sale to the Singaporean market, you will almost undoubtedly find several trade fairs to suit your purposes. Singapore is also a major site for international trade fairs for the nations of Southeast Asia, and attracts exhibitors and buyers from many countries.

The listing of trade fairs in this section is designed to acquaint you with the scope, size, frequency, and length of the events held in Singapore and to give you contact information for the organizers. While every effort has been made to ensure that this information is correct and complete as of press time, the scheduling of such events is in constant flux. Announced exhibitions can be canceled; dates and venues are often shifted. If you're interested in attending or exhibiting at a show listed here, we urge you to contact the organizer well in advance to confirm the venue and dates and to ascertain whether it is appropriate for you. (*See* Tips for Attending a Trade Fair, following this introduction, for further suggestions on selecting, attending, and exhibiting at trade fairs.) The information in this volume will give a significant head start to anyone who has considered participating in a trade fair as an exhibitor or attendee.

In order to make access to this information as easy as possible, fairs have been grouped alphabetically by product category, and within product category, alphabetically by name. Product categories, with cross references, are given following this introduction in a table of contents. Note that the first and last headings listed are out of alphabetical order. Trade fairs listed under Comprehensive do not focus on a single type of product but instead show a broad range of goods that may be from one geographic area or centered around a particular theme. The final category, Others, is a miscellaneous listing of fairs that do not fit easily into one of the other categories. When appropriate, fairs have been listed in more than one category. The breadth of products on display at a given fair means that you may want to investigate categories that are not immediately obvious. Many exhibits include the machinery, tools, and raw materials used to produce the products associated with the central theme of a fair; anyone interested in such items should consider a wide range of the listings.

The list gives the names and dates of both recent and upcoming events, together with site and contact information and a description of the products to be exhibited. Many shows take place on a regular basis. Annual or biennial schedules are common. When we were able to confirm the frequency of a show through independent sources, it has been indicated. Many others on the list may also be regular events. Some are one-time events. Because specifics on frequency are sometimes difficult to come by and because schedules for some 1994 and many 1995 shows were not available at press time, we have given both recent and future dates. It is quite possible that a fair listed for 1993 will be held again in 1994 or 1995, so it would be worthwhile getting in touch with the contact listed for any show that looks interesting. Even if we were not able to confirm the frequency, you can infer a likely time cycle if several dates are given for a fair.

As you gather further information on fairs that appeal to you, do not be surprised if the names are slightly different from those listed here. Some large trade fairs include several smaller exhibits, and some use short names or acronyms. Dates and venues, of course, are always subject to change.

For further information The Singapore Convention Bureau publishes a very complete calendar twice a year, in January and July, of exhibitions and conventions held in Singapore. Events are arranged by date, and each listing includes the name of the trade fair, the venue, number of exhibitors, and the organizer; often, there is also a detailed list of the products to be exhibited. Singapore has become a major center for international trade exhibits, so anyone interested in doing business in Asia, particularly in Southeast Asia, may want to be aware of upcoming trade fairs in Singapore. For a copy of the calendar, contact the Singapore Trade Development Board main office or one of its 25 offices worldwide. (Refer to "Important Addresses" chapter for a listing of offices.)

Singapore Trade Development Board
1 Maritime Square #10-40 (Lobby D)
World Trade Centre, Telok Blangah Rd.
Singapore 0409
Tel: [65] 2719388 Fax: [65] 2740770, 2782518
Tlx: 28617 TRADEV

Other valuable sources of information include the commercial sections of Singapore's diplomatic missions, chambers of commerce, and other business organizations dedicated to trade between your country and Singapore, and the embassy or consulate of your own country located in Singapore. Professional and trade organizations in Singapore involved in your area of interest may also be worth contacting. (Refer to "Important Addresses" chapter for Singaporean embassies and consulates, Singaporean chambers of commerce and business organizations, diplomatic missions located in Singapore, and trade organizations.)

While the annual directory *Trade Shows Worldwide* (Gale Research Inc., Detroit, Michigan) is far from comprehensive, it may provide further information on some trade fairs in Singapore, and it is worth seeking out at your local business library.

TRADE FAIRS
TABLE OF CONTENTS

Comprehensive .... p. 84
    Trade fairs exhibiting a wide range of goods
    See also: Export Fairs
Aerospace & Oceanic .... p. 84
Agriculture, Forestry & Aquaculture .... p. 85
    Includes horticulture
    See also: Food, Beverages & Food Processing
Automobiles & Automotive Parts .... p. 87
Communications & Networks .... p. 87
    See also: Computer & Information Industries,
    Electronic & Electric Equipment, Multimedia & Audiovisual Equipment
Computer & Information Industries .... p. 89
    See also: Communications & Networks, Electronic & Electric Equipment,
    Multimedia & Audiovisual Equipment
Construction & Housing .... p. 91
    See also: Furniture & Housewares
Education & Books .... p. 94
    See also: Hobbies, Recreation & Travel
Electronic & Electric Equipment .... p. 95
    See also: Communications & Networks, Computer & Information Industries,
    Multimedia & Audiovisual Equipment
Environmental & Energy Industries .... p. 98
    See also: Petroleum & Gas Industries
Export Fairs .... p. 101
    Trade fairs exhibiting a wide range of goods from specific countries
    or regions
Factory Automation .... p. 103
    See also: Computer & Information Industries, Machines & Instruments
Food, Beverages & Food Processing .... p. 105
    See also: Agriculture, Forestry & Aquaculture
Furniture & Housewares .... p. 106
    See also: Construction & Housing

TRADE FAIRS
TABLE OF CONTENTS

Gifts, Jewelry & Stationery .... p. 108
    Includes art, crafts, and timepieces
Health & Safety .... p. 110
    Includes security
    See also: Medicine & Pharmaceuticals
Hobbies, Recreation & Travel .... p. 111
    See also: Sporting Goods
Industrial Materials & Chemicals .... p. 112
    See also: Metal & Metal Finishing, Petroleum & Gas Industries
Investment & Business Services .... p. 114
Machines & Instruments .... p. 116
    See also: Tools: Precision & Measuring, and other categories which
    may include exhibitions with machines specific to those industries
Medicine & Pharmaceuticals .... p. 117
    See also: Health & Safety
Metal & Metal Finishing .... p. 124
    See also: Industrial Materials & Chemicals
Multimedia & Audiovisual Equipment .... p. 125
    See also: Communications & Networks, Computer & Information
    Industries, Electronic & Electric Equipment
Packaging, Printing & Paper .... p. 127
    Includes graphics and handling
Petroleum & Gas Industries .... p. 129
    See also: Environmental & Energy Industries, Industrial Materials
    & Chemicals
Sporting Goods .... p. 130
    See also: Hobbies, Recreation & Travel
Textiles & Apparel .... p. 131
Tools: Precision & Measuring .... p. 132
    See also: Tools & Instruments, and other categories which
    may include exhibitions with tools specific to those industries
Others .... p. 133
    Miscellaneous trade fairs

Tips for Attending a Trade Fair

Overseas trade fairs can be extremely effective for making face-to-face contacts and sales or purchases, identifying suppliers, checking out the competitors, and finding out how business really works in the host country. However, the cost of attending such fairs can be high. To maximize the return on your investment of time and money, you should be very clear about your goals for the trip and give yourself plenty of time for advance research and preparation. You should also make sure that you are aware of the limitations of trade fairs. The products on display probably do not represent the full range of goods available on the market. In fact, some of the latest product designs may still be under wraps. And while trade fairs give you an opportunity to make face-to-face contact with many people, both exhibitors and buyers are rushed, which makes meaningful discussions and negotiations difficult. These drawbacks can easily be minimized if you have sufficient preparation and background information. Allow at least three months for preparation—more if you also need to identify the fair that you will attend. Under ideal circumstances, you should begin laying the groundwork nine to 12 months in advance.

Tips for Attending a Trade Fair (cont'd.)

Selecting an appropriate trade fair

Consult the listings of trade fairs here to find some that interest you. Note the suggestions for finding the most current calendars of upcoming fairs. Once you have identified some fairs, contact their organizers for literature, including show prospectus, attendee list, and exhibitor list. Ask plenty of questions. Do not neglect trade organizations in the host country, independent show-auditing firms, and recent attendees. Find out whether there are "must attend" fairs for your particular product group. Fairs that concentrate on other but related commodities might also be a good match. Be aware that there may be preferred seasons for trade in certain products. Your research needs to consider a number of points.

Audience • Who is the intended audience? Is the fair open to the public or only to trade professionals? Are the exhibitors primarily foreigners looking for local buyers or locals looking for foreign buyers? Many trade fairs are heavily weighted to one or the other. Decide whether you are looking for an exposition of general merchandise produced in one region, a commodity-specific trade show, or both.

Statistics • How many people attended the fair the last time it was held? What were the demographics? What volume of business was done? How many exhibitors were there? How big is the exhibition space? What was the ratio of foreign to domestic attendees and exhibitors?

Specifics • Who are the major exhibitors? Are particular publications or organizations associated with the fair? On what categories of products does the fair focus? Are there any special programs, and do they require additional fees? Does the fair have particular themes that change each time? How long has the fair been in existence? How often is it held? Is it always in the same location, or does it move each time? How much does it cost to attend? To rent space?

Before you go

- If you have not already spoken with someone who attended the fair in the past, make sure to seek someone out for advice, tips, and general information.
- Make your reservations and travel arrangements well in advance, and figure out how you are going to get around once you get there. Even if the fair takes place in a large city, do not assume that getting around will be easy during a major trade fair. If the site is a small city or less-developed area, the transportation and accommodation systems are likely to be saturated even sooner than in metropolitan areas.
- Will you need an interpreter for face-to-face business negotiations? A translation service to handle documents? Try to line up providers well in advance of your need for their services.
- Do you need hospitality suites and/or conference rooms? Reserve them as soon as you can.
- Contact people you'd like to meet before you go. Organize your appointments around the fair.
- Familiarize yourself with the show hours, locations (if exhibits and events are staged at several different venues), and schedule of events. Then prioritize.

While you are there

- Wear businesslike clothes that are comfortable.
- Immediately after each contact, write down as much information as you can. Do not depend on remembering it.

After the fair

- Within a week after the conclusion of the fair, write letters to new contacts and follow up on requests for literature. If you have press releases and questionnaires, send them out quickly as well.
- Write a report evaluating the experience while it is still fresh in your mind. Even if you don't have to prepare a formal report, spend some time organizing your thoughts on paper for future reference and to quantify the results. Did you meet your goals? Why or why not? What would you do differently? What unforeseen costs arose?
- With your new contacts and your experience in mind, start preparing for your next trade fair.

If you are selling

- Set specific goals for sales leads, developing product awareness, selling and positioning current customers, and gathering industry information; for example, number of contacts made, orders written, leads converted into sales, visitors at presentations, brochures or samples distributed, customers entertained, seminars attended. You can also set goals for total revenue from sales, cost-to-return benefit ratio, amount of media coverage, and amount of competitor information obtained.
- Review your exhibitor kit, paying particular attention to show hours and regulations, payment policies, shipping instructions and dates, telephone installation, security, fire regulations, union regulations, and extra-cost services. Is there a show theme that you can tie into?
- Gear your advertising and product demonstrations to the audience. Should you stress certain aspects of your product line? Will you need brochures and banners in different languages? Even if you do not need to translate the materials currently in use into another language, do you need to re-write them for a different culture? Consider advertising in publications that will be distributed at the fair.
- Plan the display in your booth carefully; you will have only a few seconds to grab the viewer's attention. Secure a location in a high-traffic area—for example, near a door, restroom, refreshment area, or major exhibitor. Use banner copy that is brief and effective. Focus on the product and its benefits. Place promotional materials and giveaways near the back wall so that people have to enter your area, but make sure that they do not feel trapped. If you plan to use videotapes or other multimedia, make sure that you have enough space. Such presentations are often better suited to hospitality suites, because lights are bright and noise levels high in exhibition halls.
- Do not forget about the details. Order office supplies and printed materials that you will need for the booth. If you ordered a telephone line, bring your own telephone or arrange to rent one. Have all your paperwork—order forms, business cards, exhibitor kit and contract, copies of advance orders and checks, travel documents, and so on—in order and at hand. Draw up a schedule for staffing the booth.
- Plan and rehearse your sales pitch in advance, preferably in a space similar to the size of your booth.
- Do not sit, eat, drink, or smoke while you are in the booth.
- If you plan to return to the next show, reserve space while you're still at the fair.
- Familiarize yourself with import regulations for products that you wish to exhibit at the fair.

If you are buying

- Set specific goals for supplier leads and for gathering industry information; for example, number of contacts made, leads converted to purchases, seminars and presentations attended, booths visited. Other goals might be cost-to-return benefit ratio, amount of competitor information gathered, and percentage of projected purchases actually made.
- List all the products that you seek to purchase, their specifications, and the number of units you plan to purchase of each.
- Know the retail and wholesale market prices for the goods in your home country and in the country where you will be buying. List the highest price you can afford to pay for each item and still get a worthwhile return.
- List the established and probable suppliers for each of the products or product lines that you plan to import. Include their addresses and telephone numbers and your source for the information. Contact suppliers before you go to confirm who will attend and to make appointments.
- Familiarize yourself with customs regulations on the products that you seek to purchase and import into your own country or elsewhere. Be sure to include any products that you might be interested in.

Trade Fair	Site	Exhibition Profile	Organizer
COMPREHENSIVE Trade fairs exhibiting a wide range of goods See also Export Fairs			
Asia Supermarket Last held: September 22-25, 1993	International Merchandise Mart	Products, equipment and services for sale or use by wholesalers and retailers in the supermarket industry. Estimated exhibitors: 100	Tokai Agency (S) No. 2 Jurong East Street 21 5th Level IMM Building Singapore 2260 Tel: 5682437/8 Fax: 5682435 Contact: Mr. Eric Tan, Ms. Elizabeth Ng
Asian International Gift Fair June 1-4, 1993 May 10-13, 1994	World Trade Centre	Corporate gifts, character merchandise, toys, games, consumer electronics, arts and crafts, fashion accessories, festive decorations, home interiors, leather goods, pictures and frames, sports and leisure goods, stationery, watches and clocks. Estimated exhibitors: 350	Singapore Exhibition Services 11 Dhoby Ghaut #15-09 Cathay Building Singapore 0922 Tel: 3384747 Fax: 3395651 Tlx: 23597 Contact: Ms. Rae Lo
Utech Asia Exhibition Last held: March 2-4, 1993	World Trade Center	Automation, aerospace, general engineering, furnishing and bedding, building and construction, electrical and electrotechnical footwear, textile, mining and offshore industries. Domestic and medical/surgical applications of PV technology also included. Estimated exhibitors: 75	Industrial Promotions International 3 Pemimpin Drive #08-01 Lip Hing Industrial Building Singapore 2057 Tel: 3539091 Fax: 3537834

AEROSPACE & OCEANIC

Asian Aerospace incorporating Asian Airport Equipment & Technology and Asia Defence Technology Exhibition February 19-22, 1994 February 6-11, 1996	Changi International Exhibition & Convention Centre	Commercial and military aircraft, equipment and systems. Equipment, products and services for the aerospace retrofitting and manufacturing industries, flying displays and supporting industry conferences. Estimated exhibitors: '94-1150, '96-1180	Asian Aerospace 1 Maritime Square #12-01 World Trade Centre Singapore 0409 Tel: 2711013 Fax: 2744666 Tlx: 39200 Contact: Mr. Jimmy Lau, Mr. Ong Lin

Boats & Aquasports
Last held: April 8-11, 1993

World Trade Center

Held concurrently with the International Aquasports Exhibition. Boat equipment and accessories for both powered and sail, viz, speed boats, luxury yachts, work boats, sail boats, dinghies, navigation equipment, marina developers and designers. Full range of aquasports products for fishing, skiing and fashion apparel.
Estimated exhibitors: 120

Reed Exhibition
1 Maritime Square #12-01
World Trade Centre
Singapore 0409
Tel: 2711013 Fax: 2744666 Tlx: 39200
Contact: Mr. David Low

International High Speed Surface Craft Conference & Exhibition
Last held: March 9-11, 1993

Westin Stamford & Westin Plaza

High speed ferries.
Estimated exhibitors: 250

Fast Ferry International
69 Kings Road
Kingston upon Thames
Surrey KT2 5JB U.K.
Tel: [44] (081) 5491077 Fax: [44] (081) 5472893
Contact: Mr. David Woodgate

SingaPort
March 22-25, 1994

World Trade Centre

Maritime equipment, technology and services, viz ship operations and management, vessels, marine services, port/harbor operations and management. Endorsed by the STDB & the SCB.
Estimated exhibitors: 450

Port of Singapore Authority
c/o Times Conferences & Exhibitions
Times Centre
1 New Industrial Road
Singapore 1953
Tel: 2848844 Fax: 2865754
Contact: Mr. Patrick Soh, Ms. Christine Chew-Ng

AGRICULTURE, FORESTRY & AQUACULTURE Includes horticulture
See also Food, Beverages & Food Processing

Agrotechnology in the Commonwealth: Focus in the 21st Century Exhibition
May 24-26, 1994

National University of Singapore

Scientific and agrotechnology equipment.
Estimated exhibitors: 50

The Singapore Institute of Biology
c/o Department of Biotechnology
Ngee Ann Polytechnic
535 Clementi Road
Singapore 2159
Tel: 4606920 Fax: 7795671
Contact: Dr. Hedy Goh

Note: Country codes for telephone and fax numbers are not displayed unless they are *outside* of Singapore. All country codes have square brackets around them, while city codes have parentheses. The country code for Singapore is [65]. Singapore has no city codes.
Additional Note: Endorsed by the STDB & the SCB denotes trade fairs endorsed by the Singapore Trade Development Board and the Singapore Convention Bureau, indicating a high proportion of international trade visitors.

Trade Fair	Site	Exhibition Profile	Organizer
Aquarama Last held: June 24-27, 1993	World Trade Centre	Ornamental fishes, accessories, fish care products and medication. Estimated exhibitors: 150	Expoconsult Singapore 46A Horne Road Singapore 0820 Tel: 2999273 Fax: 2999782 Contact: Mr. John Neo
Asia-Pacific Bonsai & Suiseki Exhibition May/June, 1995	Chinese Garden	Bonsai plants and Suiseki stones. Estimated exhibitors: 50	Singapore Bonsai Society & Singapore Chinese Garden c/o Singapore Leisure Industries Jurong Park Yuan Ching Road Singapore 2261 Tel: 2643455 Fax: 2658133 Contact: Mr. Lim Sip Li
Fish Asia October 26-29, 1994	World Trade Centre	Aquaculture, seafood, ornamental fish, deep sea and marine fishing products, technology and services Estimated exhibitors: 160	ITP Services 2 Jurong East Street 21 #05-19/22 IMM Building Singapore 2260 Tel: 2913238 Fax: 2965384 Contact: Ms. May Loo
Forest TechAsia Last held: November 23-27, 1993	World Trade Centre	Forestry management, timber supply, handling and storage. Estimated exhibitors: 340	Singapore Exhibition Services 11 Dhoby Ghaut #15-09 Cathay Building Singapore 0922 Tel: 3384747 Fax: 3395651 Tlx: 23597 Contact: Mr. Brendan Kelly
International Orchid Show Singapore Last held: August 26-29, 1993	World Trade Centre	International manufacturers and suppliers of cut flowers, potted plants, plantlets and seedlings, nursery and floristry supplies, show and display fittings, packaging materials, landscaping products, laboratory equipment, chemicals and fertilizers, transportation and support services. Estimated exhibitors: 80	Singapore Trade Development Board c/o Expoconsult Singapore 46A Horne Road Singapore 0820 Tel: 2999273 Fax: 2999782 Contact: Ms. Jean Oh

WoodmacAsia
Last held: November 23-27, 1993

World Trade Centre

Logging, sawmilling, timber processing and woodworking, plant machinery, furniture components and fittings. Endorsed by the STDB & the SCB.
Estimated exhibitors: 570

Singapore Exhibition Services
11 Dhoby Ghaut #15-09
Cathay Building
Singapore 0922
Tel: 3384747 Fax: 3395651 Tlx: 23597
Contact: Mr. Brendan Kelly

AUTOMOBILES & AUTOMOTIVE PARTS

Asian Auto Parts & Accessories
Last held: August 26-29, 1993

World Trade Centre

International manufacturers and suppliers of automobiles, including passenger and commercial vehicles, spare parts and accessories.
Estimated exhibitors: 100

United Exhibition Services
c/o Expoconsult Singapore
46A Horne Road
Singapore 0820
Tel: 2999273 Fax: 2999782
Contact: Ms. Tan Seok Hoon

Asian Automotive & Accessories Exhibition (AAAE), co-located with the Singapore International Motor Show
Every 2 years
November 11-14, 1993
November, 1995

World Trade Centre

Automotive components, aftermarket products, test and service equipment, service, garage and machine shop equipment and products. Equipment and technology for car and truck assembly and a range of the latest cars and vehicles.
Estimated exhibitors: 350

Reed Exhibition
1 Maritime Square #12-01
World Trade Centre
Singapore 0409
Tel: 2711013 Fax: 2744666 Tlx: 39200
Contact: Mrs. Rosalind Seah

COMMUNICATIONS & NETWORKS
See also Computer & Information Industries; Electronic & Electric Equipment; Multimedia & Audiovisual Equipment

Asia TELECOM Exhibition
Last held: May 17-22, 1993

World Trade Centre

International Telecommunication Union exhibition featuring the latest developments and applications in the telecommunications and computer industries.
Estimated exhibitors: 150

International Telecommunication Union
c/o Singapore Telecom & Telecommunication Authority of Singapore
31 Exeter Road #24-00 Comcentre
Singapore 0923
Tel: 8383278 Fax: 2351650
Contact: Mrs. Marian Ling

Note: Country codes for telephone and fax numbers are not displayed unless they are *outside* of Singapore. All country codes have square brackets around them, while city codes have parentheses. The country code for Singapore is [65]. Singapore has no city codes.
Additional Note: Endorsed by the STDB & the SCB denotes trade fairs endorsed by the Singapore Trade Development Board and the Singapore Convention Bureau, indicating a high proportion of international trade visitors.

Trade Fair	Site	Exhibition Profile	Organizer
BroadcastAsia June 1-4, 1994 1996 (Dates to be announced)	World Trade Centre	Broadcast, video and film production equipment, HDTV, ENG/EFP/SNG, studio equipment, microwave transmitter and receivers, test and measurement, lighting, power, cable and accessories. Estimated exhibitors: 400	Singapore Exhibition Services 11 Dhoby Ghaut #15-09 Cathay Building Singapore 0922 Tel: 3384747 Fax: 3395651 Tlx: 23597 Contact: Mr. Nat Wong
CITY-IQ **(Advanced Technology for Infrastructure Development)** July 7-10, 1993 July, 1995	World Trade Centre	Road management systems, estate management systems, information network systems, environment management city planning technology, and telecommunications. Estimated exhibitors: 120	Asia-Pacific Exhibitions & Conventions #08-09 Midlink Plaza 122 Middle Road Singapore 0718 Tel: 3343866 Fax: 3343323 Contact: Ms. Annie Wong
CommunicAsia June 1-4, 1994	World Trade Centre	Electronic communication and information technology. Endorsed by the STDB & the SCB. Estimated exhibitors: 500	Singapore Exhibition Services 11 Dhoby Ghaut #15-09 Cathay Building Singapore 0922 Tel: 3384747 Fax: 3395651 Tlx: 23597 Contact: Ms. Ruby Tan
MobilCommAsia June 1-4, 1994	World Trade Centre	Asian international cellular, radio and satellite communications show. Estimated exhibitors: 80	Singapore Exhibition Services 11 Dhoby Ghaut #15-09 Cathay Building Singapore 0922 Tel: 3384747 Fax: 3395651 Tlx: 23597 Contact: Mr. Nat Wong
NetworkAsia June 1-4, 1994	World Trade Centre	LANs, WANs, MANs, network architecture, OSI modems/MUX, bridges, routers, gateways, cabling media, EDDIX 25. Estimated exhibitors: 100	Singapore Exhibition Services 11 Dhoby Ghaut #15-09 Cathay Building Singapore 0922 Tel: 3384747 Fax: 3395651 Tlx: 23597 Contact: Mr. Nat Wong
NOW June 16-18, 1993	Westin Stamford & Westin Plaza	Open system networking and work-stations technology. Estimated exhibitors: 100	HQ Link 150 South Bridge Road #13-01 Fook Hai Building Singapore 0105 Tel: 5343588 Fax: 5342330 Tlx: 24603 Contact: Ms. Dilys Yong

TRADE FAIRS

OFFICE-IQ **(Office Management Technology and Systems)** July 7-10, 1993 July, 1995	World Trade Centre	Application and customized software, communication systems, computers and information services, data communication and networking, office automation systems and power supply, interior architectural and design, personal environment, space and facility design and office system furniture. Estimated exhibitors: 120	Asia-Pacific Exhibitions & Conventions #08-09 Midlink Plaza 122 Middle Road Singapore 0718 Tel: 3343866 Fax: 3343323 Contact: Ms. Annie Wong
Singapore Informatics October 14-17, 1993 October 13-16, 1994 October 12-15, 1995	World Trade Centre	Computer systems, communication technology, office and organization systems, application systems, application development software, supporting services, multimedia. Estimated exhibitors: '94-1200, '95-1400	Singapore Fed. of the Computer Industry c/o Times Conferences & Exhibitions Times Centre 1 New Industrial Road Singapore 1953 Tel: 2848844 Fax: 2865754 Contact: Mr. Benjamin Ng

COMPUTER & INFORMATION INDUSTRIES
See also Communications & Networks; Electronic & Electric Equipment; Multimedia & Audiovisual Equipment

APCIC **(Automatic Processors, Computers & Instrumentation in Construction)** Last held: July 7-10, 1993	World Trade Centre	Construction automation systems and robotics, computer aided design and drafting, computer aided engineering, software for construction management, measurement and calibration equipment and instruments, diagnostic and testing tools and systems. Estimated exhibitors: 120	Asia-Pacific Exhibitions & Conventions #08-09 Midlink Plaza 122 Middle Road Singapore 0718 Tel: 3343866 Fax: 3343323 Contact: Ms. Annie Wong
AUTO-ID Asia May 18-21, 1994	World Trade Centre	Automatic identification and bar code technology for Southeast Asia. Estimated exhibitors: 45	Business and Industrial Trade Fairs 1 Maritime Square #09-07 World Trade Centre Singapore 0409 Tel: 2783900 Fax: 2718108 Contact: Ms. Faris Yee

Note: Country codes for telephone and fax numbers are not displayed unless they are *outside* of Singapore. All country codes have square brackets around them, while city codes have parentheses. The country code for Singapore is [65]. Singapore has no city codes.

Additional Note: Endorsed by the STDB & the SCB denotes trade fairs endorsed by the Singapore Trade Development Board and the Singapore Convention Bureau, indicating a high proportion of international trade visitors.

Trade Fair	Site	Exhibition Profile	Organizer
Comex Last held: August 19-22, 1993	World Trade Centre	Computer technology, hardware, software, peripherals, systems, services and other related manufacturing technology, equipment and products. Estimated exhibitors: 500	ITP Services 2 Jurong East Street 21 #05-19/22 IMM Building Singapore 2260 Tel: 2913238 Fax: 2965384
International Symposium on IC Technology Systems & Applications Exhibition (ISIC) Last held: September 15-17, 1993	Hyatt Regency Singapore	Equipment and systems from suppliers and companies. Estimated exhibitors: 10	School of Electrical & Electronic Engineering Nanyang Technological University Nanyang Avenue Singapore 2263 Tel: 7794943 Fax: 7916178 Tlx: 38851 NTU Contact: Dr. A.M. Alvarez
Manusoft October, 1995	Singapore International Convention & Exhibition Centre	Manufacturing solution and process software. Estimated exhibitors: 70	Singapore Industrial Automation Association c/o Asia-Pacific Exhibitions & Conventions #08-09 Midlink Plaza 122 Middle Road Singapore 0718 Tel: 3343866 Fax: 3343323 Contact: Ms. Annie Wong
NOW Last held: June 16-18, 1993	Westin Stamford & Westin Plaza	Open system networking and work-stations technology. Estimated exhibitors: 100	HQ Link 150 South Bridge Road #13-01 Fook Hai Building Singapore 0105 Tel: 5343588 Fax: 5342330 Tlx: 24603 Contact: Ms. Dilys Yong
OFFICE-IQ (Office Management Technology and Systems) July 7-10, 1993 July, 1995	World Trade Centre	Application and customized software, communication systems, computers and information services, data communication and networking, office automation systems and power supply, interior architectural and design, personal environment, space and facility design and office system furniture. Estimated exhibitors: 120	Asia-Pacific Exhibitions & Conventions #08-09 Midlink Plaza 122 Middle Road Singapore 0718 Tel: 3343866 Fax: 3343323 Contact: Ms. Annie Wong
Pre Press Asia / Multi-Media Asia Exhibition March 30-April 2, 1994	World Trade Centre	Production systems, design tools, services and electronic publishing. Estimated exhibitors: 100	Conference & Exhibition Management 1 Maritime Square #09-43/56 World Trade Centre Singapore 0409 Tel: 2788666 Fax: 2784077 Tlx: 35319 CEMS Contact: Mr. Wilson Liu

TRADE FAIRS

Singapore Informatics
October 14-17, 1993
October 13-16, 1994
October 12-15, 1995

World Trade Centre

Computer systems, communication technology, office and organization systems, application systems, application development software, supporting services, multimedia.
Estimated exhibitors: '94-1200, '95-1400

Singapore Federation of the Computer Industry
c/o Times Conferences & Exhibitions
Times Centre
1 New Industrial Road
Singapore 1953
Tel: 2848844 Fax: 2865754
Contact: Mr. Benjamin Ng

Singapore Information Technology Exposition (SITEX)
Annual
Last held: December 9-12, 1993

World Trade Centre

IT solutions, international IT showcase, technolab and education/multimedia.
Estimated exhibitors: 1200

Technofairs Corporation
1 Colombo Court #06-01
Singapore 0617
Tel: 3376265/3381789 Fax: 3360718
Contact: Ms. Sophia Yen

The PC Show Singapore
June 24-27, 1993
June 23-26, 1994

World Trade Centre

Computer hardware and software.
Estimated exhibitors: 300

Lines Exposition & Management Services
318-B King George's Avenue
Singapore 0820
Tel: 2998611 Fax: 2998633
Contact: Mr. Jimmy Bong

CONSTRUCTION & HOUSING
See also Furniture & Housewares

**APCIC
(Automatic Processors, Computers & Instrumentation in Construction)**
Last held: July 7-10, 1993

World Trade Centre

Construction automation systems and robotics, computer aided design and drafting, computer aided engineering, software for construction management, measurement and calibration equipment and instruments, diagnostic and testing tools and systems.
Estimated exhibitors: 120

Asia-Pacific Exhibitions & Conventions
#08-09 Midlink Plaza
122 Middle Road
Singapore 0718
Tel: 3343866 Fax: 3343323
Contact: Ms. Annie Wong

Baucon Asia Exhibition
Every 2 years
March 30-April 2, 1993
September 26-29, 1995

World Trade Center

Building materials, prefabricated parts, renovation, construction equipment and building material machinery.
Estimated exhibitors: 500

Conference & Exhibition Management
1 Maritime Square #09-43/56
World Trade Centre
Singapore 0409
Tel: 2788666 Fax: 2784077 Tlx: 35319 CEMS
Contact: Mr. Cheah Wai Hong

Note: Country codes for telephone and fax numbers are not displayed unless they are *outside* of Singapore. All country codes have square brackets around them, while city codes have parentheses. The country code for Singapore is [65]. Singapore has no city codes.

Trade Fair	Site	Exhibition Profile	Organizer
BURO-IQ **(Building Automation & Telecommunications for Commercial Buildings)** July 7-10, 1993 July, 1995	World Trade Centre	Advanced building services and automation systems, access floors, building automation/integration systems, cables, cable systems, computer-aided design, consultancy and management services, energy management systems, environment planning and design, fire and security systems, HVAC systems, lift and movers, monitoring and control systems. Estimated exhibitors: 120	Asia-Pacific Exhibitions & Conventions #08-09 Midlink Plaza 122 Middle Road Singapore 0718 Tel: 3343866 Fax: 3343323 Contact: Ms. Annie Wong
China Build Last held: November 3-6, 1993	World Trade Centre	Building technology materials and equipment from China. Estimated exhibitors: 50	Business and Industrial Trade Fairs 1 Maritime Square #09-07 World Trade Centre Singapore 0409 Tel: 2783900 Fax: 2718108 Contact: Mr. James Kwek
City Trans Asia Last held: September 2-5, 1993	World Trade Centre	Latest developments in city planning and transport technologies. Estimated exhibitors: 150	Ministry of National Development c/o Meeting Planners 2nd Floor PICO Creative Centre 20 Kallang Avenue Singapore 1233 Tel: 2972822 Fax: 2962670/ 2927577 Contact: Ms. Magdalene Sik
CITY-IQ **(Advanced Technology for Infrastructure Development)** July 7-10, 1993 July, 1995	World Trade Centre	Road management systems, estate management systems, information network systems, environment management city planning technology, and telecommunications. Estimated exhibitors: 120	Asia-Pacific Exhibitions & Conventions #08-09 Midlink Plaza 122 Middle Road Singapore 0718 Tel: 3343866 Fax: 3343323 Contact: Ms. Annie Wong
HOME-IQ **(Automation for the Home)** July 7-10, 1993 July, 1995	World Trade Centre	Advanced products and technology for the home: push-button, remote-control and programmable appliances for the kitchen and bathroom, control and management systems for home convenience, comfort, entertainment, communications, security. Estimated exhibitors: 120	Asia-Pacific Exhibitions & Conventions #08-09 Midlink Plaza 122 Middle Road Singapore 0718 Tel: 3343866 Fax: 3343323 Contact: Ms. Annie Wong

TRADE FAIRS   93

Homemakers Last held: April 24-May 2, 1993	World Trade Centre Hall 2 & 3	Furnishing, furniture, housewares and home appliances. Estimated exhibitors: 200	Reed Exhibition 1 Maritime Square #12-01 World Trade Centre Singapore 0409 Tel: 2711013 Fax: 2744666 Tlx: 39200 Contact: Mr. Jimmy Lau, Ms. Jorinda Tan
Interlight (incorporating Pro Light) March 18-21, 1993 October 26-29, 1994	World Trade Center	Lighting equipment and solutions for architects, engineers, interior designers and specifiers. Estimated exhibitors: 100	Lines Exposition & Management Services 318B King George's Avenue Singapore 0820 Tel: 2998611 Fax: 2998633 Contact: Mr. Jimmy Bong
International Kerensky Conference: "Global Trends in Structural Engineering" Exhibition July 20-22, 1994	Orchard Hotel	Computer software for design and analysis of structure, structural systems and construction technologies, as well as materials and products used in buildings and structures. Estimated exhibitors: 20	Institution of Engineers Singapore/ Institution of Structural Engineers, UK Joint Committee - Kerensky Conference Organizing Committee c/o 150 Orchard Road #07-14 Orchard Plaza Singapore 0923 Tel: 7332922 Fax: 2353530 Contact: Mr. John Tan
New Homes April 30-May 8, 1994	World Trade Centre	New and refurbished apartments, bungalows, landed properties and related products. Estimated exhibitors: 70	INTERFAMA Fairs & Exhibitions 1 Maritime Square #09-36 World Trade Centre Singapore 0409 Tel: 2766933 Fax: 2768111 Contact: Ms. Jorinda Tan
OFFICE-IQ (Office Management Technology and Systems) July 7-10, 1993 July, 1995	World Trade Centre	Application and customized software, communication systems, computers and information services, data communication and networking, office automation systems and power supply, interior architectural and design, personal environment, space and facility design and office system furniture. Estimated exhibitors: 120	Asia-Pacific Exhibitions & Conventions #08-09 Midlink Plaza 122 Middle Road Singapore 0718 Tel: 3343866 Fax: 3343323 Contact: Ms. Annie Wong

Note: Country codes for telephone and fax numbers are not displayed unless they are *outside* of Singapore. All country codes have square brackets around them, while city codes have parentheses. The country code for Singapore is [65]. Singapore has no city codes.

Trade Fair	Site	Exhibition Profile	Organizer
SIBEX - Southeast Asian International Building & Construction Exposition Annual May 12-15, 1993 May 11-14, 1994	World Trade Centre	Materials, equipment, systems for the building and construction industry, fire protection and security, M & E and building automation systems, real estate and town planning, retrofitting, renovation, interior design products. Estimated exhibitors: 800	Reed Exhibition 1 Maritime Square #12-01 World Trade Centre Singapore 0409 Tel: 2711013 Fax: 2744666 Tlx: 39200 Contact: Ms. Chua Bee Hong, Ms. Jamie Tan
Valves & Piping Asia Exhibition July 20-23, 1994	World Trade Centre	Valves, pipes, hardware and related products and services. Estimated exhibitors: 250	HQ Link 150 South Bridge Road #13-01 Fook Hai Building Singapore 0105 Tel: 5343588 Fax: 5342330 Tlx: 24603 Contact: Mr. David Chow, Director for Project Planning

EDUCATION & BOOKS
See also Hobbies, Recreation & Travel

Trade Fair	Site	Exhibition Profile	Organizer
Children's Book and Educational Fair Annual (Dates not available)	Site information not available	Children's books, educational products and services.	Academic Associates Kitchener Complex, French Rd. 03-185 Singapore 0820 Tel: 2926166 Contact: Florence Fam
Singapore International Festival of Books and Book Fair Last held: September 4-12, 1993	World Trade Centre	Book exhibition and electronic publishing. CD-I, CD-TV, video and on-line database systems. Estimated exhibitors: 220	Festival of Books Singapore 865 Mountbatten Road #05-28 Singapore 1543 Tel: 3441495 Fax: 3440180 Contact: Mr. N.T.S. Chopra, Book Fair Director
Singapore Language Fair Last held: December 2-5, 1993	World Trade Centre	Exhibits on foreign language courses. Estimated exhibitors: 60	International Trade Fairs (Singapore) 1 Maritime Square #09-20 World Trade Centre Singapore 0409 Tel: 2789166 Fax: 2748670 Tlx: 55171 Contact: Ms. Wong Yoke Lin

Stationery & Book Fair
Last held: August 6-10, 1993

World Trade Centre

Stationery and books.
Estimated exhibitors: 100

Interconex Management
Blk 261 Waterloo Street
#03-26 Waterloo Centre
Singapore 0718
Tel: 3368988 Fax: 3386826
Contact: Mr. Ernest Chan

World Chinese Book Fair
Last held: May 29-June 6, 1993

World Trade Centre

Books, stationery, art, paintings and accessories.
Estimated exhibitors: 100

Lianhe Bao
News Centre
82 Genting Lane
Singapore 1334
Tel: 7438800 Fax: 7466878 Tlx: 55653

ELECTRONIC & ELECTRIC EQUIPMENT
See also Communications & Networks; Computer & Information Industries; Multimedia & Audiovisual Equipment

Asia Electronics
September 21-24, 1994

World Trade Centre

Part of Globatronics. Electronic components, assemblies and parts used in microelectronics and printed circuit board production.
Total estimated exhibitors: 1060

Reed Exhibition
1 Maritime Square #12-01
World Trade Centre
Singapore 0409
Tel: 2711013 Fax: 2744666 Tlx: 39200
Contact: Mr. Leong Choong Cheng, Mrs. Anu Ghosh

BroadcastAsia
June 1-4, 1994
1996 (Dates to be announced)

World Trade Centre

Broadcast, video and film production equipment, HDTV, ENG/EFP/SNG, studio equipment, microwave transmitter and receivers, test and measurement, lighting, power, cable and accessories.
Estimated exhibitors: 400

Singapore Exhibition Services
11 Dhoby Ghaut #15-09
Cathay Building
Singapore 0922
Tel: 3384747 Fax: 3395651 Tlx: 23597
Contact: Mr. Nat Wong

DISKCON Singapore
March 2-3, 1994

Westin Stamford & Westin Plaza

Equipment, materials and components used to manufacture magnetic and optical rigid disk drives.
Estimated exhibitors: 80

International Disk Drive Equipment & Materials Association (IDEMA)
710 Lakeway, Suite 140
Sunnyvale, CA 94186 USA
Tel: [1] (408) 720-9352
Fax: [1] (408) 720-9380
Contact: Mr. Blaine Carmen

Note: Country codes for telephone and fax numbers are not displayed unless they are *outside* of Singapore. All country codes have square brackets around them, while city codes have parentheses. The country code for Singapore is [65]. Singapore has no city codes.

Trade Fair	Site	Exhibition Profile	Organizer
ELECTRIC ASIA Every 2 years September 21-24, 1994	World Trade Centre	Electrical manufacturing and installation technology, distribution equipment, fittings, controls and accessories. Estimated exhibitors: 250	Reed Exhibition 1 Maritime Square #12-01 World Trade Centre Singapore 0409 Tel: 2711013 Fax: 2744666 Tlx: 39200 Contact: Mrs. Anu Ghosh, Ms. Chua Bee Hong, Ms. Jamie Tan
Electronics Subcontracting Asia September 21-24, 1994	World Trade Centre	Part of Globatronics. Subcontracting and contract manufacturing technology, including mold and die manufacturing, metal stamping and precision engineering for the electronics industry. Total estimated exhibitors: 1060	Reed Exhibition 1 Maritime Square #12-01 World Trade Centre Singapore 0409 Tel: 2711013 Fax: 2744666 Tlx: 39200 Contact: Mr. Leong Choong Cheng, Mrs. Anu Ghosh
Electrotest Asia September 21-24, 1994	World Trade Centre	Part of Globatronics. Test and measurement equipment, such as bare board testers, burn-in board testers, optical measurement, test fixtures and x-ray inspection equipment. Total estimated exhibitors: 1060	Reed Exhibition 1 Maritime Square #12-01 World Trade Centre Singapore 0409 Tel: 2711013 Fax: 2744666 Tlx: 39200 Contact: Mr. Leong Choong Cheng, Mrs. Anu Ghosh
Globatronics September 21-24, 1994	World Trade Centre	Incorporates Nepcon Asia-Pacific, International Procurement, Asia Electronics, Electronics Subcontracting Asia, Semitech Asia and Electrotest Asia. Total estimated exhibitors: 1060	Reed Exhibition 1 Maritime Square #12-01 World Trade Centre Singapore 0409 Tel: 2711013 Fax: 2744666 Tlx: 39200 Contact: Mr. Leong Choong Cheng, Mrs. Anu Ghosh
International Procurement September 21-24, 1994	World Trade Centre	Part of Globatronics. International procurement exhibition featuring leading OEM and PPO buyers. Total estimated exhibitors: 1060	Reed Exhibition 1 Maritime Square #12-01 World Trade Centre Singapore 0409 Tel: 2711013 Fax: 2744666 Tlx: 39200 Contact: Mr. Leong Choong Cheng, Mrs. Anu Ghosh

International Symposium on the Physical & Failure Analysis of Integrated Circuits Exhibition (IPFA) Last held: November 1-5, 1993	Westin Stamford & Westin Plaza	Failure analysis equipment. Estimated exhibitors: 20	Institute of Electrical & Electronics Engineers Singapore Section 59D Science Park Drive The Fleming Singapore Science Park Singapore 0511 Tel: 7731141 Fax: 7731142 Contact: Ms. Jasmine Leong
Nepcon Asia-Pacific September 21-24, 1994	World Trade Centre	Part of Globatronics. The PCB, packaging production and design, technology exhibition and conference. Total estimated exhibitors: 1060	Reed Exhibition 1 Maritime Square #12-01 World Trade Centre Singapore 0409 Tel: 2711013 Fax: 2744666 Tlx: 39200 Contact: Mr. Leong Choong Cheng, Mrs. Anu Ghosh
Rigid Disk Drive Components Technology Review Exhibition March 8-9, 1994	Hyatt Regency Singapore	Disk drive magnetic recordings, current and future magnetic technology to the year 2000, as well as components and test technologies. Estimated exhibitors: 15	Head/Media Technology Review 351 S. Hitchcock Way, B-200 Santa Barbara, CA 93105 USA Tel: [1] (805) 963-8081 Fax: [1] (805) 2512 Contact: Ms. Paulette Le Blanc
SEMICON/Singapore February 9-11, 1993 January 11-13, 1994	World Trade Centre	Equipment and materials for the semiconductor and flat panel display industries. Estimated exhibitors: 150	SEMI Singapore Secretariat Theseira & Associates c/o 63 Robinson Road #04-18 Afro Asia Building Singapore 0106 Tel: 2259586 Fax: 2259785 Contact: Ms. Anne Theseira, Mr. Sathish Jeram Naidu
Semitech Asia September 21-24, 1994	World Trade Centre	Part of Globatronics. Manufacturing of semiconductors, including semiconductor packaging and design, technology and materials. Total estimated exhibitors: 1060	Reed Exhibition 1 Maritime Square #12-01 World Trade Centre Singapore 0409 Tel: 2711013 Fax: 2744666 Tlx: 39200 Contact: Mr. Leong Choong Cheng, Mrs. Anu Ghosh

Note: Country codes for telephone and fax numbers are not displayed unless they are *outside* of Singapore. All country codes have square brackets around them, while city codes have parentheses. The country code for Singapore is [65]. Singapore has no city codes.

ENVIRONMENTAL & ENERGY INDUSTRIES
See also Petroleum & Gas Industries

Trade Fair	Site	Exhibition Profile	Organizer
ChemAsia – Asian International Chemical Process Engineering & Contracting Exhibition October 4-7, 1993 1995 (Dates to be announced)	World Trade Centre	Chemical process engineering. Endorsed by the STDB & the SCB. Estimated exhibitors: 250	Singapore Exhibition Services 11 Dhoby Ghaut #15-09 Cathay Building Singapore 0922 Tel: 3384747 Fax: 3395651 Tlx: 23597 Contact: Mr. Ivan Ng
Conference of the International Association on Water Quality Exhibition June 18-21, 1996	Singapore International Convention & Exhibition Centre	Water and wastewater engineering, collection, treatment, disposal and technology. Estimated exhibitors: 240	The Environmental Engineering Society of Singapore (EESS) c/o Times Conferences & Exhibitions Times Centre 1 New Industrial Road Singapore 1953 Tel: 2848844 Fax: 2865754 Contact: Mr. Benjamin Ng
ENEX - ASIA September 21-24, 1994	World Trade Centre	South East Asian region's electrical industry. Estimated exhibitors: 250	Reed Exhibition 1 Maritime Square #12-01 World Trade Centre Singapore 0409 Tel: 2711013 Fax: 2744666 Tlx: 39200 Contact: Mrs. Anu Ghosh, Ms. Chua Bee Hong, Ms. Jamie Tan
EnviroAsia / AquatechAsia June 23-25, 1994	World Trade Centre	Environment and water technology equipment, products and services. Estimated exhibitors: 200	Times Conferences & Exhibitions Times Centre 1 New Industrial Road Singapore 1953 Tel: 2848844 Fax: 2865754 Contact: Ms. Ninette Chan

Environmex Asia Last held: October 4-7, 1993	World Trade Centre	Equipment and services for monitoring air pollution, gas cleaning, odor abatement, dust separation, storage, collection and solid waste, street cleaning, sorting, transportation, recycling incineration, composing deposition, processing of special waste, sludge treatment, reduction of pollution excavation and processing of contaminated soil. Computer models/systems consultancy services. Estimated exhibitors: 280	Singapore Exhibition Services 11 Dhoby Ghaut #15-09 Cathay Building Singapore 0922 Tel: 3384747 Fax: 3395651 Tlx: 23597 Contact: Mr. Nat Wong
Enviroworld Last held: June 24-27, 1993	World Trade Centre	Environment management, technology, equipment and control systems. Estimated exhibitors: 250	Communication International Associates 44/46 Tanjong Pagar Road Singapore 0208 Tel: 2262838 Fax: 2262877 Contact: Ms. Marilyn Au
Filtration & Water Systems Asia Exhibition July 20-23, 1994	World Trade Centre	International exhibition on filtration systems and equipment, water purification equipment, chemicals in Asia. Estimated exhibitors: 250	HQ Link 150 South Bridge Road #13-01 Fook Hai Building Singapore 0105 Tel: 5343588 Fax: 5342330 Tlx: 24603 Contact: Mr. David Chow, Director for Project Planning
POWER GENERATION ASIA September 21-24, 1994	World Trade Centre	Power generation equipment, systems and technology. Estimated exhibitors: 250	Reed Exhibition 1 Maritime Square #12-01 World Trade Centre Singapore 0409 Tel: 2711013 Fax: 2744666 Tlx: 39200 Contact: Mrs. Anu Ghosh, Ms. Chua Bee Hong, Ms. Jamie Tan

Note: Country codes for telephone and fax numbers are not displayed unless they are *outside* of Singapore. All country codes have square brackets around them, while city codes have parentheses. The country code for Singapore is [65]. Singapore has no city codes.

Additional Note: Endorsed by the STDB & the SCB denotes trade fairs endorsed by the Singapore Trade Development Board and the Singapore Convention Bureau, indicating a high proportion of international trade visitors.

Trade Fair	Site	Exhibition Profile	Organizer
POWER-GEN ASIA Last held: September 13-15, 1993	World Trade Centre	Electric utilities, industrial facilities, cogenerators, independent power producers, waste-to-energy plants, architect/engineering firms, consultants/constructors, project developers and government agencies. Estimated exhibitors: 700	Pennwell Conferences & Exhibitions c/o Times Conferences & Exhibitions Times Centre 1 New Industrial Road Singapore 1953 Tel: 2848844 Fax: 2865754 Contact: Mr. Benjamin Ng
Pumps & Systems Asia Exhibition July 20-23, 1994	World Trade Centre	Pumps and related systems in Asia. Estimated exhibitors: 250	HQ Link 150 South Bridge Road #13-01 Fook Hai Building Singapore 0105 Tel: 5343588 Fax: 5342330 Tlx: 24603 Contact: Mr. David Chow, Director for Project Planning
Refining, LNG & PetrochemAsia December 6-9, 1994	World Trade Centre	Pipeline engineering equipment, instrument and control equipment systems, power generation equipment, safety systems and equipment, environment engineering equipment, process control systems, pumps and compressors, heat exchange systems, electrical systems, components and lighting, turnkey contractors and process technologies, tankage fabricators and equipment. Estimated exhibitors: 270	Singapore Exhibition Services 11 Dhoby Ghaut #15-09 Cathay Building Singapore 0922 Tel: 3384747 Fax: 3395651 Tlx: 23597 Contact: Ms. Ruby Tan
Waste Recycle Asia Last held: April 1-4, 1993	World Trade Center	Covers the entire spectrum for management of waste: collection, transport, recycling, composition, combustion and land-filling. Estimated exhibitors: 50	United Exhibition Services 249A Victoria Street (Bugis Village) Singapore 0718 Tel: 3386662 Fax: 3381171 Contact: Mr. Anthony Chan
WatermexAsia Last held: October 4-7, 1993	World Trade Centre	Equipment and services for monitoring waste water, pre-treatment of sewage, mechanical and physical separation, biological treatment, post-treatment of sewage, chemical treatment, sludge treatment, computer models/systems, miscellaneous accessories and consultancy services. Estimated exhibitors: 120	Singapore Exhibition Services 11 Dhoby Ghaut #15-09 Cathay Building Singapore 0922 Tel: 3384747 Fax: 3395651 Tlx: 23597 Contact: Mr. Nat Wong

EXPORT FAIRS
Trade fairs exhibiting a wide range of goods from specific countries or regions

Fair	Venue	Description	Contact
A New Link – South Africa & South East Asia Last held: August 31-September 3, 1993	Westin Stamford & Westin Plaza	South African national industries, including industrial, consumer and service organizations. Estimated exhibitors: 100	South African Foreign Trade Organization (SAFTO) c/o Conference and Exhibition Secretariat 128 Sophia Road #01-01 Singapore 0922 Tel: 3373476 Fax: 3370694 Tlx: 24554 Contact: Ms. Ann Wee
Austrian Export Fair Last held: November 25-28, 1993	World Trade Centre	Wines, machinery, garments, stationery, etc., from Austria. Estimated exhibitors: 200	Debis Marketing Services Asia & Pacific 150 Beach Road #14-01 Gateway West Singapore 0718 Tel: 2961669 Fax: 2918263 Contact: Mr. Bernd Adler
Beijing Economic Trade Fair Last held: September 16-19, 1993	World Trade Centre	Commodities, technologies products from Beijing, China. Estimated exhibitors: 50	Zhongzhan Exhibition Services 62 Cecil Street #B1-00 TPI Building Singapore 0104 Tel: 2223188 Fax: 2259328 Contact: Ms. Lui Wei Fun
China Jiangsu Export Commodities Fair Last held: July 30-August 4, 1993	World Trade Centre	Export products and commodities from Jiangsu, China. Estimated exhibitors: 120	Zhongzhan Exhibition Services 62 Cecil Street #B1-00 TPI Building Singapore 0104 Tel: 2223188 Fax: 2259328 Contact: Ms. Lui Wei Fun
China Jiangxi Provincial Export Commodities Fair & Economic & Technological Co-operation Symposium Last held: September 13-17, 1993	World Trade Centre	Commodities and technologies from Jiangxi, China. Estimated exhibitors: 130	Zhongzhan Exhibition Services 62 Cecil Street #B1-00 TPI Building Singapore 0104 Tel: 2223188 Fax: 2259328 Contact: Ms. Gan Loo Kiat, Ms. Lui Wei Fun

Note: Country codes for telephone and fax numbers are not displayed unless they are *outside* of Singapore. All country codes have square brackets around them, while city codes have parentheses. The country code for Singapore is [65]. Singapore has no city codes.

Trade Fair	Site	Exhibition Profile	Organizer
China New Technology Last held: October 26-30, 1993	World Trade Centre	Chinese international new technological systems, machinery and applications. Estimated exhibitors: 80	XPO Consultant Services 4-A Jalan Gelenggang Singapore 2057 Tel: 4551566 Fax: 4570731 Contact: Mr. Jeremy Chew
China Ningbo Export & Investment Fair Last held: August 17-22, 1993	World Trade Centre	Products from Ningbo City in China. Estimated exhibitors: 40	Skieslands Resources 150 Orchard Road #07-07 Singapore 0953 Tel: 7363988 Fax: 2856972 Contact: Ms. Nana Lin
China Showcase in Singapore (Light Industrial Products & Handicrafts) Last held: December 4-7, 1993	International Merchandise Mart	High-quality products from China, including handicrafts, pottery, fashion accessories and apparel, leather goods, shoes, sporting goods, stationery, gift items, consumer electronics, household, furniture and lighting products. Estimated exhibitors: 200	Headway Trade Fairs (Singapore) 10 Anson Road #15-20 International Plaza Singapore 0207 Tel: 2236603 Fax: 2230289 Contact: Mr. Manson Yip, Manager
Chinese Beverage Exhibition Last held: September 3-6, 1993	World Trade Centre	Different beverages from China. Estimated exhibitors: 40	Sea Sky Cultural Enterprise 300 Sin Ming Road (Singapore-Johore Express Depot) Singapore 2057 Tel: 4550887 Fax: 4523331 Contact: Mr. Seet Chee Peng
Global Business Contacts (GBC) Last held: September 23-26, 1993	World Trade Centre	Small to medium sized manufacturers and exporters of light industry machinery, products and consumer goods of Europe and Asia-Pacific region. Estimated exhibitors: 200	Richsmile Promotions & Marketing Services 1 Colombo Court #06-29 North Bridge Road Singapore 0617 Tel: 3360303 Fax: 2727928 Contact: Mr. Albert Lee
Hangzhou Trade & Investment Fair Last held: August 25-29, 1993	World Trade Centre	Hangzhou products, investment opportunities and fashion show of Hangzhou silk apparel. Estimated exhibitors: 30	Fumay 1 North Bridge Road #19-01 High Street Centre Singapore 0617 Tel: 3362203 Fax: 3343606 Contact: Mr. Tan Tji Kwang, Mr. Cai Qing Ming

TRADE FAIRS   103

Made in Indonesia Last held: August 5-9, 1993	World Trade Centre	National products and commodities from Indonesia. Estimated exhibitors: 160	PT Kompass Indonesia 4 SX-TX Jln Hayam Wuruk Jakarta Pusat 10045 Indonesia Tel: [62] (21) 3801909 Fax: [62] (21) 373707 Tlx: 46344 PASSIA IA Contact: Mr. Freddy Sutedi

FACTORY AUTOMATION
See also Computer & Information Industries, Machines & Instruments

AutomAsia November 15-19, 1994 November, 1995	World Trade Centre	Automated manufacturing technology and robotics. Estimated exhibitors: 1725	Singapore Exhibition Services 11 Dhoby Ghaut #15-09 Cathay Building Singapore 0922 Tel: 3384747 Fax: 3395651 Tlx: 23597 Contact: Mr. Brendan Kelly
IA – The Asia Pacific Industrial Automation Exhibition Every 2 years October, 1995 October, 1997 October, 1999	Singapore International Convention & Exhibition Centre	Factory automation, logistics and material handling, fluid control and power transmission and computers for manufacturing. Estimated exhibitors: 370	Singapore Industrial Automation Association c/o Asia-Pacific Exhibitions & Conventions #08-09 Midlink Plaza 122 Middle Road Singapore 0718 Tel: 3343866 Fax: 3343323 Contact: Ms. Annie Wong
ICCIM – International Exhibition on Computer Integrated Manufacturing Last held: September 6-10, 1993	Westin Stamford & Westin Plaza	Technology in the area of computer integrated manufacturing. Estimated exhibitors: 20	GINTIC Institute of CIM Institute of Manufacturing Technology c/o HQ Link 150 South Bridge Road #13-01 Fook Hai Building Singapore 0105 Tel: 5343588 Fax: 5342330 Tlx: 24603 Contact: Ms. Joyce Lim

Note: Country codes for telephone and fax numbers are not displayed unless they are *outside* of Singapore.
All country codes have square brackets around them, while city codes have parentheses. The country code for Singapore is [65]. Singapore has no city codes.

104 **SINGAPORE Business**

Trade Fair	Site	Exhibition Profile	Organizer
Industrial Automation March 23-26, 1994	World Trade Centre	Logistics, material handling, power transmission, fluid control, manufacturing, process software and factory automation. Estimated exhibitors: 100	Singapore Industrial Automation Association c/o Reed Exhibition 1 Maritime Square #12-01 World Trade Centre Singapore 0409 Tel: 2711013 Fax: 2744666 Tlx: 39200 Contact: Mrs. Anu Ghosh, Mr. Alex Eow
International Conference on Automation, Robotics & Computer Vision – ICARCV Exhibition November 8-11, 1994 November, 1996	1994: Shangri-La Hotel Singapore 1996: To be announced	Automation, robotics and computer vision. Estimated exhibitors: 20	The Institution of Electrical Engineers c/o School of Electrical & Electronic Engineering Nanyang Technological University Nanyang Avenue Singapore 2263 Tel: 7795399 Fax: 7912687 Contact: Assoc. Prof. Leonard Chin, Assoc. Prof. Dinesh P. Mital
MachineAsia Every 2 years Last held: October 26-30, 1993	World Trade Centre	Electrical and electronics in manufacturing, sub-contracting and engineering, computers in manufacturing, material handling and pneumatics and hydraulics, factory equipment maintenance and specialized machinery, plastics and rubber machines. Endorsed by the STDB & the SCB. Estimated exhibitors: 600	Singapore Exhibition Services 11 Dhoby Ghaut #15-09 Cathay Building Singapore 0922 Tel: 3384747 Fax: 3395651 Tlx: 23597 Contact: Ms. Vicki Lim
Robotics October, 1995	Singapore International Convention & Exhibition Centre	Industrial robots, service robots, machine vision systems, automation systems, peripherals including conveyors, sensors, process controls grippers and simulation systems. Estimated exhibitors: 150	Singapore Industrial Automation Association c/o Asia-Pacific Exhibitions & Conventions #08-09 Midlink Plaza 122 Middle Road Singapore 0718 Tel: 3343866 Fax: 3343323 Contact: Ms. Annie Wong

FOOD, BEVERAGES & FOOD PROCESSING
See also Agriculture, Forestry & Aquaculture

Asia Supermarket Last held: September 22-25, 1993	International Merchandise Mart	Products, equipment and services for sale or use by wholesalers and retailers in the supermarket industry. Estimated exhibitors: 100	Tokai Agency (S) No. 2 Jurong East Street 21 5th Level IMM Building Singapore 2260 Tel: 5682437/8 Fax: 5682435 Contact: Mr. Eric Tan, Ms. Elizabeth Ng
Brew Drink Tech Asia Exhibition Last held: March 24-27, 1993	World Trade Center	International exhibition on beverage technology. Estimated exhibitors: 40	Conference & Exhibition Management 1 Maritime Square #09-43/56 World Trade Centre Singapore 0409 Tel: 2788666 Fax: 2784077 Tlx: 35319 CEMS Contact: Ms. Maggie Phang
Chinese Beverage Exhibition Last held: September 3-6, 1993	World Trade Centre	Different beverages from China. Estimated exhibitors: 40	Sea Sky Cultural Enterprise 300 Sin Ming Road (Singapore-Johore Express Depot) Singapore 2057 Tel: 4550887 Fax: 4523331 Contact: Mr. Seet Chee Peng
Food & Hotel Asia Every 2 years April 12-15, 1994 1996 (Dates to be announced)	World Trade Centre	Food and drink, hotel, restaurant and catering equipment, supplies and services. Endorsed by the STDB & the SCB. Estimated exhibitors: 1400	Singapore Exhibition Services 11 Dhoby Ghaut #15-09 Cathay Building Singapore 0922 Tel: 3384747 Fax: 3395651 Tlx: 23597 Contact: Ms. Lindy Wee
Food Ingredients Asia (FIA) Last held: April 20-22, 1993	World Trade Centre	Food ingredients, food analysis equipment and services for the Asian region. Estimated exhibitors: 130	Expoconsult Singapore 46A Horne Road Singapore 0820 Tel: 2999273 Fax: 2999782 Contact: Ms. Ting Siew Mui

Note: Country codes for telephone and fax numbers are not displayed unless they are *outside* of Singapore. All country codes have square brackets around them, while city codes have parentheses. The country code for Singapore is [65]. Singapore has no city codes.

Additional Note: Endorsed by the STDB & the SCB denotes trade fairs endorsed by the Singapore Trade Development Board and the Singapore Convention Bureau, indicating a high proportion of international trade visitors.

Trade Fair	Site	Exhibition Profile	Organizer
IMFEX August 18-21, 1994	World Trade Centre	Muslim food, food ingredients, technology, Halal food preparation, equipment, packaging and processing machinery. Estimated exhibitors: 400	ITP Services 2 Jurong East Street 21 #05-19/22 IMM Building Singapore 2260 Tel: 2913238 Fax: 2965384 Contact: Mr. Justin Wong, Project Executive
ProPakAsia Every 2 years April 12-15, 1994	World Trade Centre	Food processing and packaging technology. Estimated exhibitors: 175	Singapore Exhibition Services 11 Dhoby Ghaut #15-09 Cathay Building Singapore 0922 Tel: 3384747 Fax: 3395651 Tlx: 23597 Contact: Mr. Nat Wong
Salon Culinaire April 11-14, 1994	World Trade Centre	Culinary arts competition featuring events for national and regional teams of individual participants from the food, hotel and catering industry. Estimated exhibitors: 130	Singapore Exhibition Services 11 Dhoby Ghaut #15-09 Cathay Building Singapore 0922 Tel: 3384747 Fax: 3395651 Tlx: 23597 Contact: Ms. Adrienne Pereira

FURNITURE & HOUSEWARES
See also Construction & Housing

Trade Fair	Site	Exhibition Profile	Organizer
Asia Pacific Furniture Fair Last held: July 17-25, 1993	World Trade Centre	Furniture items by members of the Singapore Furniture Manufacturers' & Traders' Association and other furniture companies. Estimated exhibitors: 100	Singapore Furniture Manufacturers' and Traders' Association 16C 4th Storey NTWU Building Lorong 37, Geylang Road Singapore 1438 Tel: 7441600 Fax: 7452917 Tlx: 20855 Contact: Mr. Fong Chee Ying
Combine Home & Appliances Last held: April 9-18, 1993	World Trade Center	Furniture and home appliances. Estimated exhibitors: 120	Combine Management 1 Maritime Square #09-57A World Trade Centre Singapore 0409 Tel: 2737818 Fax: 2702766 Tlx: 22512

TRADE FAIRS

Decoexpo October 26-29, 1994	World Trade Centre	Decorative materials, products and services. Estimated exhibitors: 80	Lines Exposition & Management Services 318B King George's Avenue Singapore 0820 Tel: 2998611 Fax: 2998633 Contact: Mr. Jimmy Bong
FurniTek Asia Last held: November 23-27, 1993	World Trade Centre	Machinery and accessories for furniture production, upholstery and furnishings. Estimated exhibitors: 340	Singapore Exhibition Services 11 Dhoby Ghaut #15-09 Cathay Building Singapore 0922 Tel: 3384747 Fax: 3395651 Tlx: 23597 Contact: Mr. Brendan Kelly
Home Creations Last held: August 14-22, 1993	World Trade Centre	Furniture, furnishings, appliances and wedding accessories. Estimated exhibitors: 72	XPO Consultant Services 4A Jalan Gelenggang Singapore 2057 Tel: 4551566 Fax: 4570731 Contact: Mr. Chester Thang
HomePride Asia April 30-May 8, 1994	World Trade Centre	Home artifacts, antiques, arts, paintings, handicrafts, home-gifts and accessories. Estimated exhibitors: 150	INTERFAMA Fairs & Exhibitions 1 Maritime Square #09-36 World Trade Centre Singapore 0409 Tel: 2766933 Fax: 2768111 Contact: Ms. Jorinda Tan
Homex Last held: November 20-28, 1993	World Trade Centre	Furniture, furnishings and home appliances. Estimated exhibitors: 100	Combine Management 1 Maritime Square #09-57A World Trade Centre Singapore 0409 Tel: 2737818 Fax: 2702766 Tlx: 22512 Contact: Ms. Molly Chua
International Furniture Fair Singapore **In conjunction with the ASEAN Furniture Show** Last held: March 9-13, 1993	World Trade Center	Wood, metal, leather, rattan and cane furniture for residential, office, outdoor and hotel use. Endorsed by STDB & SCB. Estimated exhibitors: 200	Singapore Furniture Industries Council 2 Jurong East St. 21 #02-110 IMM Building Singapore 2260 Tel: 5682626 Fax: 5682922 Contact: Ms. Jannis Seow

Note: Country codes for telephone and fax numbers are not displayed unless they are *outside* of Singapore. All country codes have square brackets around them, while city codes have parentheses. The country code for Singapore is [65]. Singapore has no city codes.

Additional Note: Endorsed by the STDB & the SCB denotes trade fairs endorsed by the Singapore Trade Development Board and the Singapore Convention Bureau, indicating a high proportion of international trade visitors.

Trade Fair	Site	Exhibition Profile	Organizer
LIGHTING ASIA September 21-24, 1994	World Trade Centre	Interior and exterior lighting systems, fittings and accessories. Estimated exhibitors: 250	Reed Exhibition 1 Maritime Square #12-01 World Trade Centre Singapore 0409 Tel: 2711013 Fax: 2744666 Tlx: 39200 Contact: Mrs. Anu Ghosh, Ms. Chua Bee Hong, Ms. Jamie Tan
Model Apartment Last held: October 23-30, 1993	World Trade Centre	Furniture, furnishings, appliances and wedding accessories. Estimated exhibitors: 90	XPO Consultant Services 4-A Jalan Gelenggang Singapore 2057 Tel: 4551566 Fax: 4570731 Contact: Mr. Chester Thang

GIFTS, JEWELRY & STATIONERY   Includes art, crafts, timepieces

Trade Fair	Site	Exhibition Profile	Organizer
Art & Watch Fair February 17-21, 1993 February 25-27, 1994	Westin Stamford & Westin Plaza	Paintings and calligraphy by famous artists, antiques, watches, innovative arts and crafts from the region. Estimated exhibitors: 20	Taisei Gallery Taisei Stamps & Coins 15 Phillip Street #01-00 Singapore 0104 Tel: 535 7255 Fax: 535 5752 Contact: Mr. Peter P.Y. Loke, Ms. Karen Chan
Art in Asia – Singapore Art Fair Last held: July 28-August 1, 1993	World Trade Centre	Paintings, sculptures, drawings and prints, ceramics, calligraphy, fine art photography, installation and performance art, antiquities, ethnic arts and tribal arts. Estimated exhibitors: 88	National Arts Council PSA Building, 35th Floor 460 Alexandra Road Singapore 0511 Tel: 2700722 Fax: 2736880 Contact: Ms. Elaine Jek
Asian International Gift Fair June 1-4, 1993 May 10-13, 1994	World Trade Centre	Corporate gifts, character merchandise, toys, games, consumer electronics, arts and crafts, fashion accessories, festive decorations, home interiors, leather goods, pictures and frames, sports and leisure goods, stationery, watches and clocks. Estimated exhibitors: 350	Singapore Exhibition Services 11 Dhoby Ghaut #15-09 Cathay Building Singapore 0922 Tel: 3384747 Fax: 3395651 Tlx: 23597 Contact: Ms. Rae Lo

TRADE FAIRS

Diamond & Jewellery Fair February 19-21, 1993 February 25-27, 1994	Westin Stamford & Westin Plaza	Hong Kong's latest designs in coin jewelry, diamonds, emeralds, sapphires, rubies and other precious stones, including jewelry pieces. Estimated exhibitors: Not available	Taisei Jewellery & Goldsmith Taisei Stamps & Coins 15 Phillip Street #01-00 Singapore 0104 Tel: 535 7255 Fax: 535 5752 Contact: Mr. Jackson Leow
Handicraft World November 18-21, 1994	World Trade Centre	Artifacts, antiques, batik, carpets, embroidery, paintings, sculptures, souvenirs, etc. Estimated exhibitors: 200	ITP Services 2 Jurong East Street 21 #05-19/22 IMM Building Singapore 2260 Tel: 2913238 Fax: 2965651 Contact: Mr. Justin Wong
HomePride Asia April 30-May 8, 1994	World Trade Centre	Home artifacts, antiques, arts, paintings, handicrafts, home-gifts and accessories. Estimated exhibitors: 150	INTERFAMA Fairs & Exhibitions 1 Maritime Square #09-36 World Trade Centre Singapore 0409 Tel: 2766933 Fax: 2768111 Contact: Ms. Jorinda Tan
International Jewellery Singapore May 6-8, 1993 May 19-21, 1994	World Trade Centre Hall 6	Manufacturers and distributors of fine jewelry, watches and clocks, targeting the ASEAN region. Estimated exhibitors: 100	Reed Exhibition 1 Maritime Square #12-01 World Trade Centre Singapore 0409 Tel: 2711013 Fax: 2744666 Tlx: 39200 Contact: Mr. Jimmy Lau,
Singapore International Watch & Gems Fair Last held: August 19-22, 1993	Dynasty Hotel	The latest jewelry pieces, gemstones and watches. Estimated exhibitors: 80	H-IMS Exhibitions & Conference (M) Sdn Bhd 1611 IGB Plaza No. 6 Jalan Kampar Kuala Lumpur, Malaysia Tel: (03) 442 0833 Fax: (03) 442 0216 Contact: Ms. Lee Maying
Stationery & Book Fair Last held: August 6-10, 1993	World Trade Centre	Stationery and books. Estimated exhibitors: 100	Interconex Management Blk 261 Waterloo Street #03-26 Waterloo Centre Singapore 0718 Tel: 3368988 Fax: 3386826 Contact: Mr. Ernest Chan

Note: Country codes for telephone and fax numbers are not displayed unless they are *outside* of Singapore. All country codes have square brackets around them, while city codes have parentheses. The country code for Singapore is [65]. Singapore has no city codes.

Trade Fair	Site	Exhibition Profile	Organizer
Tresors – The International Fine Art & Antiques Fair for Asia Last held: Sept. 29-Oct. 3, 1993	World Trade Centre	Antiques and fine art for the collector. Estimated exhibitors: 200	Bradbury (International) Limited 5 Wyvern Road Sutton Coldfield West Midlands B74 2PS U.K. Tel: [44] (21) 3545805 Fax: [44] (21) 3553455 Contact: Mr. William R. Burris
HEALTH & SAFETY Includes security See also Medicine & Pharmaceuticals			
Healthy Living/Mums, Babies & Children Last held: March 17-21, 1993	World Trade Center Hall 7	Health and parenting related products, services and activities. Estimated exhibitors: 100	XPO Consultant Services 4-A Jalan Gelenggang Singapore 2057 Tel: 4551566 Fax: 4570731 Contact: Ms. Jilyn Chew
International Airport Emergency Exhibition January 12-14, 1994	Singapore Aviation Academy	Fire protection, fire-fighting and rescue equipment and services. Also featuring manufacturers and other suppliers. Estimated exhibitors: 50	Singapore Aviation Academy No. 1 Aviation Drive Singapore 1749 Tel: 5406209 Fax: 5429890 Contact: Ms. Diana Lee
International Nutrition & Healthcare July 1-4, 1994	World Trade Centre	Nutritional food and healthcare products for hospital institutions, consumers, household. Estimated exhibitors: 180	Success Dynamics (Singapore) Blk A, 2 Leng Kee Road #04-11 Thye Hong Centre Singapore 0315 Tel: 4713218 Fax: 4713036 Contact: Ms. Elsie Kwek, Mr. Alfred Tan
Lifestyle Singapore 2000 Last held: June 25-28, 1993	World Trade Centre	Health and nutritional products, healthcare consumer products and services, over the counter drugs/vitamins, medicated cosmetics, food/soya/natural products. Estimated exhibitors: 100	Success Dynamics (Singapore) Blk A, 2 Leng Lee Road #04-11 Thye Hong Centre Singapore 0315 Tel: 4713218 Fax: 4713036 Contact: Ms. Elsie Kwek, Mr. Alvin Choon
Motherhood Last held: June 11-15, 1993	World Trade Centre	Products for babies and children. Estimated exhibitors: 100	Eastern Publishing Associates 1123 Serangoon Road, #03-01 Singapore 1232 Tel: 2966166 Fax: 2987551

TRADE FAIRS

Occupational Safety & Health Asia Last held: August 18-20, 1993	Mandarin Singapore	International manufacturers and suppliers of chains, ropes, slings, safety belts, safety nets, fire fighting equipment, fire resistant clothing, asbestos abatement, dermatitis control products, medical equipment and services, wellness and fitness programs, eye, face andbody protection, cleaning maintenance material and devices, plant operations equipment, alarms, signaling devices. Estimated exhibitors: 100	Expoconsult Singapore 46A Horne Road Singapore 0820 Tel: 2999273 Fax: 2999782 Contact: Mr. John Neo
Security Asia Last held: February 17-20, 1993	World Trade Center	International exhibition on security equipment. Estimated exhibitors: 250	Conference & Exhibition Management 1 Maritime Square #09-43/56 World Trade Centre Singapore 0409 Tel: 2788666 Fax: 2784077 Tlx: 35319 CEMS Contact: Ms. Maggie Phang

HOBBIES, RECREATION & TRAVEL
See also Sporting Goods

ABTIME Last held: April 15-18, 1993	World Trade Centre	Business travel, incentive travel & corporate hospitality	Connex Connex (Int'l Exhibition Division) 7500A Beach Road The Plaza #07-308/9 Singapore 0719 Tel: 2966961 Fax: 2935628 Tlx: 36463 Contact: Mr. Stephen S. Allen
Photo Fair Last held: September 15-19, 1993	World Trade Centre	Photography, photographic equipment, cameras and accessories. Estimated exhibitors: 120	Photo Fair 11 Tannery Road Singapore 1334 Tel: 7441366 Fax: 7477589 Contact: Mr. Richard Koong, Mr. Lim Kiew Hong
Singapore Taisei International Coin Convention & Exhibition (STICC) February 25-27, 1994	Westin Stamford & Westin Plaza	Ancient and modern coins, medals and bank notes. Public has opportunity to engage in unrestricted buying and selling on the spot. Estimated exhibitors: 100	Taisei Stamps & Coins 15 Phillip Street #01-00 Singapore 0104 Tel: 5357255 Fax: 5355752 Contact: Mr. Peter P.Y. Loke, Ms. Karen Chan

Note: Country codes for telephone and fax numbers are not displayed unless they are *outside* of Singapore. The country code for Singapore is [65]. Singapore has no city codes. All country codes have square brackets around them, while city codes have parentheses.

Trade Fair	Site	Exhibition Profile	Organizer
Singapore World Stamp Exhibition September 1-10, 1995	Singapore International Convention & Exhibition Centre	Priceless competitive and specially invited stamp collections from all over the world. Showcasing Singapore's stamp collections. Estimated exhibitors: 200	Association of Singapore Philatelists 3rd Story, Crosby House 71-75 Robinson Road Singapore 0106 Tel: 3231995 Fax: 3232395 Contact: Ms. F.L. Lee
SingPex Last held: August 21-23, 1993	Hotel New Otani	Philatelic exhibition and seminar on stamps and covers from the Asian region. Estimated exhibitors: 100	Association of Singapore Philatelists 3rd Storey, Crosby House 71-75 Robinson Road Singapore 0106 Tel: 3231995 Fax: 3232395 Contact: Ms. F.L. Lee
Travel March 19-21, 1993 March 12-14, 1994	World Trade Centre	Promotion of tour packages (open to trade and public). Travel agents, national tourist offices, hotels/resorts, airlines and tourist attractions. Estimated exhibitors: 200	National Association of Travel Agents Singapore (NATAS) 401 Havelock Road #04-06 Hotel Miramar Singapore 0316 Tel: 7324022 Fax: 7374834 Contact: Ms. Patricia Auyeong, Ms. Lee Ping Sie
TRAVEX - ASEAN Tourism Forum January 11-13, 1994	World Trade Centre	ASEAN suppliers of tourism products and services in which buyers worldwide are invited to meet by scheduled appointments with the suppliers. Estimated exhibitors: 400	Singapore Tourist Promotion Board Raffles City Tower #36-04 250 North Bridge Road Singapore 0617 Tel: 3396622 Fax: 3399423 Contact: Mr. Chew Khin Peow, Ms. Teresa Cheong

INDUSTRIAL MATERIALS & CHEMICALS
See also Metal & Metal Finishing, Petroleum & Gas Industries

Trade Fair	Site	Exhibition Profile	Organizer
Asian Industrial Coatings Last held: June 10-13, 1993	Site information not available	Range of products from raw materials to end products like architectural coatings, original equipment manufacturer coatings and special purpose coatings. Estimated exhibitors: 150	IMS Exhibitions & Conferences (Singapore) 190 Middle Road #18-06A Fortune Centre Singapore 0718 Tel: 3364611 Fax: 3366762 Tlx: 37011 Contact: Ms. Kelly Chan

TRADE FAIRS

AsiaPlas June 16-19, 1993 June 14-17, 1995	World Trade Centre	Plastic rubber technology. Endorsed by the STDB and the SCB. Estimated exhibitors: 500	Reed Exhibition 1 Maritime Square #12-01 World Trade Centre Singapore 0409 Tel: 2711013 Fax: 2744666 Tlx: 39200 Contact: Mr. Adrian Low, Mr. Rowland Tan, Ms. Chia Chui Lan
International Conference on Geotextiles, Geomembranes & Related Products Exhibition September 5-9, 1994	Westin Stamford & Westin Plaza	Geotextiles, geomembranes and related products. Estimated exhibitors: 100	Southeast Asia Chapter of the International Geotextile Society (SEAC-IGS) P.O. Box 0177 Thomson Road Singapore 9157 Tel: 3535511 Fax: 3532424 Contact: Mr. R.S. Douglas
International Rubber & Plastics Exhibition Last held: October 6-9, 1993	International Merchandise Mart	Plastics and rubber materials, additives, compounds and masterbatch, recycling, blow and injection molding machines, crushing, mixing and screening, extrusion, molds, molding tools, presses, robotics and automation systems, testing and measuring equipment, tire machinery, welding, machines for foam production and other processing machinery, trade publications. Estimated exhibitors: 150	Mass Communications (S) 289 Tanjong Katong Road Singapore 1542 Tel: 3455188 Fax: 3460797 Contact: Mr. Richard Tan
Metals, Manufacturing & Engineering Asia Exhibition Last held: November 23-25, 1993	Westin Stamford & Westin Plaza	Plant supplies and services for foundry, furnace, metallurgical and advanced materials industries. Estimated exhibitors: 100	FMJ International Publications Ltd. Queensway House 2 Queensway Redhill, Surrey RH1 1QS U.K. Tel: [44] (737) 768611 Fax: [44] (737) 761685 Tlx: 948669 TOPJNLG Contact: Mr. Brian Wilkinson, Exhibition Organizer

Note: Country codes for telephone and fax numbers are not displayed unless they are *outside* of Singapore. All country codes have square brackets around them, while city codes have parentheses. The country code for Singapore is [65]. Singapore has no city codes.

Additional Note: Endorsed by the STDB & the SCB denotes trade fairs endorsed by the Singapore Trade Development Board and the Singapore Convention Bureau, indicating a high proportion of international trade visitors.

SINGAPORE Business

Trade Fair	Site	Exhibition Profile	Organizer
Plastec Asia Last held: July 6-10, 1993	World Trade Centre	Plastics machinery, materials, production technology and ancillary equipment. Estimated exhibitors: 150	Business and Industrial Trade Fairs 1 Maritime Square #09-07 World Trade Centre Singapore 0409 Tel: 2783900 Fax: 2718108 Mr. Ben Wong
World of Concrete Asia Last held: February 16-18, 1993	World Trade Center	Products, materials, equipment and services for all aspects of concrete production, construction and rehabilitation. Estimated exhibitors: 100	World of Concrete – International Ltd. c/o CI-Premier 150 Orchard Road #07-14 Orchard Plaza Singapore 0923 Tel: 7332922 Fax: 2353530 Tlx: 33205 Contact: Mr. John S.Y. Tan

INVESTMENT & BUSINESS SERVICES

Trade Fair	Site	Exhibition Profile	Organizer
Asia Pacific Exhibition: HRM Dimensions in the 21st Century – Changes & Strategies Last held: March 9-10, 1993	Hilton International Singapore	Showcases the latest training aids, packages, services, technologies and wellness programs to raise the professionalism of HR and the healthiness of the company. Estimated exhibitors: 20	Singapore Institute of Personnel Management c/o Novo Conferences 1 Science Park Drive SISIR Building Singapore 0511 Tel: 8701136 Fax: 7774463 Tlx: 28499 SISIR Contact: Ms. Nancy Wee
Banque Asia Last held: April 13-15, 1993	Westin Stamford & Westin Plaza	Financial technology applications in the Asia/Pacific region. Estimated exhibitors: 50	Consulton 1 Colombo Court #06-01 Singapore 0617 Tel: 3376265, 3381789 Fax: 3360718 Contact: Mr. B.J. Fernandes
Exhibition on Financing, Developing & Operating Power Stations in Asia March 2-3, 1994	Hilton International Singapore	International project finances, project organizations, operating companies in Asia, Europe and America. In conjunction with Electricity International. Estimated exhibitors: 25	Theseira & Associates Marketing & Communications Network c/o 63 Robinson Road #04-18 Afro Asia Building Singapore 0106 Tel: 2263025 Fax: 2259785 Contact: Ms. Anne Theseira, Mr. Sathish Jeram Naidu

TRADE FAIRS 115

Franchise Asia Last held: May 26-28, 1993	Site information not available	Franchising. Estimated exhibitors: 30	Conference & Exhibition Management 1 Maritime Square #09-43/56 World Trade Centre Singapore 0409 Tel: 2788666 Fax: 2784077 Tlx: 35319 CEMS Contact: Ms. Jessie Goh
Global Investment Last held: May 11-14, 1993	Westin Stamford & Westin Plaza	Real estate, commodities, banking services, financial institutions, consultancy services, credit card services, government institutions, insurance companies, financial information services, publications, bank equipment and supplies, security system and services, investment, banking technology. Estimated exhibitors: 100	Times Conferences & Exhibitions Times Centre 1 New Industrial Road Singapore 1953 Tel: 2848844 Fax: 2865754 Contact: Ms. Christine Chew-Ng
HRD Asia – Premier Human Resource Development, Instructional Design and Training Technology Exhibition October 25-27, 1994	Westin Stamford & Westin Plaza	Audio-visual equipment/accessories, conference facilities, consultancy services, computer based training, correspondence courses, publications, training boards, instructional systems and training packages. Estimated exhibitors: 150	Conference & Exhibition Management 1 Maritime Square #09-43/56 World Trade Centre Singapore 0409 Tel: 2788666 Fax: 2784077 Tlx: 35319 CEMS Contact: Mr. Cheah Wai Hong
Manufacturing ServicesAsia Last held: October 26-30, 1993	World Trade Centre	Subcontracting and industrial equipment, materials and services. Estimated exhibitors: 205	Singapore Exhibition Services 11 Dhoby Ghaut #15-09 Cathay Building Singapore 0922 Tel: 3384747 Fax: 3395651 Tlx: 23597 Contact: Ms. Vicki Lim
Singapore "China Property Exhibition" Last held: July 22-25, 1993	Mandarin Singapore	Investment properties in Beijing, Shanghai and other provinces in China. Estimated exhibitors: 20	Pro & Pub Ltd. (Singapore) #06-01 Tannery Block Ruby Industrial Complex 35 Tannery Road Singapore 1334 Tel: 7423313 Fax: 7424366 Contact: Mr. Don Leong

Note: Country codes for telephone and fax numbers are not displayed unless they are *outside* of Singapore. All country codes have square brackets around them, while city codes have parentheses. The country code for Singapore is [65]. Singapore has no city codes.

116 SINGAPORE Business

MACHINES & INSTRUMENTS
See also Tools: Precision & Measuring, and other categories which may include exhibitions with machines specific to those industries

Trade Fair	Site	Exhibition Profile	Organizer
Asia-Pacific Forex Assembly Exhibition November 10-12, 1994	Westin Stamford & Westin Plaza	Dealing equipment & programs. Estimated exhibitors: 30	The Forex Association of Singapore c/o Hexa-Team Planners 26 Duxton Hill Singapore 0208 Tel: 2278110 Fax: 2278113 Contact: Mrs. Sharon Ho, Mrs. Lynn Yong
InstrumentAsia Last held: October 4-7, 1993	World Trade Centre	Electronic instruments for findustrial applications, on-line and off-line process control equipment, maintenance and repair instruments, testing, calibration and inspection equipment, computer systems and accessories. Estimated exhibitors: 450	Singapore Exhibition Services 11 Dhoby Ghaut #15-09 Cathay Building Singapore 0922 Tel: 3384747 Fax: 3395651 Tlx: 23597 Contact: Mr. Ivan Ng
Sign Asia Last held: April 1-4, 1993	World Trade Center	Sign making equipment and services Estimated exhibitors: 40	Expoconsult Singapore 46A Horne Road Singapore 0820 Tel: 2999273 Fax: 2999782 Contact: Mr. Roel Van Leeuwen
Transfluid October, 1995	Singapore International Convention & Exhibition Centre	Power transmission, pneumatics, hydraulics and fluid control technology. Estimated exhibitors: 110	Singapore Industrial Automation Association c/o Asia-Pacific Exhibitions & Conventions #08-09 Midlink Plaza 122 Middle Road Singapore 0718 Tel: 3343866 Fax: 3343323 Contact: Ms. Annie Wong
TurboMachinery Asia July 20-23, 1994	World Trade Centre	Turbo-machinery and rotating equipment, maintenance and service, condition monitoring equipment, noise and vibration alignment, balancing equipment and service, controls and instrumentation. Estimated exhibitors: 250	HQ Link 150 South Bridge Road #13-01 Fook Hai Building Singapore 0105 Tel: 5343588 Fax: 5342330 Tlx: 24603 Contact: Mr. Roland Kiew, Business Development Manager

MEDICINE & PHARMACEUTICALS
See also Health & Safety

AIDEX Last held: April 15-18, 1993	World Trade Centre	Equipment and care for the elderly and disabled. Estimated exhibitors: 100	Seatro International (Singapore) 5001 Beach Road #12-33 Golden Mile Complex Singapore 0710 Tel: 2965881 Fax: 2961171 Tlx: 36493 SEATRO Contact: Ms. Siti Studaidah
AnaLabAsia – South East Asia International Laboratory & Analytical Exhibition Every 2 years October 4-7, 1993 September, 1995	World Trade Centre	Equipment for wet and dry chemical analysis equipment, microbiological analysis equipment, biochemical analysis equipment, pharmacological analysis, medical analysis equipment, general laboratory apparatus and accessories. Estimated exhibitors: 200	Singapore Exhibition Services 11 Dhoby Ghaut #15-09 Cathay Building Singapore 0922 Tel: 3384747 Fax: 3395651 Tlx: 23597 Contact: Mr. Ivan Hg.
ASEAN Congress of Rheumatology Exhibition Last held: October 31-November 4, 1993	Pan Pacific Hotel Singapore	Medicines and pharmaceuticals, medical equipment, diagnostic aids, patient care items and furnishing and other products relevant to the subject of rheumatology. Estimated exhibitors: 10	Singapore Society of Immunolgy, Allergy and Rheumatology c/o Communication Consultants 336 Smith Street #06-302 New Bridge Centre Singapore 0105 Tel: 2279811 Fax: 2270257 Contact: Mr. Jerry Ng, Ms. Nina Sharma
Asia Pacific Cancer Society Exhibition October 17-20, 1995	Westin Stamford & Westin Plaza	High-tech equipment, latest medical technology and new drugs available in the investigation and treatment of cancer, as well as patient education. Estimated exhibitors: 40	Singapore Cancer Society 15 Enggor Street #06-03/04 Realty Centre Singapore 0207 Tel: 2219577 Fax: 2227424 Contact: Prof. Lee Hin Peng

Note: Country codes for telephone and fax numbers are not displayed unless they are *outside* of Singapore. All country codes have square brackets around them, while city codes have parentheses. The country code for Singapore is [65]. Singapore has no city codes.

Trade Fair	Site	Exhibition Profile	Organizer
Asia Pacific Congress of Endoscopic Surgery Exhibition Last held: August 6-8, 1993	Shangri-La Hotel Singapore	Endoscopic and laparoscopic equipment and instruments. Estimated exhibitors: 40	Endoscopic & Laparascopic Surgeons of Asia c/o Conference & Exhibition Management 1 Maritime Square #09-43/56 World Trade Centre Singapore 0409 Tel: 2788666 Fax: 2784077 Tlx: 35319 CEMS Contact: Ms. Maggie Phang
Asian Congress of Stereotactic, Functional and Computer Assisted December 10-14, 1994	Site information to be announced	Medical equipment and medical computers. Estimated exhibitors: 30	Department of Neurosurgery Singapore General Hospital Outram Road Singapore 0316 Tel: 3213608 Fax: 2263824 Contact: Dr. Prem Kumar Pillay
Asian Pacific Society of Respriology Exhibition Last held: October 7-10, 1993	Mandarin Singapore	Medical and pharmaceutical products. Estimated exhibitors: 40	Singapore Thoracic Society c/o Communication Consultants 336 Smith Street #06-302 New Bridge Centre Singapore 0105 Tel: 2279811 Fax: 2270257 Contact: Prof. Tan Wan Cheng
Asian-Pacific Conference on Doppler and Echocardiography Exhibition Last held: August 6-8, 1993	Mandarin Singapore	Medical equipment and drugs. Estimated exhibitors: 40	Singapore Cardiac Society c/o Academy of Medicine College of Medicine Building 16 College Road #01-01 Singapore 0316 Tel: 2238968 Fax: 225 5155 Contact: Ms. Lim Ai Choo
Basic & Advanced Techniques in Interventional Cardiology Exhibition Last held: July 26-28, 1993	College of Medicine Building	Medical equipment relating to interventional cardiology, such as balloons, catheters, stents, etc. Estimated exhibitors: 15	Department of Cardiology Singapore General Hospital Outram Road Singapore 0316 Tel: 3214026 Fax: 2273562 Contact: Ms. Janie Chan, Prof. Arthur Tan

TRADE FAIRS

Chinese Medicine & Phamacology Show Last held: October 27-30, 1993	World Trade Centre	Products, theories and tactics of traditional Chinese medicine and pharmacology. Estimated exhibitors: 75	XPO Consultant Services 4-A Jalan Gelenggang Singapore 2057 Tel: 4551566 Fax: 4570731 Contact: Mr. Jeremy Chew
Congress of the Federation of Asia & Oceania Perinatal Societies Exhibition November 10-11, 1996	Site information to be announced	Medical products in reproductive medicine. Estimated exhibitors: 50	Perinatal Society of Singapore c/o Ken-Air Skylinks Tours 35 Selegie Road #09-19 Parklane Shopping Mall Singapore 0718 Tel: 3368857/8, Fax: 3363613 Contact: Ms. Felicia Teng
Congress of the International Society of University Colon & Rectal Surgeons Trade Exhibition July 2-6, 1994	Westin Stamford & Westin Plaza	Medical products and services relating to general and colorectal surgery. Estimated exhibitors: 50	International Society of University Colon & Rectal Surgeons (ISUCRS) c/o Department of Colorectal Surgery Singapore General Hospital Outram Road Singapore 0316 Tel: 3214677 Fax: 2262009 Contact: Dr. Y.H. Ho, Dr. Goh Hak Su
Dentistry February 18-20, 1994	World Trade Centre	Dental health care, equipment, supplies and techniques. Estimated exhibitors: 70	Lines Exposition & Management Services 318-B King George's Avenue Singapore 0820 Tel: 2998611 Fax: 2998633 Contact: Mr. Jimmy Bong
Expomed Singapore 2000 Last held: August 12-15, 1993	World Trade Centre	Product development/ equipment for general and specialist application, information technology in healthcare, interacting software and hospital/medical products and services. Estimated exhibitors: 100	Success Dynamics (Singapore) Blk A, 2 Leng Kee Road #04-11 Thye Hong Centre Singapore 0315 Tel: 4713218 Fax: 4713036 Contact: Ms. Elsie Kwek, Mr. Alfred Tan

Note: Country codes for telephone and fax numbers are not displayed unless they are *outside* of Singapore. All country codes have square brackets around them, while city codes have parentheses. The country code for Singapore is [65]. Singapore has no city codes.

Trade Fair	Site	Exhibition Profile	Organizer
Hands-on Course on Total Joint Replacement Exhibition Last held: December 1-4, 1993	Hyatt Regency Singapore	Orthopaedic implants and pharmaceutical products. Estimated exhibitors: 10	Department of Orthopaedic Surgery National University Hospital c/o Communication Consultants 335 Smith Street #06-302 New Bridge Centre Singapore 0105 Tel: 2279811 Fax: 2270257 Contact: Ms. Nina Sharma
International Association for Dental Research Exhibition (IADR) June 28-July 1, 1995	Westin Stamford & Westin Plaza	Dental products including laboratory equipment, clinical instruments and material. Estimated exhibitors: 70	International Association for Dental Research Southeast Asian Division c/o Faculty of Dentistry National University of Singapore 5 Lower Kent Ridge Road Singapore 0511 Tel: 7724944 Fax: 7732602 Contact: Prof. Teo Choo Soo, Dr. Keson Tan
International Conference on Biomedical Engineering (ICBME) Exhibition December 7-10, 1994	Site information to be announced	Medical instruments and equipment, medical devices, implants, computers in medicine and medical books. Estimated exhibitors: 10	Department of Orthopedic Surgery National University Hospital Lower Kent Ridge Road Singapore 0511 Tel: 7724424 Fax: 7780720 Contact: Dr. James Goh, Assistant Manager
International Congress of Radiology Exhibition January 23-28, 1994	World Trade Centre	Radiology and related equipment. Estimated exhibitors: 5000	Singapore Radiological Society c/o The Secretariat International Congress of Radiology Department of Diagnostic Imaging National University Hospital Lower Kent Ridge Road Singapore 0511 Tel: 7761981 Fax: 7762081 Contact: Dr. Lenny Tan
International Congress on Lasers in Dentistry Exhibition August 6-10, 1994	Pan Pacific Hotel Singapore	Laser equipment related to dentistry and medicine. Estimated exhibitors: 20	International Society for Lasers in Dentistry c/o Faculty of Dentistry National University of Singapore Lower Kent Ridge Road Singapore 0511 Tel: 7724933 Fax: 7732600 Contact: Prof. Loh Hong Sai

TRADE FAIRS

International Congress on Transplantation in Developing Countries Exhibition May 10-13, 1995	Westin Stamford & Westin Plaza	Hospital supplies, pharmaceuticals, dialysis, transplantation, medical, healthcare and surgical care products. Estimated exhibitors: 40	National Kidney Foundation 705 Serangoon Road Singapore 1232 Tel: 2990200/3905730 Fax: 2993164/2998711 Contact: Ms. Esther Ho
International Exhibition of the International Society of Haematology August 25-29, 1996	Singapore International Convention & Exhibition Centre	New technology and trends in medical and laboratory equipment, computer systems and innovative medical products related to hematology. Estimated exhibitors: Not available.	International Society of Haematology / Singapore Society of Haematology c/o National Blood Centre Outram Road Singapore 0316 Tel: 2290600 Fax: 2238682 Contact: Dr. Ong Yong Wan
International Symposium on Limb Salvage – ISOLS Last held: August 22-27, 1993	Shangri-La Hotel Singapore	Medical equipment. Estimated exhibitors: 30	Department of Orthopaedic Surgery National University Hospital 5 Lower Kent Ridge Road Singapore 0511 Tel: 7734340 Fax: 7780720 Contact: Prof. Robert Pho
International Union of Biochemistry & Molecular Biology Exhibition April 17-22, 1995	Site information to be announced	Scientific instrumentation and accessories. Estimated exhibitors: 30	Singapore Society for Biochemistry & Molecular Biology c/o Department of Biochemistry National University of Singapore Kent Ridge Road Singapore 0511 Tel: 7723682/7723247 Fax: 7791453 Contact: Dr. Raymond Yuen, Assoc. Prof. Lee Kay Hoon
IUVDT World STD/AIDS Exhibition March 19-23, 1995	Site information to be announced	Pharmaceuticals and diagnostic aids. Estimated exhibitors: 30	Singapore Society of Infectious Disease c/o Communication Consultants 336 Smith Street #06-302 New Bridge Centre Singapore 0105 Tel: 2279811 Fax: 2270257 Contact: Ms. Nina Sharma

Note: Country codes for telephone and fax numbers are not displayed unless they are *outside* of Singapore. All country codes have square brackets around them, while city codes have parentheses. The country code for Singapore is [65]. Singapore has no city codes.

Trade Fair	Site	Exhibition Profile	Organizer
Optics February 18-20, 1994	World Trade Centre	Eye wear, technology and equipment for optometry and ophthalmology. Estimated exhibitors: 70	Lines Exposition & Management Services 318-B King George's Avenue Singapore 0820 Tel: 2998611 Fax: 2998633 Contact: Mr. Jimmy Bong
Pan Ophthalmologica, Singapore Last held: April 15-16, 1993	Site information not available	Ophthalmic products. Estimated exhibitors: 15	Department of Ophthalmology National University Hospital 5 Lower Kent Ridge Road Singapore 0511 Tel: 7725318 Fax: 7777161 Contact: Dr. Ronald Yeoh
Post-Congress Meeting of the International Society of Urology (SIU) Exhibition September 18-22, 1994	Rasa Sentosa Resort	Medical instruments and pharmaceutical products in the field of urology. Estimated exhibitors: 20	Singapore Urological Association c/o Ken-Air Skylinks Tours 35 Selegie Road #09-19 Parklane Shopping Mall Singapore 0718 Tel: 3368857/8, Fax: 3363613 Contact: Ms. Nancy Lee
Regional Conference on Dermatology Exhibition May 21-24, 1994	Shangri-La Hotel Singapore	Dermatology products and equipment. Estimated exhibitors: 40	Dermatological Society of Singapore c/o Conference & Exhibition Management 1 Maritime Square #09-43/56 World Trade Centre Singapore 0409 Tel: 2788666 Fax: 2784077 Tlx: 35319 CEMS Contact: Ms. Maggie Phang
Renal Update Last held: September 4-5, 1993	The Oriental Singapore	Equipment and products related to nephrology, dialysis and transplantation. Estimated exhibitors: 250	Singapore Society of Nephrology Department of Renal Medicine Singapore General Hospital Outram Road Singapore 0316 Tel: 3214436 Fax: 2202308 Contact: Prof. Evan Lee
Singapore National Eye Centre (SNEC) International Meeting Last held: April 16-18, 1993	Shangri-La Hotel Singapore	Ophthalmic instruments and equipment. Estimated exhibitors: 30	Singapore National Eye Centre 11 Third Hospital Avenue Singapore 0316 Tel: 2277255 Fax: 2277290 Tlx: 22842 Contact: Ms. Charity Wai

Singapore Obstetrics & Gynaecology Exhibition Last held: November 20-22, 1993	Shangri-La Hotel Singapore	Medical instruments related to the field of obstetrics and gynecology. Estimated exhibitors: 25	Obstetrical & Gynaecological Society of Singapore c/o Dept. of Obstetrics & Gynaecology National University Hospital Lower Kent Ridge Road Singapore 0511 Tel: 7770313 Fax: 7794753 Contact: Dr. Clement Chan
Singapore-Malaysia Congress of Medicine Exhibition August 4-7, 1994	Site information to be announced	Medical and surgical equipment. Estimated exhibitors: 30	Academy of Medicine College of Medicine Building 16 College Road #01-01 Singapore 0316 Tel: 2238968 Fax: 2255155 Contact: Ms. Lim Ai Choo
The Singapore International Dental Exhibition (SIDEC) June 2-5, 1994	Mandarin Singapore	Technological advances in material science and equipment supplies. Estimated exhibitors: 35	Orchard Dental Centre 268 Orchard Road #05-07 Singapore 0923 Tel: 7343162 Fax: 7321979 Contact: Ms. Evelyn Soh
World Meeting on Impotence Exhibition September 12-16, 1994	Shangri-La Hotel Singapore	Implants (prosthesis) pharmaceuticals and various equipment related to the diagnosis and treatment of impotence. Estimated exhibitors: 20	International Society for Impotent Research/Obstetrical & Gynaecological Society of Singapore c/o Dept. of Obstetrics & Gynaecology National University Hospital Lower Kent Ridge Road Singapore 0511 Tel: 7770313 Fax: 7794753 Contact: Dr. P. Ganesan Adaikan

Note: Country codes for telephone and fax numbers are not displayed unless they are *outside* of Singapore. All country codes have square brackets around them, while city codes have parentheses. The country code for Singapore is [65]. Singapore has no city codes.

METAL & METAL FINISHING
See also Industrial Materials & Chemicals

Trade Fair	Site	Exhibition Profile	Organizer
Asia Weldex/Asia Fastener & Hardware Last held: June 10-13, 1993	Site information not available	Fastener, production machinery, hardware and tool accessories, material and technology exhibition, welding, NDT, surface treatment and metal fabrications industries exhibition. Estimated exhibitors: 200	IMS Exhibitions & Conferences (Singapore) 190 Middle Road #18-06A Fortune Centre Singapore 0718 Tel: 3364611 Fax: 3366762 Tlx: 37011 Contact: Mr. Henry Lee
MetalAsia Every 2 years November 15-19, 1994	World Trade Centre	Machine tool and metalworking. Endorsed by the STDB & the SCB. Estimated exhibitors: 1725	Singapore Exhibition Services 11 Dhoby Ghaut #15-09 Cathay Building Singapore 0922 Tel: 3384747 Fax: 3395651 Tlx: 23597 Contact: Mr. Brendan Kelly
Metals, Manufacturing & Engineering Asia Exhibition Last held: November 23-25, 1993	Westin Stamford & Westin Plaza	Plant supplies and services for foundry, furnace, metallurgical and advanced materials industries. Estimated exhibitors: 100	FMJ International Publications Ltd. Queensway House 2 Queensway Redhill, Surrey RH1 1QS U.K. Tel: [44] (737) 768611 Fax: [44] (737) 761685 Tlx: 948669 TOPJNLG Contact: Mr. Brian Wilkinson
Mould & Die Asia Last held: July 6-10, 1993	World Trade Centre	Die-cast molds, CAD/CAM/CAE systems, surface treatment, mold and die processing machines, components and supplies, parts and components manufacturing machines, measuring/testing and control equipment and auxiliary equipment. Estimated exhibitors: 200	Business and Industrial Trade Fairs 1 Maritime Square #09-07 World Trade Centre Singapore 0409 Tel: 2783900 Fax: 2718108 Contact: Mr. Ben Wong
Stamptec Asia Last held: July 6-10, 1993	World Trade Centre	Metal stamping machinery, production technology and ancillary equipment. Estimated exhibitors: 150	Business and Industrial Trade Fairs 1 Maritime Square #09-07 World Trade Centre Singapore 0409 Tel: 2783900 Fax: 2718108 Mr. Ben Wong

TRADE FAIRS

The International Advanced Materials & Manufacturing Technology Exhibition (AMT) Last held: November 17-20, 1993	International Merchandise Mart	Advanced metals and materials, metal forming and components, metallurgical equipment and plants. Estimated exhibitors: 80	HQ Link 150 South Bridge Road #13-01 Fook Hai Building Singapore 0105 Tel: 5343588 Fax: 5342330 Tlx: 24603 Contact: Mr. Roland Kiew, Business Development Manager
WeldtechAsia Last held: October 26-30, 1993	World Trade Centre	Welding equipment and surface treatment. Estimated exhibitors: 205	Singapore Exhibition Services 11 Dhoby Ghaut #15-09 Cathay Building Singapore 0922 Tel: 3384747 Fax: 3395651 Tlx: 23597 Contact: Ms. Vicki Lim

MULTIMEDIA & AUDIOVISUAL EQUIPMENT

See also Communications & Networks, Computer & Information Industries; Electronic & Electric Equipment

Asia Pacific Imaging and Information Management Exhibition (AIM) Last held: June 16-18, 1993	World Trade Centre	Systems, equipment, accessories and supplies related to image processing and multimedia technology. Estimated exhibitors: 150	HQ Link 150 South Bridge Road #13-01 Fook Hai Building Singapore 0105 Tel: 5343588 Fax: 5342330 Tlx: 24603 Contact: Ms. Dilys Yong
HRD Asia – Premier Human Resource Development, Instructional Design and Training Technology Exhibition October 25-27, 1994	Westin Stamford & Westin Plaza	Audio-visual equipment/accessories, conference facilities, consultancy services, computer based training, correspondence courses, publications, training boards, instructional systems and training packages. Estimated exhibitors: 150	Conference & Exhibition Management 1 Maritime Square #09-43/56 World Trade Centre Singapore 0409 Tel: 2788666 Fax: 2784077 Tlx: 35319 CEMS Contact: Mr. Cheah Wai Hong

Note: Country codes for telephone and fax numbers are not displayed unless they are *outside* of Singapore. All country codes have square brackets around them, while city codes have parentheses. The country code for Singapore is [65]. Singapore has no city codes.

Additional Note: Endorsed by the STDB & the SCB denotes trade fairs endorsed by the Singapore Trade Development Board and the Singapore Convention Bureau, indicating a high proportion of international trade visitors.

Trade Fair	Site	Exhibition Profile	Organizer
International Conference on Multi-Media Modeling Exhibition (MMM) Last held: November 8-12, 1993	National University of Singapore	Multimedia products, systems and solutions. Estimated exhibitors: 10	Department of Information Systems and Computer Science National University of Singapore Lower Kent Ridge Singapore 0511 Tel: 7722505 Fax: 7794580 Contact: Dr. Chua Tat Seng
Multi-Media 2000 Exhibition Last held: March 19-21, 1993	Pan Pacific Hotel Singapore	Interactive multi-media technologies, uses/functions in all areas of work and play. Estimated exhibitors: 60	Success Dynamics (Singapore) Blk A, 2 Leng Kee Road #04-11 Thye Hong Centre Singapore 0315 Tel: 4713218 Fax: 4713036 Contact: Ms. Elsie Kwek, Mr. Alvin Choon
Pre Press Asia / Multi-Media Asia Exhibition March 30-April 2, 1994	World Trade Centre	Production systems, design tools, services and electronic publishing. Estimated exhibitors: 100	Conference & Exhibition Management 1 Maritime Square #09-43/56 World Trade Centre Singapore 0409 Tel: 2788666 Fax: 2784077 Tlx: 35319 CEMS Contact: Mr. Wilson Liu
Pro Audio & Light Asia July 7-10, 1993 July 6-8, 1994	World Trade Centre	Professional recording, sound reinforcement and duplication, lighting, lasers, special effects and associated equipment for the leisure, presentation, entertainment and related industries in Asia. Estimated exhibitors: 275 (1994)	Business and Industrial Trade Fairs 18/F First Pacific Bank Centre 51-57 Gloucester Road Wanchai, Hong Kong Tel: [852] 8652633 Fax: [852] 8661770, 8655513 Tlx: 64882 Contact: Mr. Alan Suen
Singapore Information Technology Exposition (SITEX) Annual Last held: December 9-12, 1993	World Trade Centre	IT solutions, international IT showcase, technolab and education/multimedia. Estimated exhibitors: 1200	Technofairs Corporation 1 Colombo Court #06-01 Singapore 0617 Tel: 3376265/3381789 Fax: 3360718 Contact: Ms. Sophia Yen
Singapore International Festival of Books and Book Fair Last held: September 4-12, 1993	World Trade Centre	Book exhibition and electronic publishing. CD-I, CD-TV, video and on-line database systems. Estimated exhibitors: 220	Festival of Books Singapore 865 Mountbatten Road #05-28 Singapore 1543 Tel: 3441495 Fax: 3440180 Contact: Mr. N.T.S. Chopra, Book Fair Director

World TV, Cable & Satellite
Last held: May/June, 1993

Site information not available

Film and program market for television, video, cable, satellite.
Estimated exhibitors: 100

Asian Exhibitions Management
1 Marine Parade Central #12-01
Parkway Builders Centre
Singapore 1544
Tel: 4473680 Fax: 4473690
Contact: Mr. Zainal Abidin Shah

PACKAGING, PRINTING & PAPER Includes graphics and handling

Asian Paper
April 20-22, 1994

World Trade Centre

International suppliers and manufacturers of paper and board machines, pulping digesters, bleaching systems, pulp refiners, deinking and recycling systems, paper sheeters and converting, box-making and tissue converting, computer process automation, specialty chemicals and starches, measurement and sensor devices, stock preparation equipment, energy and chemical recovery.
Estimated exhibitors: 120

Expoconsult Singapore
46A Horne Road
Singapore 0820
Tel: 2999273 Fax: 2999782
Contact: Ms. Jean Oh

AsiaPack
June 16-19, 1993
June 14-17, 1995

World Trade Centre

Packaging machinery, materials and food processing machinery and systems.
Endorsed by the STDB and the SCB.
Estimated exhibitors: 500

Reed Exhibition
1 Maritime Square #12-01
World Trade Centre
Singapore 0409
Tel: 2711013 Fax: 2744666 Tlx: 39200
Contact: Ms. Daisy Chua

AsiaPrint
June 16-19, 1993
June 14-17, 1995

World Trade Centre

Printing, graphic arts equipment and supplies. Endorsed by the STDB and the SCB.
Estimated exhibitors: 500

Reed Exhibition
1 Maritime Square #12-01
World Trade Centre
Singapore 0409
Tel: 2711013 Fax: 2744666 Tlx: 39200
Contact: Mr. Adrian Low, Mr. Rowland Tan, Ms. Chia Chui Lan

Note: Country codes for telephone and fax numbers are not displayed unless they are *outside* of Singapore. All country codes have square brackets around them, while city codes have parentheses. The country code for Singapore is [65]. Singapore has no city codes.

Additional Note: Endorsed by the STDB & the SCB denotes trade fairs endorsed by the Singapore Trade Development Board and the Singapore Convention Bureau, indicating a high proportion of international trade visitors.

Trade Fair	Site	Exhibition Profile	Organizer
Covertech Asia Last held: April 1-4, 1993	World Trade Center	Paper, boardfilm and foil converting exhibition. Provides platform for suppliers of initial substrates, paper, board, film and foils ancillary equipment to meet converters of S.E. Asia. Estimated exhibitors: 100	United Exhibition Services 249A Victoria Street (Bugis Village) Singapore 0718 Tel: 3386662 Fax: 3381171 Contact: Mr. Anthony Chan
International Packaging Exhibition Last held: October 6-9, 1993	International Merchandise Mart	Packaging machinery and equipment, packaging materials and supporting materials, packaging services, food processing equipment, additives/ingredients, trade publications on packaging and food processing. Estimated exhibitors: 400	Mass Communications (S) 289 Tanjong Katong Road Singapore 1542 Tel: 3455188 Fax: 3460797 Contact: Mr. Richard Tan
International Printing Exhibition Last held: October 6-9, 1993	International Merchandise Mart	Equipment for photocomposing, typesetting, photographic, reproduction, darkrooms, graphic arts. Printing, cutting, collating, laminating and binding machines, and paper and ink publications and other related services. Estimated exhibitors: 200	Mass Communications (S) 289 Tanjong Katong Road Singapore 1542 Tel: 3455188 Fax: 3460797 Contact: Mr. Richard Tan
LabelexpoAsia Last held: June 16-19, 1993	World Trade Centre	For label printers and those involved within the field of narrow web printing production. Estimated exhibitors: 300	Reed Exhibition 1 Maritime Square #12-01 World Trade Centre Singapore 0409 Tel: 2711013 Fax: 2744666 Tlx: 39200 Contact: Mr. Rowland Tan
Logismat October, 1995	Singapore International Convention & Exhibition Centre	Logistic distribution and materials handling technology. Estimated exhibitors: 160	Singapore Industrial Automation Association c/o Asia-Pacific Exhibitions & Conventions #08-09 Midlink Plaza 122 Middle Road Singapore 0718 Tel: 3343866 Fax: 3343323 Contact: Ms. Annie Wong

Material HandlingAsia
Last held: October 26-30, 1993

World Trade Centre

Material handling and storage systems.
Estimated exhibitors: 205

Singapore Exhibition Services
11 Dhoby Ghaut #15-09
Cathay Building
Singapore 0922
Tel: 3384747 Fax: 3395651 Tlx: 23597
Contact: Ms. Vicki Lim

ProPakAsia
Every 2 years
April 12-15, 1994

World Trade Centre

Food processing and packaging technology.
Estimated exhibitors: 175

Singapore Exhibition Services
11 Dhoby Ghaut #15-09
Cathay Building
Singapore 0922
Tel: 3384747 Fax: 3395651 Tlx: 23597
Contact: Mr. Nat Wong

ProPakAsia
Every 2 years
April 12-15, 1994

World Trade Centre

Food processing and packaging technology.
Estimated exhibitors: 175

Singapore Exhibition Services
11 Dhoby Ghaut #15-09
Cathay Building
Singapore 0922
Tel: 3384747 Fax: 3395651 Tlx: 23597
Contact: Mr. Nat Wong

Screen Printing Asia
Last held: April 1-4, 1993

World Trade Center

Screen printing technologies and techniques exhibited by the region's leading screen printers and suppliers.
Estimated exhibitors: 100

United Exhibition Services
249A Victoria Street (Bugis Village)
Singapore 0718
Tel: 3386662 Fax: 3381171
Contact: Mr. Anthony Chan

PETROLEUM & GAS INDUSTRIES
See also Environmental & Energy Industries, Industrial Materials & Chemicals

Asia Pacific Oil and Gas Exhibition
Last held: February 8-10, 1993

Westin Stamford & Westin Plaza

Exploration, evaluation, drilling, environment and safety, pipeline and transportation, new technology for production, well completion and other aspects of engineering.
Estimated exhibitors: 40

Society of Petroleum Engineers
c/o Conference and Exhibition Secretariat
128 Sophia Road #01-01
Singapore 0922
Tel: 3373476 Fax: 3370694
Contact: Ms. Ann Wee

Note: Country codes for telephone and fax numbers are not displayed unless they are *outside* of Singapore. All country codes have square brackets around them, while city codes have parentheses. The country code for Singapore is [65]. Singapore has no city codes.

Trade Fair	Site	Exhibition Profile	Organizer
International Exhibition on Safety, Health and Loss Prevention in the Oil, Chemical and Process Industries Last held: February 15-19, 1993	Pan Pacific Hotel Singapore	Equipment related to the loss prevention of oil, chemical and process industries. Estimated exhibitors: 12	Society of Loss Prevention in the Oil, Chemical and Process Industries c/o Mansfield International 71 Robinson Road #04-02 Crosby House Singapore 0106 Tel: 2280940 Fax: 2263733 Tlx: 22485 Contact: Ms. Tan Bee Nah
Offshore South East Asia Exhibition & Conference Every 2 years December 6-9, 1994	World Trade Centre	Machinery, vessels, equipment and services used for oil and gas exploration and production, offshore and onshore. Endorsed by the STDB & the SCB. Estimated exhibitors: 1200	Singapore Exhibition Services 11 Dhoby Ghaut #15-09 Cathay Building Singapore 0922 Tel: 3384747 Fax: 3395651 Tlx: 23597 Contact: Mr. Ivan Ng
Petro-Safe Asia September 14-16, 1993 September 20-22, 1994	Westin Stamford & Westin Plaza	Environmental, safety and health technology of the oil, gas and petrochemical industries. Estimated exhibitors: 130	Times Conferences & Exhibitions Times Centre 1 New Industrial Road Singapore 1953 Tel: 2848844 Fax: 2865754 Contact: Mr. Patrick Soh

SPORTING GOODS
See also Hobbies, Recreation & Travel

Trade Fair	Site	Exhibition Profile	Organizer
BoatAsia Last held: June 3-7, 1993	World Trade Centre	Pleasure craft, marine/boating equipment and accessories, watersports equipment, accessories, nautical wear, marina/resorts, marina consultants. Estimated exhibitors: 120	Marine Recreation 50 Jalan Sultan #10-01 Jalan Sultan Centre Singapore 0719 Tel: 2938811 Fax: 2981298 Contact: Mr. Kenneth Pereira
Boats & Aquasports Last held: April 8-11, 1993	World Trade Center	Held concurrently with the International Aquasports Exhibition. Boat equipment and accessories for powered and sail, viz, speed boats, luxury yachts, work boats, sail boats, dinghies, navigation equipment, marina developers and designers. Full range of products for fishing, skiing, fashion apparel. Estimated exhibitors: 120	Reed Exhibition 1 Maritime Square #12-01 World Trade Centre Singapore 0409 Tel: 2711013 Fax: 2744666 Tlx: 39200 Contact: Mr. David Low

GCSAA Pacific Rim Golf Course Show March 19-21, 1993 March 21-27, 1994	International Merchandise Mart	Manufacturers, suppliers and distributors of golf course management equipment, supplies, services golf course developers, architects and builders. Estimated exhibitors: 200	Golf Course Superintendents Association of America (Singapore) 2 Jurong East Street 21 #04-216 IMM Building Singapore 2260 Tel: 5682224 Fax: 5682473 Contact: Ms. Linda Fortunato
Golf Asia March 25-28, 1993 March 31-April 3, 1994	World Trade Center	Golfing apparel and accessories, golf antiques and arts, literature, holidays, courses, properties, clubs, schools, shop retail, club management, indoor golf, tournaments, sponsorship, course planning, equipment and maintenance, associations and golf tours. Estimated exhibitors: 275	Connex (Int'l Exhibition Division) 7500A Beach Road The Plaza #07-308/9 Singapore 0719 Tel: 2966961 Fax: 2935628 Tlx: 36463 Contact: Mr. Stephen S. Allen
Sports Asia Last held: June 8-11, 1993	World Trade Centre	Technology, services and education aspects of the sports industry. Exhibits range from upstream construction sector of sports centers to downstream suppliers of sporting goods. Estimated exhibitors: 100	Times Conferences & Exhibitions Times Centre 1 New Industrial Road Singapore 1953 Tel: 2848844 Fax: 2865754 Contact: Mr. Garrick Sim, Ms. Pauline Lee

TEXTILES & APPAREL

Fashion Asia Annual Last held: September, 1993	Changi International Exhibition & Conference Center	Fashions and accessories.	Cahners Exposition Group (Singapore) 1 Maritime Sq., 12-01 World Trade Centre Singapore 0409 Tel: 2711013 Fax: 2744666 Tlx: 39200
International Trade Fair for Footwear, Leather Goods, Bags & Travel Goods (IBF) Last held: October 6-9, 1993	World Trade Centre	Footwear, bags, leather products and travel goods, manufacturing materials, components and parts, chemicals, machinery and technology. Estimated exhibitors: 100	HQ Link 150 South Bridge Road #13-01 Fook Hai Building Singapore 0105 Tel: 5343588 Fax: 5342330 Tlx: 24603 Contact: Ms. Kate Chia, Marketing Manager

Note: Country codes for telephone and fax numbers are not displayed unless they are *outside* of Singapore. All country codes have square brackets around them, while city codes have parentheses. The country code for Singapore is [65]. Singapore has no city codes.

Additional Note: Endorsed by the STDB & the SCB denotes trade fairs endorsed by the Singapore Trade Development Board and the Singapore Convention Bureau, indicating a high proportion of international trade visitors.

Trade Fair	Site	Exhibition Profile	Organizer
Wedding Festival Last held: April 8-11, 1993	World Trade Center	Wedding gowns and accessories. Estimated exhibitors: 70	Full House Marketing & Entertainment 50 Jalan Sultan #16-06 Jalan Sultan Centre Singapore 0719 Tel: 2965320 Fax: 2984764

TOOLS: PRECISION & MEASURING
See also Machines & Instruments, and other categories which may include exhibitions with tools specific to those industries

Trade Fair	Site	Exhibition Profile	Organizer
China Exhibition for Measuring Instruments (CESMI) Last held: July 27-30, 1993	World Trade Centre	Measuring, systems and equipment metrology, weighing, testing and inspection, calibration equipment, etc. Estimated exhibitors: 80	China National Institute of Metrology c/o HQ Link 150 South Bridge Road #13-01 Fook Hai Building Singapore 0105 Tel: 5343588 Fax: 5342330 Tlx: 24603 Contact: Ms. Dilys Yong
International Electro-Optics, incorporating Optics Asia Last held: November 11-14, 1993	International Merchandise Mart	Optics, laser and electro-optics exposition, encompassing commercial industrial, marine, scientific and laboratory, as well as laser technology systems and products. Estimated exhibitors: 380	ITP Services 2 Jurong East Street 21 #05-19/22 IMM Building Singapore 2260 Tel: 2913238 Fax: 2965384 Contact: Mr. William Lim
Measuring Asia – Asia's Premier Exhibition & Conference on Measurement Technology Last held: July 27-30, 1993	World Trade Centre	Measuring systems and equipment and related technologies. Estimated exhibitors: 150	HQ Link 150 South Bridge Road #13-01 Fook Hai Building Singapore 0105 Tel: 5343588 Fax: 5342330 Tlx: 24603 Contact: Ms. Kate Chia, Marketing Manager
Metrology Asia November 15-19, 1994	World Trade Centre	Test forms, standards and gauges, length scanners, angle and inclination measuring equipment, workshop length testing equipment, optical, co-ordinate and other measuring equipment and testing devices, measuring robots, software, materials testing and machine monitoring. Estimated exhibitors: 90	Singapore Exhibition Services 11 Dhoby Ghaut #15-09 Cathay Building Singapore 0922 Tel: 3384747 Fax: 3395651 Tlx: 23597 Contact: Mr. Brendan Kelly

Photo Fair Last held: September 15-19, 1993	World Trade Centre	Photography, photographic equipment, cameras and accessories. Estimated exhibitors: 120	Photo Fair 11 Tannery Road Singapore 1334 Tel: 7441366 Fax: 7477589 Contact: Mr. Richard Koong, Mr. Lim Kiew Hong
Quality Asia Every 2 years July 12-15, 1995	Singapore International Convention & Exhibition Centre	Metrology, test and measuring equipment, systems and consultancy service on quality. Estimated exhibitors: 80	Reed Exhibition 1 Maritime Square #12-01 World Trade Centre Singapore 0409 Tel: 2711013 Fax: 2744666 Tlx: 39200 Contact: Mrs. Anu Ghosh

OTHERS Miscellaneous trade fairs

City Trans Asia Last held: September 2-5, 1993	World Trade Centre	Latest developments in city planning and transport technologies. Estimated exhibitors: 150	Ministry of National Development c/o Meeting Planners 2nd Floor PICO Creative Centre 20 Kallang Avenue Singapore 1233 Tel: 2972822 Fax: 2962670/ 2927577 Contact: Ms. Magdalene Sik
COMMIT Peace Exhibition August, 1994	World Trade Centre	Military equipment for conversion to commercial use. Estimated exhibitors: 120	Seatro International (S) 5001 Beach Road #12-33 Golden Mile Complex Singapore 0719 Tel: 2965881 Fax: 2961171 Tlx: 36943 Contact: Ms. Sant Kaur-Jayaram
Interclean Asia January 14-16, 1993 March 30-April 1, 1995	World Trade Centre	Industrial cleaning and maintenance. Estimated exhibitors: 120	Conference and Exhibition Secretariat 128 Sophia Road #01-01 Singapore 0922 Tel: 3373476 Fax: 3370694 Tlx: 24554 Contact: Ms. Ann Wee

Note: Country codes for telephone and fax numbers are not displayed unless they are *outside* of Singapore. All country codes have square brackets around them, while city codes have parentheses. The country code for Singapore is [65]. Singapore has no city codes.

Trade Fair	Site	Exhibition Profile	Organizer
The Singapore Green Plan Exhibition November 9-14, 1995	World Trade Centre	Publicize the Green Plan, the national vision of Singapore as a model Green City by the year 2000. Estimated exhibitors: 50	Ministry of the Environment Environment Building, 40 Scotts Road #11-00 Singapore 0922 Tel: 7327733 Fax: 7319922 Contact: Mr. Yeo Boon Leng

Note: Country codes for telephone and fax numbers are not displayed unless they are *outside* of Singapore. All country codes have square brackets around them, while city codes have parentheses. The country code for Singapore is [65]. Singapore has no city codes.

Business Travel

Business travel in Singapore is nearly effortless. The entire city-state is modern, compact, efficiently run, and clean. Telecommunications, transportation, and public health and safety are the equal of any in the world. Hotels leave nothing to chance. The only things you have to concentrate on in Singapore are business and where and what to eat next.

NATIONAL TRAVEL OFFICES WORLDWIDE

The Singapore Tourist Promotion Board (STPB) has offices in cities around the world to answer questions and steer prospective travelers in the right direction with maps and brochures.

Auckland Singapore Tourist Promotion Board, c/o Walshes World, 2nd Floor, Dingwall Building, 87 Queen Street, Box 279, Auckland 1, New Zealand; Tel: (9) 79-3708 Tlx: 21437.

Chicago Suite 818, 333 North Michigan Avenue, Chicago, IL 60601, USA; Tel: (312) 220-0099 Fax: (312) 220-0020.

Frankfurt Fremdenverkehrsburo von Singapur, Postrasse 2-4, Frankfurt, Germany 6000; Tel: (69) 231-456/7 Fax: (69) 233-924 Tlx: 4189742.

Hong Kong Suite 1402, 1-13 D'Aguilar Street, Central, Century Square, Hong Kong; Tel: 224-0523 Fax: 810-6694 Tlx: 86630 ETBHK HX.

London Singapore Tourist Promotion Board, 1st Floor, Carrington House, 126-130 Regent Street, London W1R 5FE, UK; Tel: (71) 437-0033 Fax: (71) 734-2191 Tlx: 893491.

Los Angeles Suite 510, 8484 Wilshire Boulevard, Beverly Hills, CA 90211, USA; Tel: (213) 852-1901 Fax: (213) 852-0129 Tlx: 278141.

New York NBR, 12th Floor, 590 Fifth Avenue, New York, NY 10036, USA; Tel: (212) 302-4861 Fax: (212) 302-4801.

Osaka Singapore Tourist Promotion Board, 1F Sumito Seimei Nishi-Honmachi Building, 1-chome, 6-5 Nishi Honmachi, Nishi-ku, Osaka 550, Japan; Tel: (6) 538-3389 Fax: (6) 538-3384.

Paris L'Office National du Tourisme de Singapour, Centre d'Affaires le Louvre, 2 place du Palais Royal, 75044 Paris, France; Tel: (1) 42-97-16-16 Fax: (1) 42-97-16-17 Tlx: SINGPAR 213593F.

Perth Singapore Tourist Promotion Board, c/o Forum Organisation, Suite 1202, Level 12, Westpac Plaza, 60 Margaret Street, Perth, WA 6000, Australia; Tel: (6) 241-3771/2 Fax: (6) 252-3586 Tlx: 127775.

Seoul Singapore Tourist Promotion Board, c/o Nara Corporation, Dongsan Building, 28-1, Jamwon-dong Seocho-ku, Seoul 137-030, CPO Box 1894, Seoul, Republic of Korea; Tel: (2) 549-0691 Tlx: 29956.

Taipei Singapore Tourist Promotion Board, 9th Floor, TFIT Tower, 85 Jen Ai Road, Section 4, Taipei, Taiwan; Tel: (2) 721-0644 Tlx: 24974.

Tokyo Singapore Tourist Promotion Board, 1st Floor, Yamato Seimei Building, 1-chome, 1-7 Uchisaiwai-cho, Chiyoda-ku, Tokyo 100, Japan; Tel: (3) 3593-3388 Fax: (3) 3591-1480.

Toronto Suite 1112, North Tower, 175 Bloor Street East, Toronto, ON M4W 3R8, Canada; Tel: (416) 323-9139 Fax: (416) 323-3514.

Zurich Hochstrasse 48, Zurich CH 8044, Switzerland; Tel: (1) 252-5454 Fax: (1) 252-5303.

STPB OFFICES IN SINGAPORE

The headquarters in Singapore is located at: 36-04 Raffles City Tower, 250 North Bridge Street; Tel: 3396622 Fax: 3399423 Tlx: 33375.

Information Centres are located at:
Ground Floor, Raffles City Tower, 250 North Bridge Road, Singapore 0617; Tel: 3396622 Fax: 3399423 Tlx: 33375; *and* 6 Scotts Road, 02-02/03 Scott's Shopping Centre, Singapore 0922; Tel: 7383778/9.

VISA AND PASSPORT REQUIREMENTS

Singapore's entry requirements are easy for travelers who aren't planning to become employed in Singapore. Citizens of most Western nations, all British Commonwealth nations, Japan, Korea, and Pakistan don't need visas for stays up to 90 days. When you arrive, you'll get a 14 day visitor permit—which you can extend for another 14 days if you can prove you have enough money to live on and a plane ticket to leave. After that, it gets a little more complex to extend your stay, but usually you'll have no real problem staying up to 90 days. The visitor permit is all you'll need to conduct business with your Singapore contacts as long as you're not earning money in some way. Holders of British or US passports don't even need permits.

You'll need a variety of other permits if you want to work, set up a business, undergo training, or exhibit at a trade fair, among other business activities. Some can be granted upon arrival; others you'll need to get before you arrive. Check with your local Singapore consulate. (Refer to "Important Addresses" chapter for a listing of Singaporean embassies and consulates.)

IMMUNIZATION

You need no proof of immunizations unless you've passed through an infected area, such as South America (cholera and yellow fever) or Africa (yellow fever), within the past six days.

CLIMATE

Lying just north of the equator, Singapore has the sunny, hot, humid kind of climate that you'd expect in the tropics. The greater part of the annual 200 cm. (78 inches) of rain falls from November through February, when the average high temperature cools down a bit—30°C (86°F) in January compared to 31°C (88°F) in June. In fact, the winter months are famous for their spectacular thunderstorms and drenching rains, although the average low temperature never falls below 23°C (74°F). However, the humidity creates tropical showers all year long. You'll be glad that the entire city is air-conditioned.

BUSINESS ATTIRE

The year-round humid warmth of Singapore's tropical climate mandates a common sense balance between saving face and heat prostration. The city is renowned for its tailors, and business folk are expected to wear the finest tropical-weight suits, but only long enough to strip down to a first-rate shirt and tie—or, in a woman's case, to the finest, most tastefully conservative blouse and skirt. Men also favor the ubiquitous safari-style suits. Women, be forewarned: perspiration-proof and rainproof makeup is de rigueur. If you don't care to be refreshed by the copious showers, umbrellas are a necessity. Combat the heat with a raincoat that breathes.

AIRLINES

At least 53 airlines fly into Singapore's Changi Airport. Among them are Air China, Air India, Air New Zealand, Alitalia, All Nippon, British Airways, Cathay Pacific, China Airlines, Japan Airlines, Japan Airlines, Japan Air System, KLM, Korean Air, Lufthansa, Malaysia Airlines, Northwest, Philippine Airlines, Qantas, SAS, Singapore Airlines, South African Airways, Swissair, Thai Airways, Turk Hava Yollari, and United.

Air Travel Time*

- From Auckland nonstop on Air New Zealand: 6 hours 30 minutes
- From Bangkok nonstop on Thai Airways: 2 hours 10 minutes
- From Beijing am-nonstop on Air China or Dragonair to Hong Kong, pm-nonstop on China Airlines: 7 hours 30 minutes, including 1-hour layover in Hong Kong
- From Frankfurt nonstop on Lufthansa: 12 hours 10 minutes
- From Hong Kong nonstop on Singapore Airlines: 3 hours 30 minutes
- From Jakarta nonstop on Garuda Indonesia: 3 hours
- From Kuala Lumpur nonstop on Malaysia Airlines: 55 minutes
- From London nonstop on British Airways: 13 hours 10 minutes
- From Manila nonstop on Philippine Airlines: 3 hour 15 minutes
- From New York City direct on Singapore Airlines via Amsterdam: 20 hours 35 minutes
- From San Francisco direct on Singapore Airlines via Hong Kong: 19 hours
- From Seoul direct on Korean Airlines via Bangkok: 7 hours
- From Sydney nonstop on Qantas: 8 hours
- From Taipei nonstop on China Airlines: 3 hours 30 minutes
- From Tokyo nonstop on Japan Airlines: 6 hours 45 minutes

*Direct-flight times include time on the ground in layover cities

TIME CHANGES

Singapore shares a time zone 8 hours ahead of Greenwich Mean Time with China, Taiwan, Hong Kong, the Philippines, Malaysia, central Indonesia, and Western Australia. When you're in Singapore, you can determine what time it is in any of the cities listed here by adding or subtracting the number shown to or from Singapore time.

Auckland	+4
Beijing	0
Bangkok	-1
Frankfurt	-7
Hong Kong	0
Jakarta	-1
Kuala Lumpur	0
London	-8
Manila	0
New York City	-13
San Francisco	-16
Seoul	+1
Sydney	+2
Taipei	0
Tokyo	+1

CUSTOMS ENTRY

Singapore's customs entry is similar to that of most countries in East Asia, except that the city-state is more worried about chewing gum than guns. And its *Welcome to Singapore—Your Customs Guide* brochure is succinct enough: "Warning: Death for Drug Traffickers Under Singapore Law." You must declare only dutiable items.

Duty-free for personal use only
- 1 liter of alcohol
- Up to S$50 (about US$31) worth of chocolate, cookies, cakes, candy
- Personal effects
- ATA carnet items—professional equipment, commercial samples, and advertising material

Cash
No limit on foreign currencies, checks or bank drafts.

Non-dutiable goods
- Cameras and electronic gear
- Cosmetics
- Jewelry, precious metals, precious stones

Dutiable goods (non-personal)
- More than 1 liter of alcohol
- Tobacco
- Clothing and accessories
- Leather bags and wallets
- Imitation jewelry
- More than S$50 worth of chocolate, cookies, cakes, candy

Restricted—entry by permit only
- Arms, ammunition, and dangerous weapons including knives
- Live or stuffed animals
- Meat
- Plants or seeds
- Vaccines, poisons, controlled drugs
- Prerecorded cartridges, cassettes, videotapes, and audio and computer disks
- Newspapers, books, and magazines
- Telecommunications equipment

Prohibited
- Chewing gum
- Pornography
- Illegal narcotics
- Firecrackers

FOREIGN EXCHANGE

The Singapore dollar is divided into 100 cents, and comes in coin denominations of 1, 5, 20, and 50 cents, and S$1. Notes come in denominations of S$1, S$5, S$10, S$20, S$50, S$100, S$500, S$1,000, and S$10,000 (it's unlikely you will ever see a S$10,000 note). At the end of 1993 the exchange rate was S$1.605 = US$1. The Singapore dollar is slowly strengthening against the US dollar.

The first and best place to change money is at the Changi Airport money-changing counters. You can get equally good rates at money-changing booths in shopping centers, and foreign exchange houses such as Thomas Cook will try to match them. Banks give somewhat lesser rates, while hotels and shops give the poorest rates. As in all other countries, you get the best rates when changing traveler's checks.

TIPPING

The Singaporean government discourages the practice of tipping. Tipping is forbidden at the airport and discouraged at hotels and restaurants that add a 10 percent service charge. However, you can tip hotel porters S$1-S$2 (about US$0.62 to US$0.25) per bag up to the front desk, S$2-S$3 up to your room. Room service gets S$1-S$2. Taxi drivers also welcome tips from foreigners, partly because they don't get them from Singaporeans.

ACCESS TO CITY FROM AIRPORT

Changi Airport is about 18 km (11 miles) northeast of the city. It's one of the best airports in the world—in layout, services (from haircuts to information), cleanliness (of course—this is Singapore), efficiency, and facilities (including showers and day rooms). There's an expressway from the terminal directly to the city and buses, taxis, and limousines to carry you to and fro.

Buses run about every 10 minutes, take about 30-45 minutes to get to city center, and cost about S$1.20 (about US$0.75). At last count, there were 12,705 taxis on Singapore streets, and a fair percentage of them are at the airport. The typical metered fare runs S$10-S$15, plus a S$3 surcharge (about US$11.50 total) on rides from the airport (rides to the airport don't have this surcharge.)

ACCOMMODATIONS

A hotel room in Singapore will cost you about what an equivalent room does in New York, a little less than it would in Hong Kong, and somewhat less than a lesser room would in Tokyo. For the high price, you'll get a large, clean (this is Singapore) room, excellent service, and a host of amenities to choose from (for a fee). Most of the major hotels have fully equipped business centers, and many have health clubs. And yet even in expensive Singapore there are still budget hotels, which also have good service.

It's very easy to book a room through your travel agent, but you can even book a room on arrival through the Singapore Hotel Association counters at Changi Airport. The room rates quoted here range from the lowest-priced single to the highest-priced double. Note: The word is that Singapore hotels were planning to boost their room rates by 15 percent in late 1993.

Top-end

Goodwood Park 22 Scotts Road, Singapore 0922; in shopping district. Circa 1900, renovated, refined, classy, no glitz. Business center with laptop computers, conference facilities, meeting rooms, personalized service, pools, restaurants, shops. Rates: S$400-S$450 (about US$250 to US$280). Tel: 7377411 Fax: 7328558 Tlx: 24377.

Oriental 5 Raffles Avenue, #01-200, Singapore 0103; in Marina Square-Raffles City area. Best service in Singapore. Business center, computers, conference facilities, meeting rooms, health club, pools, tennis, travel desk, restaurants, shops. Rates: S$370-S$590 (about US$230 to US$368). Tel: 3380066 Fax: 3399537 Tlx: 29117.

Expensive

Dynasty 320 Orchard Road, Singapore 0923; in shopping district. Business center, computers, conference facilities, meeting rooms, health club, pool, restaurants, shops. Rates: S$300-S$430 (about US$187 to US$268). Tel: 7349900 Fax: 7335251 Tlx: 36633.

Omni Marco Polo Tanglin Road, Singapore 1024; near shopping district, next to Botanic Gardens. Business center, computers, meeting rooms, health club (best in town), pool, restaurants, shops. Rates: S$350-S$900 (about US$218 to US$560). Tel: 4747141, (800) 223-5652 from US Fax: 4710521 Tlx:21476.

Westin Plaza 2 Stamford Road, Singapore 0617; in Raffles City. Business center, computers, convention facilities, meeting rooms, health club, pool, tennis, restaurants, shops. Rates: S$340-S$480 (about US$212 to US$299). Tel: 3388585 Fax: 3382862 Tlx: 22206.

Moderate

Inn of the Sixth Happiness 33-35 Erskine Road, Singapore 0106; Chinatown. Small, Chinese decor, friendly service. Meeting rooms, restaurant. Rates: S$230-S$270 (about US$143 to US168). Tel: 2233266 Fax: 2237951.

Ladyhill 1 Ladyhill Road, Singapore 1025; in residential area, 10-minute walk from Orchard Road. Small, quiet, relaxed. Conference rooms, pool, restaurant. Rates: S$200-S$450 (about US$125 to US$280). Tel: 7372111 Fax: 7374606 Tlx: 23157.

Plaza 7500A Beach road, Singapore 0719; in Little Araby. Business center, conference facilities, meeting rooms health club, pool, free shuttle to downtown. Rates: S$270-S$550 (about US$168 to US$343). Tel: 2980011 Fax: 2963600 Tlx: 22150.

Budget

Mitre 145 Killiney Road, Singapore 0923; near Orchard Road and Singapore Telecommunications Centre. 1930s Charlie Chan crime mysteries atmosphere. Overhead fans, shared baths, no credit cards, no restaurant. Rates: S$40 and up (about US$25). Tel: 7373811.

YMCA International House 1 Orchard Road; Singapore 0923; near Raffles City, National Museum, National Library. Hotel-like; direct-dial phones, private baths, excellent gym, rooftop pool, McDonald's. Rates: S$20 dorm; S$45 single; S$70 double (about US$12.50 to US$44). Tel: 3366000.

YMCA Metropolitan Tanglin Centre 60 Stevens Road, Singapore 1025; near shopping district. Conference rooms, private baths, suites, pool, gym, good restaurant; no credit cards. Rates: S$50 (about US$31) and up. Tel: 7377755 Fax: 2355528.

EATING

Singapore's hotels may not have much variety, but its restaurants more than make up for that. It's worth going to Singapore just to eat. You can choose from several regional Chinese cuisines and three from India, plus Thai, Malay, Nyona, Sumatran, Japanese, French, Italian, Mexican, English, American *nouvelle cuisine* or fast food, or that ubiquitous and inexplicably popular franchise *Hard Rock Café*, and even a steak-and-salad-bar place called *Ponderosa*.

To sample Singapore's incredible variety of fresh seafood, go to the UDMC Seafood Centre on *East Coast Parkway*, where you'll find a selection of moderately priced restaurants. Or try *Choon Hoon Seng*, on the northern shore 25 minutes by cab from downtown, where you'll find local color and great seafood in a popular, moderately priced restaurant overlooking the Straits of Johore (892 Punggol Road; Tel: 288-3472; no credit cards, no reservations).

In fact, Singapore doesn't have to be an expensive place to eat: simply go to one of the dozens of hawkers' centers scattered throughout the city. These are the Singapore version of pushcart food. To keep things hygienic, the government put them all under roof and strictly regulates cleanliness (this is Singapore, after all). At each center, you'll find dozens of hawkers, tables and chairs; you can sample the wares of each one and watch the activity—all for about S$2 to S$3 (less than US$2) per dish. The best is the *Satay Club* (evenings only), on Connaught Drive at Stamford Road between Raffles City and Marina Square. *Satay* is a Malaysian dish of marinated meat barbecued on coconut sticks and served with a spicy peanut sauce. Also try the *Newton Circus Hawker Centre* in the Scotts Shopping Centre at the corner of Scotts and Orchard roads—touristy, extremely lively at night (but avoid the overpriced seafood); *Empress Place* (lunch and dinner) and *Boat Quay* (lunch only) on the Singapore River in the business district off Bridge Road; *Marina South* with its hundreds of hawkers of Chinese and Malay food (get there either by taxi or by shuttle bus from major hotels to Marina Village); and *Albert Centre* on Albert Street between Waterloo and Bencoolen streets near Rochor Canal Road and not far from Raffles City.

With such a variety of restaurants, it's hard to draw a line, so we've decided on a few of the very best and most popular places to entertain and impress your guests with fine food and excellent service. All advise reservations and all accept major credit cards. Dress is casual or dressy-casual, unless otherwise noted.

Cherry Garden Specialties: minced-pigeon broth with scallops steamed in a bamboo tube, camphor-smoked duck in bean-curd (tofu) crust. Dinner: S$30 (about US$19) and up. In Oriental Hotel, 6 Raffles Boulevard; Tel: 3380066.

Harbour Grill Continental cuisine. Elegant, executives' favorite. Specialties: blanquette of salmon with mushrooms and baby onions, venison in sauce of Noilly Prat, cognac, game stock, butter, and juniper. Dinner: S$50 (about US$31) and up. Jacket and tie. In Hilton Hotel, 581 Orchard Road; Tel: 7372233.

La Brasserie Traditional French cuisine. Relaxed, country-style. Specialties: sliced veal with mushrooms in cream sauce, fluffy lemon pancakes with ice cream. Dinner: S$30 (about US$19) and up. In Marco Polo Hotel, Tanglin Road; Tel: 4747141.

Latour Continental cuisine. Quiet elegance, good for business. Specialties: wine list, lobster ragout, fillet of pomfret, blueberry soufflé. Dinner: S$45 (about US$28) and up. Jacket and tie. In Shangri-La Hotel, 22 Orange Grove Road; Tel: 7373644.

Li Bai Cantonese cuisine. Pavilion around courtyard. Specialties: fried lobster in black bean sauce, shark's fin with Chinese wine and -ham. Dinner: S$45 (about US$28) and up. In Sheraton Towers Hotel, 39 Scotts Road; Tel: 7376888.

Pete's Place Italian American cuisine. Informal and lively. Dinner: S$25 (about US$16) and up. In Hyatt Hotel, 10/12 Scotts Road; Tel: 7331188.

Suntory Japanese cuisine. Best in town. Teppanyaki, sushi. Dinner: S$50 (about US$31) and up. 06-01/02 Delfi Orchard, 402 Orchard Road; Tel: 7325111.

Tandoor Indian cuisine. Moghul-court ambience. Specialties: roast leg of lamb, garlic naan (bread), tandoori fowl and fish. Dinner: S$30 (about US$19) and up. In Holiday Inn Park View, 11 Cavenagh Road; Tel: 7338333.

LOCAL CUSTOMS OVERVIEW

Singaporean society has a Confucian mind, a communalist body, and a capitalist heart. In Singapore, the saying "the government means business," has two meanings, and neither should be ignored. On the one hand, the government is in the business of business, and it's done a superb job, as shown by the high standard of living, rampant entrepreneurism, and booming international trade. On the other hand, longtime leader and father figure Lee Kuan Yew has wedded the patriarchal traits of Confucianism to the collectivist traits of Communism (the leftist but non-Communist Lee was leader of the Communist-dominated People's Action Party party in the years preceding independence and preceding the party's purge of leftist elements). From this harmonious marriage he has fashioned a uniquely stern and unforgiving authoritarian regime that keeps a close watch over the daily lives and most intimate habits of a people already inclined to conformity. Writing in *Wired* magazine, author William Gibson dubbed Singapore "Disneyland with the Death Penalty."

This conformity is at the heart of the Chinese sector of Singaporean society, and it is the Chinese who dominate business in the city-state. Confucianism came out of China and spread its concepts of the common good throughout East Asia. Any East Asian society that can be called Confucian—China, Taiwan, Singapore, Japan, Korea, Hong Kong—can also be called collectivist. The community reigns supreme; the individual is subservient. Yet the individual is important enough that his or her wrongdoing undermines the community. A foreign businessperson who wants to be successful must understand and adapt to this social construct. Following are its most prominent characteristics:

- Family and community are paramount. The father, the political leader, the president of the company—all are revered and obeyed, and their authority increases with age. The family-community-business is a closed social system; strangers gain entrance only by introduction through a respected intermediary. Business success depends on developing close personal relationships with business contacts.
- Only the patriarchal figure assumes responsibility; all other decision-making is by consensus. Thus, Singaporean business is marked by meetings, negotiations, and consultations. The Westerner must call on all his or her comparatively sparse reserves of patience, politeness, and equanimity.
- Because the community depends on the integrity of the individual, that integrity must never be undermined. Singaporean government and business alike are known for their incorruptibility. Although gift giving is appreciated later in a relationship, if done too early it may be viewed as an attempt at bribery. One strengthens the relationship gradually and circumspectly with small favors, dinners, visits, and careful revealing of personal family information—all given with sincerity and as signs of respect.
- If the individual is seen as losing respect, the community also loses. This notion is central to the concept of face. One can have face, create it, lose it, or save it. Status, age, and experience give you face; hard work and accomplishment, or praising another in front of his or her peers, create face; one loses face by being criticized in front of peers. Thus, typical Western directness in criticism is counterproductive, as is self-praise. Courtesy, humility, and deference are prized.
- You may never hear the word no. It causes loss of face. The Westerner must learn to avoid asking yes-or-no questions and to read between the lines of Chinese responses. Any response but a direct yes could mean no.
- Business cards are indicators of status. You exchange cards at your first meeting with a business contact. Although Singaporeans can read and understand English, the reverse of your card should be printed in Chinese characters. Both sides should give your name, your company's name and your title—and your title should be clearly descriptive of your status and duties; for example, Chief Financial Officer says much more than Vice President.
- While your business is most likely to take place with Chinese, it may include the two other major segments of Singaporean society: Malay and Indian. The Malays also have a collectivist society, but with an Islamic bent, while the Indians vary widely in cultural values. A successful business relationship requires knowledge and understanding.
- Avoid political discussions, especially criticism of the Singaporean government. Such talk can only do harm to your efforts. Singapore's citizens are both proud and wary of their leaders.

(Refer to "Business Culture" chapter for a detailed discussion.)

DOMESTIC TRANSPORTATION

Getting around Singapore is easy. Whether you go by subway, taxi, or bus, you'll find it quick, convenient, and user-friendly.

Subway Singapore's superb subway is called the MRT. It's well-lighted and clean. One line runs north-south, the other east-west; they cross at Raffles Place and City Hall. In rush hour trains run every 4 to 5 minutes, and every 6 to 8 minutes at other times. You can buy magnetic tickets from vending machines or at a booth, either for one ride, or a S$10 (about US$6.25) ticket for several. Easy-to-read maps show you where to go, which train to catch, where to get off, and how much it costs. If you underpay, there's a S$2 fine. Insert the ticket in the turnstile to get on and off the platform. The minimum fare is S$.60, the maximum, S$1.10 (about US$0.37 to US$0.69). For example, the fare between the Raffles Place and Orchard Road stations is S$.70. The Orchard Road hotel district is served by the Somerset, Orchard and Newton stations. The business district is served by the Raffles Place station, while Chinatown is served by the Outram Park station.

Bus In contrast to many East Asian countries, Singapore has an efficient bus system that is easy for foreigners to use. Buses are even faster than taxis during rush hour because they ride in special bus lanes. Buses run everywhere and often. Their routes and numbers are clearly marked; bus stops list the destinations and numbers of the lines that serve them. The handy Bus Guide costs only S$.75 (about US$0.47) at any bookstore or hotel.

Buses are very cheap. Fares range from S$.40 to S$.80 (about US$0.25 to US$0.50) for unair-conditioned buses and S$.50 to S$1.00 (about US$0.31 to

US$0.62) for air-conditioned buses. You need exact change, which you deposit in the fare box when you get on the bus. Or you can get a Singapore Explorer Bus Pass—S$5 (about US$3.10) for a 1-day pass or S$12 (about US$7.50) for a 3-day pass—which you show the driver as you board. With the pass you get a map complete with routes, bus line numbers and destinations. There's also the Farecard, which costs S$12 or S$22 (about US$7.50 to US$13.75) at MRT stations and bus interchanges. The card is a multitrip card; each trip deducts from the balance until it's all used up.

Taxi Except for rush hour, taxis may be the most convenient way for a foreigner to get around Singapore, although they're not inexpensive. The first 1.5 km (0.9 mile) costs S$2.20 (about US$1.37); every 275 meters (900 feet) thereafter costs S$.10. After 10 km (6 miles), the rate rises to S$.10 for every 250 meters (820 feet). And the meter clicks up another S$.10 for every 45 seconds of waiting time. There are also surcharges for luggage, each additional person, trips between midnight and 6 am, and trips into and out of the central business district.

It's fairly easy to hail a taxi from the street, except during rush hour. You can also call for one (452-5555, 4747707, or 2500700), or just catch one from in front of your hotel.

The government strictly regulates taxis, so you won't find any of the "broken" meters so common in some other East Asian countries. If you honestly think a driver is trying to cheat you, merely threatening to call the Singapore Tourist Promotion Board is enough to make him back off. Nor do you have to tip the drivers, although far too many visitors do anyway, to the chagrin of Singaporeans.

HOLIDAYS/BANK HOLIDAYS

Singapore has 11 public holidays (indicated by an asterisk in the list that follows; Chinese New Year is two days) and several additional festivals that fill in the gaps between them. Although Singapore's official calendar is the Gregorian (Western), Chinese, Hindus, and Muslims all follow lunar calendars, which means their holidays fall on different Western-calendar dates each year.

Chinese New Year in particular is not a good time to try to conduct business (but try to take advantage of businesses' tradition of clearing their debts). It goes on for 15 days, with parades, feasts, open houses, shows, and fireworks. During this time, shops and businesses close for about a week, but banks and government agencies close only for two days.

BUSINESS HOURS

Government agencies open between 7:30 am and 9:30 am and close between 4 pm and 6 pm Monday through Friday. In addition, they are open Saturdays until 11:30 am to 1 pm. Banks are open weekdays

Singaporean Holidays and Bank Holidays

Holiday	Date	Significance
New Year's Day*	January 1	Western New Year
Ponggal	January	South Indian harvest festival; 4 days
Chinese New Year*	January-February	Lunar New Year; 15 days
Thaipusam	January-February	Hindu penance
Hari Raya Puasa*	March-April	Muslim; celebrates the end of Ramadan (month of fasting)
Good Friday*	March-April	Christian; church services
Labor Day*	May 1	Traditional workers' day
Vesak Day*	April-May	Celebration of the Buddha's birth, enlightenment, and death
Hari Raya Haji*	June-July	Muslim celebration of the pilgrimage to Mecca
Dragon Boat Festival	May-June	boat races
National Day*	August 9	Singapore independence day
Hungry Ghosts Festival	August-September	Chinese; 1 month
Thimith	September-October	Hindus honor goddess Duropadai
Deepavali*	October-November	Hindu celebration of victory of good over evil
Christmas*	December 25	Birth of Christ

from 10 am to 3 pm and Saturday from 9:30 to 11:30 am, although the Changi Airport bank is open whenever flights are arriving. Department store hours generally run from 10 am to 9 pm or later seven days a week, while shops usually close on Sunday.

COMMUNICATIONS

Telecommunications, mail and courier services, and English-language media are all first-rate.

Telephones Most hotel rooms have international direct-dial (IDD) phones, but hotels tack an 80 percent markup onto overseas calls. Your best bet is to use the international phones at Changi Airport or the General Post Office, or a Singapore Telecom phone card in a public phone—fewer and fewer public phones accept anything but this card. You can buy a phone card at post offices, 7-Eleven stores, Guardian drugstores, and Telecom's Customer Service outlets. They come in denominations of S$10, S$20 and S$50 (about US$6.25, US$12.50, or US$31). You insert the card into the pay phone, which punches holes in the card to denote your remaining balance. Or you can use your MasterCard, Visa, American Express, or Diners Club charge cards on phones at the airport and several post offices.

USEFUL TELEPHONE NUMBERS
- Local directory assistance 103
- International operator 104
- International access 005
- AT&T USADirect .... (800) 0111-111
- Singapore Tourist Promotion Board
  .... 3300431, 3396622, 7383778/9
- Changi Airport Flight Information 5425680
- Taxi (24 hours) 4525555, 2500700
- Immigration Department 5322877
- Lost or stolen credit cards
  (International collect calls only to US regardless of the country your card was issued in)
  Amex .... (919) 333-3211
  Diners Club .... (303) 799-1504
  MasterCard .... (314) 275-6690
  Visa .... (410) 581-7931

To direct dial internationally Dial the international access number—005—then the country code, then the area code (if there is one), and the phone number. For example, to call World Trade Press direct, dial 005-1-415-454-9934.

Fax, telex, and telegram Most major hotels have fax and telex services, available to guests and nonguests alike. You can send telegrams and cables from the General Post Office or Changi Airport.

Post Office Singapore's postal service is fast and efficient. The General Post Office at Fullerton Square off Collyer Quay (Tel: 5336234) and the Customer Services Centre at Exeter Road are both open 24 hours a day, while the Changi Airport and Orchard Point branches are open from 8 am to 8 pm every day. All 87 branches have friendly and helpful staff, who can also sell you packing cartons.

Most likely, you'll use your hotel's services for stamps and mailings. Local letters cost S$.10; aerograms cost S$.35; and airmail letters up to 10 grams cost S$.35 within Asia, S$.50 to Australia, S$.75 to the United Kingdom, and S$1.00 to the US (between about US$0.06 for local to US$0.62 for overseas).

English-language media English is the official language, and the only hindrance to news in English or any other language is often heavy-handed government censorship. It is criminal to sell or even possess certain books. Articles seen as unfavorable to or critical of Singapore or its government are suppressed and the offending publications suspended until they see the light.

Straits Times and *Business Times* are daily news-

Country Codes of Major Countries

Australia	61
Brazil	55
Canada	1
China	86
France	33
Germany	49
Hong Kong	852
India	91
Indonesia	62
Italy	39
Korea	82
Malaysia	60
Mexico	52
New Zealand	64
Pakistan	92
Philippines	63
Russia	7
Singapore	65
South Africa	27
Spain	34
Taiwan	886
Thailand	66
United Kingdom	44
United States	1

papers that report the official line. Such international dailies as the *Asian Wall Street Journal* and the *International Herald Tribune* are available unless they've been banned for offending the government's sensibilities; that happened to the AWSJ recently for saying a judge had been biased when he found that the magazine *Far Eastern Economic Review* had criminally criticized Lee Kuan Yew. The same can happen to other normally available publications, such as *Asiaweek, Newsweek,* and *Time.*

Radio and television broadcast in English and three other languages, and listings are in English as well. BBC's World Service is at 88.9 FM, Radio One at 90.5 FM, and Radio Five at 92.4 FM. Nearly every hotel room has a TV, and most can receive CNN.

Courier services

The large international courier services operate in Singapore.

Federal Express Federal Express (S) PTE Ltd., 3 Kaki Bukit Road 2, Block A Unit 3E, Eunos Warehouse Complex, Singapore 1441; Tel: 7432626 Fax: 7414225 Tlx: 22454.

TNT TNT Express Worldwide, 140 Paya Lebar Road, #02-10, A-Z Building, Singapore 1440; Tel: 7429000, 7453122, 5423321 (airport office) Fax: 7422214, 7476349, 5424534 (airport office) Tlx: 38102 RS TNTSKY, 23525 SKYPAK (airport office)

UPS United Parcel Service Singapore Pte Ltd., 3 Killiney Road, Winsland House #06-01, Singapore 0923; Tel: 7383388 Fax: 7382683/2883.

Local services

Just about any kind of business service you'll need while traveling is available in Singapore. Just ask at your hotel desk or look in the yellow pages.

Audiovisual rentals AV-Science Marketing, 22 Pasir Panjang Road, #10-28 PSA Multi-Story Complex; Tel: 2788283.

Business centers Most major hotels have business centers with services ranging from secretarial and translation service to telecommunications and computers. Many of the hotels make these centers available to non-guests. The following hotels have business centers:

Allson, ANA, Avant, Boulevard, Cairnhill, Carlton, Concorde, Crown Prince, Dai-Ichi, Duxton, Dynasty, Goodwood Park, Hilton, Holiday Inn Park View, Hyatt Regency, Imperial, Mandarin, Marina Mandarin, Melia Scotts, Meridien, New Otani, Omni Marco Polo, Orchard, Oriental, Pan Pacific, Plaza, Raffles, Regent, River View, Royal Holiday Inn Crowne Plaza, Shangri-La, Sheraton Towers, Tai-Pan Ramada, Westin Plaza, Westin Stamford.

Computer rentals Bizland Systems, 25 Delta Road, #10-01 Seiclene House; Tel: 2781771.

Messenger services On-Call Couriers, Tel: 5325088; Steiner Courier, Tel: 3374214.

Office space rentals Jones Lang Wooten, #45-00 Shell Tower, 50 Raffles Place; Tel: 220-3888; Richard Ellis, #12-01 Hongkong Bank Bldg., 21 Collyer Quay; Tel: 2248181; Servcop Serviced Offices, #36-00 Hong Leong Bldg., 16 Raffles Quay; Tel: 2255535.

Secretarial services Bez Secretarial Services, #03-26 Furama Hotel and Shopping Center, 60 Eu Tong Seng Street; Tel: 5353988; Drake Training and Personnel, #16-02 Hong Leong Building, 16 Raffles Quay; Tel: 2243488.

Translation services Interlingua Language Services, 141 Cecil Street, 06-01 Tung Ann Association Building; Tel: 2223755; Worldwide Translation Services, 64 Lloyd Road; Tel: 7377672.

STAYING SAFE AND HEALTHY

You will not get sick in Singapore. The government forbids it.

Unlike most East Asian nations, Singapore has relatively clean air, so bronchitis is not a problem. The tap water is safe to drink, and fresh fruits and vegetables are safe to eat. All the personal care products and medicines you might need are readily available.

Singapore isn't totally free of crime, but believe us when we say the government is working on it. The worst you might encounter is a pickpocket in some crowded areas. In fact, it's just as likely that you will be the criminal if you commit such high-fine (S$500—about US$312—and up) violations as jaywalking, spitting, chewing gum, or forgetting to flush in a public restroom.

EMERGENCY INFORMATION

Police 999
Ambulance and fire 995

Hospitals and clinics

Singapore's health-care system is first-rate. Most doctors, nurses, and hospital staff speak English, and many major hotels either have their own staff or on-call doctors. Doctors and hospitals expect immediate payment in cash. Among the hospitals and clinics accustomed to treating foreigners are:

Ming Clinic, 19 Tanglin Road, #12-02 Tanglin Shopping Centre; Tel: 2358166, 5358833 (after hours).

Paul's Clinic, 435 Orchard Road, 11-01 Wisma Atria; Tel: 2352511.

Alexandra Hospital, Alexandra Road; Tel: 4735222.

Kadang Kerbau Hospital, Hampshire Road; Tel: 2934044.

Singapore General Hospital, Outram Road; Tel: 2223322.

Dentists

International Plaza Dental Surgeons, 10 Anson Road, #02-39 International Plaza; Tel: 2206230, 5358833 (after hours).

Claymore Dental Surgeons, 05-11B Orchard Towers, 400 Orchard Road; Tel: 7325226, 5330088 (after hours).

DEPARTURE FORMALITIES

Singapore has simple exit formalities. There are no export duties, but you do need export permits for gold, platinum, precious stones set in jewelry, arms and ammunition, explosives, and animals.

BEST TRAVEL BOOKS

For practical information, insight, and thoroughness, three travel books on Singapore stand out from the crowd. Keep in mind that Singapore, like the rest of East Asia, is changing rapidly, so the reliability of such volatile information as prices and telephone numbers cannot be guaranteed.

Fodor's Singapore, edited by Craig Seligman. New York: Fodor's Travel Publications, 1993. ISBN 0-679-02340-2. 204 pages, US$13.00. Oriented towards the tourist and shopper but strong on transportation, hotels, and restaurants. Good maps.

The Four Dragons Guidebook, by Fredric M. Kaplan. Boston: Houghton Mifflin, 1991. ISBN 0-395-58577-5. 688 pages, US$18.95. 160 pages devoted to Singapore (the rest to Hong Kong, Taiwan, and Thailand), including 24 pages on doing business in Singapore. Good basic information on travel, although in need of updating.

Singapore City Guide, by Peter Turner and Tony Wheeler. Hawthorn, Victoria, Australia: Lonely Planet Publications, 1991. ISBN 0-86442-109-5. 191 pages, US$8.95. The usual Lonely Planet guide for the independent, budget-minded traveler—thorough, insightful, irreverent. Very good maps. Needs updating.

Typical Daily Expenses in Singapore

All prices are in Singapore dollars (S$) unless otherwise noted.
At press time, the exchange rate was US$1 = S$1.60

Expense	LOW	MODERATE	HIGH
Hotel	70	250	400
Local transportation *	5	10	25
Food	25	80	150
Local telephone †	1	0	0
Tips	2	5	10
Personal entertainment **	7	25	40
TOTAL	**110**	**370**	**625**
One-way airport transportation	1.20	9	9

US Government per diem allowance as of December, 1993 was US$117 for lodgings and US$84 for meals and incidentals (approximately S$322).

* Based on a 1-day bus pass for low cost, 2 medium length taxi rides for moderate cost and 4 longer taxi rides for high cost.
† Based on 2 telephone calls from pay phones for low cost. Local calls from better hotels are generally free.
** Based on a visit to a cultural site and coffee shop for low cost, a visit to a disco with a drink for moderate cost and a live floor show for high cost.

Business Culture

Singapore is a multicultural city-state where Chinese, Malay, and Indian traditions coexist beneath a veneer of Western cosmopolitan culture. As the major crossroads of international trade in Southeast Asia, its cosmopolitan inhabitants are used to dealing with people from all walks of life. Singapore is one of the easiest places in Asia for foreign businesspeople to work.

To a casual observer, Singapore's modern skyline, clean streets, and efficient transportation may give the impression of a totally Westernized society with little or no trace of indigenous culture. This impression is incorrect. Although Singaporean society is in the midst of great changes, including Westernization, urbanization, and industrialization, the people remain Asian, and their underlying values stem from traditional beliefs and customs.

Singapore's modern history began in 1965 when the city became independent from Malaysia. After several years of ethnic strife, mostly between Malays and ethnic Chinese, the city emerged as a harmonious environment in which different races could live and work side by side. While there is little ethnic tension today, Singapore is not really a melting pot where people adopt habits from a variety of cultures. Instead, people of differing cultural backgrounds preserve their own ethnic identities while living lifestyles befitting a modern industrial nation. To understand the culture of a particular Singaporean, it is necessary to know whether he is of Chinese, Malay, or Indian background and to understand the traditional beliefs and social practices associated with that background.

A MAN NAMED LEE

Modern Singapore is largely the result of one man's leadership and vision. That man is Lee Kuan Yew, who reigned as the prime minister and paramount leader of Singapore for 31 years. Although Lee has officially retired, he still exerts tremendous influence over policy and government affairs.

Lee's vision of how Singapore should operate is based largely on the Confucian principles of hard work and obedience to superiors. He is the father of modern Singapore in a very real sense. Lee's government policies are paternalistic; people are provided with housing, health care, education, and a host of human services at low cost. In return, Lee receives support, praise and unquestioning obedience from the majority of Singapore's citizens.

But just like a Confucian father, Lee is a strict disciplinarian. Opponents to his People's Action Party (PAP) are seen as disobedient and as working to undermine the collective good of society. Over the years, Lee's opponents have been continually harassed and sometimes imprisoned. Although Lee's successor as prime minister, Goh Chok Tong, nominally supports increased democracy, Lee's policies are expected to continue to guide Singapore for the foreseeable future.

In conjunction with strict political control, Singapore's government has seen fit to institute what some observers consider excessive social controls. As caretaker of the people, the government has severely restricted activities deemed contrary to the common good, and violators are subject to harsh penalties. Spitting, dropping a cigarette butt on the ground, or leaving a public toilet unflushed can be punished with a fine running into hundreds of dollars. And, unbelievably to most Westerners, Singapore has now totally banned the importation and sale of chewing gum! It seems the government got tired of scraping old gum off bathroom floors and from underneath chairs.

Despite government authoritarianism, it appears that most people in Singapore are content to exchange some personal liberties for a stable and clean society and—most of all—continued economic success.

SINGLISH

The lingua franca of modern Singapore is English, although the vast majority of Singaporeans speak it as a second language. As a result, common English in Singapore incorporates Chinese and Malay grammatical structures as well as much non-English slang. This odd mixture of languages has come to be called Singlish.

For visitors to Singapore, Singlish can seem bizarre and almost incomprehensible. Accents are thick, and in conversation it is necessary to pay close attention to what is being said. But by paying attention and remembering a few characteristics of the language, a visitor will soon be able to communicate freely. Here are a few points to remember about Singaporean English:

- **"r" becomes "l" at the beginning of a word.** For example, the word *rice* may sound more like *lice* when it comes from the lips of a Chinese Singaporean. Don't laugh!
- **"r" and "l" are dropped at the end of words.** *Bar stool* in Singlish will sound more like "ba stoo."
- **"la" is used at the end of a sentence.** This syllable emphasizes a point made, a change in condition, or a completed action, such as in the following phrases:
  - "No way, la!"
  - "The movie's over, la."
  - "I bought it, la."

  The use of "la" is derived from Chinese, where it is used for emphasis exactly as it is in Singlish.
- **Saying the same adjective twice for emphasis** For example, "It's hot hot outside!" means that it is very hot. This feature also is derived from Chinese.

These are only a few of the idiosyncracies of Singlish, and there are many more. However, in a business setting your Singaporean associates should know that you are unfamiliar with the local slang, and they will make an effort to speak standard English. If you don't understand someone's accent or grammar, politely ask him to repeat what he has just said.

CHINESE SINGAPOREANS AND BUSINESS

Singapore's business environment is dominated by people of Chinese descent, who comprise almost 80 percent of the population. However, Chinese Singaporeans are not a completely homogeneous group. Many are first- or second-generation immigrants from China. Others are descended from Chinese who came to Malaysia several centuries ago. Those with recent ties to China may speak Cantonese, Fujianese, Hokkien, or Mandarin in the home, but the Malaysian Chinese often speak Malay but no Chinese dialects. Thus English is the standard language of business, science, and education.

Singapore has the largest population of ethnic Chinese in Southeast Asia. Since it also has the most pro-business environment and modern communications and infrastructure, it serves as the nerve center of business in Southeast Asia. Ethnic Chinese control most businesses in almost every Southeast Asian nation, and establishing business relationships with Singaporean Chinese is one of the best means of establishing business relations throughout the region.

Although Chinese Singaporeans are removed from their ancestral homeland, they retain many of the cultural characteristics of Chinese in mainland China. In contrast to most mainlanders, Chinese Singaporeans are more conscious of non-Chinese habits and have a distinctly broader worldview. However, their social philosophies and rules of etiquette remain essentially Chinese. By understanding a few of the major concepts of Chinese life, you will be more capable of succeeding in business relationships in Singapore.

CONFUCIANISM

The mentality of Chinese Singaporeans is still shaped largely by the teachings of Confucius, who lived more than 2,500 years ago. Confucianism is not so much a religion as it is a code for social conduct. Its influence is so pervasive that Chinese function unconsciously in a Confucian manner. The basic tenets of Confucian thought are obedience to and respect for superiors and parents, duty to family, loyalty to friends, humility, sincerity, and courtesy.

Age and rank are respected in Singapore, and young people are expected to obey their elders unquestioningly. In the workplace, respect and status increase with age. Older foreign businesspeople have an advantage in this regard, and they are likely to receive more serious attention than younger people.

The family is the preeminent institution among Chinese. One's first duty is to the welfare of one's family, and working family members often pool their financial resources. In many ways, Chinese view themselves more as parts of the family unit than as free individuals. This is in sharp contrast to the Western emphasis on individualism. Grown children typically live with their parents even if they are married (usually the couple lives with the husband's parents).

Humility and courtesy are honored traits. Chinese are seldom overly boastful or self-satisfied, even if their accomplishments are laudable. When Chinese are being polite, they can seem excessively self-deprecating. Chinese are among the most courteous people in the world toward their friends. Every detail of a guest's stay with a Chinese friend may be prearranged, and the guest may not be allowed to spend money on even the smallest items. For individualists from the West, this form of courtesy can be overwhelming.

The effect of more than two millennia of indoctrination in Confucian thought has led the Chinese to be an extremely conformist people. The word *individualism* has a decidedly negative connotation in Chinese, and people can create resentment by standing out from the crowd. To function as an individual invites criticisms of selfishness and opportunism. Conservatism, group consciousness, and unquestioning acceptance of the status quo are norms of Confucian behavior.

The Chinese in Singapore favor increased modernization and a steady improvement in living standards. They look to the West for new ideas concerning technology and business management as well as culture. Western influences have caused some traditional values to erode, but you can expect the basic tenets of proper Confucian behavior to be observed among Chinese Singaporeans for a long time.

FACE VALUE

No understanding of the Chinese mentality is complete without a grasp of the concept of face, *mianzi* in Chinese. Having face means having a high status in the eyes of one's peers, and it is a mark of personal dignity. Chinese are acutely sensitive to having and maintaining face in all aspects of social and business life. Face can be likened to a prized possession: it can be given, lost, taken away, or earned. You should always be aware of the face factor in your dealings with Chinese Singaporeans and never do or say anything that could cause someone to lose face. Doing so could ruin business prospects and even invite retribution.

The easiest way to cause someone to lose face is to insult the individual or to criticize him or her harshly in front of others. Westerners can offend Chinese unintentionally by making fun of them in the good-natured way that is common among friends in the West. Another way to cause someone to lose face is to treat him or her as an underling when his or her official status in an organization is high. People must always be treated with proper respect. Failure to do so makes them and the transgressor lose face for all others aware of the situation.

Just as face can be lost, it can also be given by praising someone for good work in front of peers or superiors or by thanking someone for doing a good job. Giving someone face earns respect and loyalty, and it should be done whenever the situation warrants. However, it is not a good idea to praise others too much, as it can make you appear to be insincere.

You can also save someone's face by helping him to avoid an embarrassing situation. For example, in playing a game you can allow your opponent to win even if you are clearly the better player. The person whose face you save will not forget the favor, and he will be in your debt.

A person can lose face on his own by not living up to other's expectations, by failing to keep a promise, or by behaving disreputably. Remember in business interactions that a person's face is not only his own but that of the entire organization that he represents. Your relationship with the individual and the respect accorded him is probably the key to your business success with Chinese Singaporeans.

IT'S NOT WHAT YOU KNOW...

The premium placed on close personal connections is one main element of doing business in Singapore. As in China, Taiwan, and Hong Kong, little or no distinction is made between business and personal relationships. For a foreign businessperson to succeed in Singapore, he must cultivate close personal ties with business associates and earn their respect and trust.

Money Talks The one area of business in Singapore where the importance of personal connections is minimal is export trade. Because Singaporean businesspeople have been exporting to the West for decades, they understand that foreigners have little understanding of their society and traditions. Selling goods to foreigners is now itself a tradition, and because it usually does not involve close cooperation between the two sides, it can be conducted impersonally. But to market products or engage in cooperative ventures in Singapore, foreign businesspeople must cultivate connections.

All in the Family Any successful businessperson in Singapore will belong to a loose network of personal friends, friends of friends, former classmates, relatives, and associates with shared interests. These people do favors for one another and always maintain a rough running tally of the help that they have given and received. The importance of personal connections has its roots in the traditional concept of family. For the Chinese, individuals are parts of the collective family whole. The family is the source of identity, protection, and strength. In times of hardship, war, or social chaos, the Chinese family structure was a bastion against the brutal outside world, in which no one and nothing could be trusted. As a result, trust and cooperation were reserved for family members and extremely close friends. Moreover, ancient China was a land ruled by decree. Officials could act with impunity, and innocent people could get hurt unless they had powerful friends to protect them. By establishing close connections with other households and persons of higher prestige, Chinese could survive and perhaps even prosper.

Chinese Singaporeans today live in a safe and modern society free of the dangers that their ancestors faced, but the tradition of personal connections is as strong as ever. In essence the Chinese possess a clan mentality under which those inside the clan work cooperatively and those outside the clan are seen as either inconsequential or as potential threats. To be accepted into a network of personal or business relationships is an honor for foreigners. It also entails responsibility for and commitment to the members of the network.

In Singapore, executives and entrepreneurs work constantly to maintain and expand their networks of connections. Networks extend to other companies and individuals in Southeast Asia, to the PRC, Hong Kong, Taiwan, and even to Europe and the United States. While the purpose of such contacts is often mutual financial profit, the criteria are the same as for personal networks: trustworthiness and loyalty.

CULTIVATING RELATIONSHIPS

For the foreign businessperson with little or no understanding of Chinese behavior, cultivating solid relationships can be a big obstacle to success in Singapore. A Chinese Singaporean who does not already know a potential associate will be hesitant to do business with him until he has had time to get acquainted and size up the potential associate's character and intentions.

The best way to make contact with potential business associates in Singapore is to have a mutual friend serve as an intermediary and introducer. If the third party has close relationships with both sides, that alone may constitute solid grounds for the conduct of business. Anyone who has worked in Singapore or who has cooperated with Singaporeans in the past could be a key source of business contacts. There are also many business consultants who can provide assistance for a fee. Chambers of commerce, small business associations, and international government trade offices may also help you to find contacts.

If finding a third party for introductions in your home country proves impossible, consider making a fact-finding trip to Singapore. A trade show that allowed you to display your goods or services would give you a good opportunity to gauge your business prospects, or you could spend time meeting potential contacts in your area of business. Before leaving for Singapore, fax the businesses in which you are interested and try to arrange a visit. However, don't be surprised if some businesses ignore your request. They don't know you, and may therefore consider you to be a non-entity.

On your first trip to Singapore you may accomplish nothing more than getting to know several possible candidates for business relationships. Learn the lesson that rushing into business before you have established a personal relationship is an invitation to failure. After drawing up a list of potential associates, take time to evaluate each party carefully. Weigh each one's strengths and weaknesses you decide who to follow up on. Don't underestimate your gut feelings or your comfort level with various individuals, but also appraise their business abilities realistically and, of course, whom they know.

After making your first contacts with businesspeople in Singapore, be prepared to spend a lot of time deepening and strengthening these relationships through visits, dinners, gift giving, and the bestowing and receiving of many small favors. While this process can be costly and time consuming, Chinese appreciate all sincere efforts in this area, and no favor done goes unnoticed. Likewise, keep a running account of all favors done for you, all small gifts presented, and the like. The odds are good that you will be expected to reciprocate in the future. Remember this aspect of Singaporean business culture whenever someone offers you a favor, dinner, or gift. If you absolutely do not want to be in the person's debt, be creative and find some polite excuse to decline the offer. And decline it only if you have no intention of having a relationship, because declining an offer can be insulting to Singaporeans.

Foreign businesspeople will benefit from the process of cultivating relationships by keeping in mind that it gives them an opportunity to learn about the people with whom they are dealing. Getting to know your business associates is useful regardless of your culture. Learning about the personality of an associate can make communication and understanding smoother, and the resulting knowledge can be critical when it comes time to decide how far to take the business relationship.

GIVING AND RECEIVING GIFTS

Chinese are inveterate gift givers. Gifts express friendship, and they can symbolize hopes for good future business, the successful conclusion of an endeavor, or appreciation for a favor done. Foreign businesspeople should spend some time choosing appropriate presents before embarking on a trip to Singapore. Chinese consider the Western habit of simply saying thank you for a favor glib and perhaps less than sincere. Favors should be rewarded materially, although gifts can have more symbolic than monetary value. Avoid very expensive gifts unless the recipient is an old associate who has proved to be particularly important in business dealings. Singapore's government is very concerned that gift giving can lead to corruption and bribery. Thus, you should not give a gift to a government official unless a solid personal relationship exists. Never give expensive gifts to government officials, and be aware that even small gifts may be declined.

To be on the safe side, a foreigner visiting a place of business may present a single large gift to the company as a whole. In this way, no single individual will be the recipient, and therefore the issue of bribery cannot be raised. Gifts may vary in value from a nice piece of art to a coffee-table book. Gifts to individuals should be of lesser value, somewhere in the range of US$10 to US$15. If only one gift is to be given, it should be presented to the head of the group at a dinner banquet or on the conclusion of a business meeting. After a short speech expressing your appreciation, you may present the gift. As with other objects, the polite way to give and receive a gift is to hold it with both hands.

If gifts are to be given to several individuals, be sure that each person concerned receives a gift of roughly equal value. The gifts may be placed on the table at a dinner banquet or presented during an appropriately relaxed time. If you give many gifts, do not omit anyone present or anyone who has shown hospitality during your stay. Bring along extras just in case.

If you are invited to a Chinese person's home, it is courteous to arrive with a small gift. Suitable presents include a basket of fruit, coffee, tea, sweets, or any memento from your home country that the Singaporeans can associate with your relationship. Picture books of your home area make good presents. Presenting a wife with perfume or children with toys is likely to be appreciated. Such presents show that you are concerned about the family as a whole and not simply the business relationship. Foreign liquor is another much appreciated gift. French cognac is the most prized, although it can be rather expensive, and it should only be given to those with whom you already have a close relationship.

It is polite for a recipient to refuse a gift two or three times before finally accepting it. For Westerners, the process can be tricky. In initially refusing a gift, a Chinese Singaporean may appear embarrassed and say that he cannot possibly receive such a nice item. The proper thing to do is insist your gift is only a small token and you would be honored if it were accepted. As a rule, after some hemming and hawing, the person will accept the present graciously. The custom of at first refusing something offered extends to drinks, cigarettes, and small items that Westerners would not usually consider to be gifts. Be prepared to insist on acceptance.

However, if your attempts to give a present are rejected several times and it is clear that the intended recipient is serious about not wishing to accept it, it may be that he is really sincere and that your offer should be withdrawn. He may refuse your gift for any number of reasons. If he views the gift as too valuable, he may be concerned that accepting it would open him to the charge of bribery. Or he may not want to be in your debt, or he has no intention of having a relationship with you.

When a gift is offered to you, it is not necessary for you to refuse it ceremonially as the Chinese do. Humble acceptance and a few choice words of appreciation are enough. A gift from a businessperson may simply be a courtesy that he accords to all visitors, but it can also be an acknowledgment that a relationship with you exists. Or it may indicate that you will be asked for a favor. In any case, if someone presents you with a gift, you are expected to reciprocate eventually in kind or through a favor.

If the gift is wrapped, it is considered impolite to open it in front of the giver unless he encourages you to do so. Tearing the wrapping off hastily is a sign of greediness. Any gift that you give wrapped to an individual should be wrapped in traditional lucky colors of gold or red. Avoid giving gifts wrapped in white, black, or dark blue, because these colors are associated with mourning.

MEETING THE CHINESE SINGAPOREANS

When meeting Chinese Singaporean businesspeople, foreigners should display sincerity and respect. Handshaking, imported from the West, is generally the accepted form of salutation. However, Chinese handshakes differ in two ways from those common in the West. First, Chinese tend to shake hands very lightly, without the Western custom of gripping the hand firmly and pumping vigorously. Second, a handshake can last as long as ten seconds, instead of the brisk three-second contact common in the West.

The handshake is always followed by a ritualistic exchange of business cards. Foreigners should always carry an ample supply of business cards, preferably with English text on one side and Chinese on the other. Seek the advice of a knowledgeable person on the choice of characters for your name and company, as some characters have better meanings than others.

The proper procedure for exchanging business cards is to give and receive cards with both hands, holding the card corners between thumb and forefinger. When receiving a card, do not simply pocket it immediately, but take a few moments to study the card and what it says. The name card represents the person who presents it, and it should be given respect accordingly.

Presenting letters of introduction from well-known business leaders, overseas Chinese, or former government officials who have dealt with Singapore is an excellent way of showing both that you are a person of high standing and that you mean business. Singaporean Chinese are very concerned about social standing, and anything that you can do to enhance their regard for you is a plus. But be careful not to appear arrogant or haughty, as Confucian morality condemns such behavior.

Women in the Workplace

More and more women are entering management and technical positions in Singapore. While they may appear to be quite liberated in the Western sense (especially Chinese women), there are some important rules to remember for interactions with members of the opposite sex in the workplace:

- Keep all interaction businesslike and relatively formal. Flirtatiousness is taboo and could cause a woman to lose her job.
- Any compliments should concern a woman's work, not how she looks. Comments of the latter type could be misinterpreted as flirting.
- Never hug or kiss a person of the opposite sex in a business environment. Most physical contact is taboo.
- When a man meets a woman for the first time, he should wait for her to initiate the handshake. If none is forthcoming, a smile and a nod will do.

Body Language

Singaporeans often use body language that can be incomprehensible to Westerners, and some Western body language or positions can be misunderstood. This section reviews a few key examples to keep in mind when you visit Singapore. Ethnic groups which adhere to each behavior are in parentheses.

- When Singaporeans want someone to approach, they extend the hand palm down and curl the fingers as if scratching an imaginary surface (Chinese, Malay).
- Holding one's hand up near the face and slightly waving means no, or it can be a mild rebuke (Chinese).
- Pointing at someone with the forefinger is an accusatory motion considered rude or hostile. When you point, use your whole open hand (all Singaporeans).
- The meaning of laughter or smiles among Chinese depends on the situation. When they are nervous or embarrassed, Chinese often smile or laugh nervously and cover their mouths with their hands. This may be responding to an inconvenient request, a sensitive issue which has been brought up, or a social faux pas committed by the smiler or another person near by.
- When Chinese are embarrassed, they cover their faces with their hands.
- Hissing is a sign of difficulty or uncomfortableness (Chinese).
- When Chinese yawn, cough, or use a toothpick, they cover their mouths.
- It is impolite to point one's feet at another person (all Singaporeans). Chinese sit upright in chairs with both feet on the floor. Malay men sit cross-legged on a mat, and Malay women tuck their legs beneath their skirt.
- In Chinese and Malay homes, people remove their shoes when they enter. Be sure to wear clean socks!
- Singaporeans are not a touchy people and they rarely hug in public. Lightly touching another person's arm when speaking is a sign of close familiarity. Men and women rarely hold hands in public, but it is not uncommon for Chinese friends of the same sex to hold hands or to clasp each other by the shoulders.
- When eating with Chinese, never stick your chopsticks upright into a bowl. It reminds them of the incense burned at a funeral.

Saying No

When asked for a favor, Chinese will usually avoid saying no, as to do so causes embarrassment and loss of face. If a request cannot be met, Chinese may say it is inconvenient or under consideration. This generally means no. Another way of saying no is to ignore a request and pretend it wasn't made. Unless a request is really urgent, its best to respect these subtleties and not to press the issue.

Sometimes a Chinese will respond to a request by saying, "Yes, but it will be difficult." To a Westerner, this response may seem to be affirmative, but in Singapore it may well mean no or probably not. If a person says yes to a question and follows by making a hissing sound of sucking breath between his teeth, the real answer could be no.

The Chinese also have the habit of telling a person whatever they believe he or she wants to hear, whether or not it is true. They do this as a courtesy, rarely with malicious intent, although it can be a real problem in the workplace. If bad news needs to be told, Chinese will be reluctant to break it. Sometimes they will use an intermediary for communication, or perhaps they will imply bad news without being blunt. To cut through such murkiness, it is best to explain to your Singaporean coworkers that you appreciate direct communication, and that you will not be upset at bearers of bad news.

Names and Forms of Address

In Chinese an individual's family name precedes his or her personal name. The family name is almost always monosyllabic, and the personal name usually has two syllables, although one-syllable personal names are not uncommon. For example, Lee Kuan Yew's family name is *Lee*, and his personal name is *Kuan Yew*. Lu Xun, a famous author of the early 20th century, has a one-syllable personal name: *Xun*. Each syllable is the pronunciation of a single written character.

Chinese Singaporeans often adopt English names, since they know it is difficult for foreigners to remember Chinese names. Sometimes they invert their name: so that the family name follows the personal name, Western-style. This can be confusing, since you may not know if a name has been inverted. As a general rule, assume that a name has not been inverted.

People outside the family almost never call each other by their personal names, often even if they are very close. Westerners should use *Mister*, *Miss*, or *Mrs.* when addressing Chinese, just as they do in Western society. Although a woman does not take her husband's family name when she marries, it is acceptable for Westerners to address a married woman in the Western form, such as *Mrs. Hu* if the woman's husband is Mr. Hu.

Another common form of address is to use a person's designated position in society. For example, a teacher with the family name *Yuan* can be referred to as *Teacher Yuan*. This system of address also applies to company managers, directors, and higher-ranking officials.

The Chinese Banquet

It is fair to say that the number one pastime among Chinese Singaporeans is eating. If you like Chinese food, going to a traditional banquet may be among your most pleasant experiences in Singapore. The form of the meal is ancient, and thus there are rules of etiquette which should be followed. Although your host will not expect you to know everything about proper banquet behavior, he will greatly appreciate your displaying some knowledge of the subject, because it shows that you have respect for Chinese culture and traditions.

Arrival Banquets are usually held in restaurants or clubs in private rooms that have been reserved for the purpose. All members of your delegation should arrive together and on time. You will be met at the door and escorted to the banquet room, where the hosts are likely to have assembled. Traditionally, and as in all situations, the head of your delegation should enter the room first. Do not be surprised if your hosts greet you with a loud round of applause. The proper response is to applaud back.

Seating and Settings The banquet table is large and round and can seat up to twelve people. If there are more than twelve people, guests and hosts will have been divided equally among tables. Seating arrangements, which are based on rank, are stricter than in the West. This is another reason why you should give your host a list of delegation members that clearly identifies their rank. The principal host is seated facing the entrance and farthest from the door, usually with his back to the wall. The principal guest sits to the host's immediate right. If there are two tables, the second-ranking host and guest sit at the other table facing the principal host and guest. Interpreters sit to the right of the principal and second-ranking guests if there are two tables. Lower-ranking delegation members are seated in descending order around the tables, alternating with Singaporean hosts. Guests should never assume that they may sit where they please and should wait for hosts to guide them to their places.

Each place setting at the table contains a rice bowl, a dish for main courses, a dessert dish, a spoon, and chopsticks on a chopstick rest; usually there is a napkin. Two glasses are customary: a larger glass for beer or soda and a small thin glass for hard liquor. In the middle of the table is a revolving tray on which entrees are placed. During the meal it can be spun at will to gain access to the dishes that it holds.

Chopsticks Your host may politely ask if you are able to use chopsticks or if silverware would be more convenient. It is advisable to learn how to use chopsticks before you come to Singapore. One good method of learning is to practice picking up peanuts from a bowl. If you are able to pick up a bowlful of peanuts with relative ease, then you should have no trouble at a banquet. If you absolutely cannot master chopsticks, silverware may be made available. Asking for silverware entails some loss of face, but the loss is less than it would be if you are truly inept and messy with chopsticks.

Drinks Beer is the standard drink of choice at banquets, but you may feel free to substitute soda. Hard liquor, usually rice wine or perhaps brandy, is served ceremonially and is reserved for toasts. It is impolite to drink liquor alone.

Beginning the Feast The first round of food at a banquet consists of small plates of coldcuts. They may already be on the revolving tray when you sit down. The dishes may contain pork, chicken, pickled vegetables, codfish, scallops, tofu, or any number of different foods. It is polite, but certainly not mandatory, to try a taste of each dish. It is better not to partake of foods that you cannot eat than to gag at the table, but if you find something on your plate that you dislike, you may simply push it around on your plate to make it look as if you have at least tasted it.

It is the host's responsibility to serve the guests, and at very formal banquets people do not begin to eat until the principal host has broken into the dishes by serving a portion to the principal guest. Or, the host may simply raise his chopsticks and announce that eating has begun. After this point, one may serve oneself any food in any amount, although it is rude to dig around in a dish in search of choice morsels. Proper etiquette requires that serving spoons or a set of large chopsticks be used to transport food to one's dish, but in fact many Chinese will use the wide end of their eating chopsticks for this purpose. Watch your host to determine which procedure to use.

After the first course of coldcuts comes a succession of delicacies. Waiters will constantly remove and replace dishes as they are soiled or emptied, so that it is hard to tell exactly how many courses are served through the event. Some banquets can include more than twelve courses, but ten are more likely. Remember to go slow on eating. Don't fill yourself up when five courses are left to go. To stop eating in the middle of a banquet is rude, and your host may incorrectly think that something has been done to offend you.

Manners Table manners often have no relation to manners in the West. There are no prohibitions on putting one's elbows on the table, reaching across the table for food, or making loud noises when eating. Usually it is impolite to touch one's food with anything except chopsticks, but when eating chicken, shrimp, or other hard-to-handle food, Chinese use their hands. Bones and shells are usually placed directly on the tablecloth next to the eating dish. Waiters periodically come around and unceremoniously rake the debris into a bowl or small bucket. Although banquets have their prescribed methods for behavior, manners conform more to practicality than they do in the West. In fact, banquet time is when businesspeople tend to be the most relaxed and comfortable.

Liquor and Toasts One reason is that drinking figures prominently in Chinese banquets. Toasting is mandatory, and the drinking of spirits commences only after the host has made a toast at the beginning of the meal. It is likely that he will stand and hold his glass out with both hands while saying a few words to welcome the guests. When he says the words *yam seng*, which means bottoms up (literally, *dry glass*), all present should drain their glasses. After this initial toast, drinking and toasting are open to all, but the head of the visiting group will be expected to toast the well-being of his hosts in return. Subsequent toasts can be made from person to person or to the group as a whole. No words are needed to make a toast, and it is not necessary to drain your glass, although to do so is considered more respectful.

Remember that hard liquor should never be drunk alone. If you are thirsty, you can sip beer or a soft drink individually, but if you prefer to drink hard liquor, be sure to catch the eye of someone at your table, smile and raise your glass, and drink in unison. Beer or soft drinks can also be used for toasting, but do not switch from alcohol to a soft drink in the middle of the banquet lest the host think that something has offended you.

Also, it is impolite to fill your own glass without first filling the glasses of all others. This applies to all drinks and not just to alcohol. If your glass becomes empty and your host is observant, it is likely that he will fill it for you immediately. When filling another's glass, it is polite to fill it as full as you can without having the liquid spill over the rim. This symbolizes full respect and friendship.

It is a matter of courtesy for the host to try to get his guests drunk. If you do not intend to drink alcohol, make it known at the very beginning of the meal to prevent embarrassment. Even then, the host may good-naturedly try to goad you into drinking. One way to eliminate this pressure is to tell your host that you are allergic to alcohol.

In the course of drinking at banquets, it is not unusual for some Chinese to become quite inebriated, although vomiting or falling down in public entails loss of face. After a few rounds of heavy drinking, you may notice your hosts excusing themselves to the bathroom, from whence they often return a bit lighter and rejuvenated for more toasting! Also, many Asians are unable to metabolize alcohol as fast as Westerners. The result is that they often get drunk sooner, and their faces turn crimson, as if they were blushing.

The Main Dish The high point of a Chinese banquet is often the presentation of a large whole cooked fish. In formal situations, the fish is placed on the revolving tray with its head pointing toward the principal guest. The guest should accept the first serving, after which everyone helps himself.

The final rounds of food follow, usually a soup followed by rice, concluding with fresh fruit. Chinese consider soup to be conducive to digestion. Rice is served at the end so that guests can eat their fill, as if the preceding courses had not been enough. It is polite to leave some rice and other food on your plate; to finish everything implies that you are still hungry and that you did not have enough to eat. Fruit is served to cleanse the palate.

Concluding the Banquet When the fruit is finished, the banquet has officially ended. There is little ceremony involved with its conclusion. The host may ask if you have eaten your fill, which you undoubtedly will have done. Then, without further ado, the principal host will rise, signaling that the banquet has ended. Generally, the principal host will bid good evening to everyone at the door and stay behind to settle the bill with the restaurateur. Other subordinate hosts usually accompany guests to their vehicles and remain outside waving until the cars have left the premises.

Singapore

Legend:
- City or Village
- Urban Area
- Expressway
- Primary Road
- MRT System (Fast train)
- Railroad

Locations:

Pulau Tekong Besar
Pulau Ubin
Changi
Singapore Changi Airport
Serangoon Harbor
Pasir Ris
Tampines
Bedok
Katong
Punggol
Hougang
Serangoon
Bishan
Geylang
Yio Chu Kang
Ang Mo Kio
Thomson Village
Toa Payoh
Downtown
Pulau Brani
Queenstown
Tiong Bahru
Sentosa
Nee Soon
Singapore Island
Clementi
Pasir Panjang
Pulau Semakau
Yishun
Sembawang
Woodlands
Bukit Panjang
Bukit Batok
Jurong East
Pulau Bukum
Selat Pandan
Pulau Ayer Chawan
Choa Chu Kang
Jurong West
Tuas

Surrounding areas (Malaysia):
Sungai Johor
Masai
Selat Johor
Johor Baharu
Kangkar Pendas
Selat Johor

Straits of Singapore

Scale: 0 – 5 – 10 km / 0 – 5 mi

© 1993 Magellan Geographix℠

Business District

1. Amara Hotel and Shopping Center
2. 6 Battery Road
3. Carlton Hotel
4. Century Tower 21
5. City Hall
6. Clifford Center
7. Clifford Pier
8. Colombo Court
9. CPF Building
10. Crasco Building
11. DBS Building
12. Empress Place Building
13. Gateway
14. General Post Office Fullerton Building
15. GMG Building
16. Golden Shoe Car Park
17. Hong Leong Building
18. IBM Towers
19. International Exhibition and Convention Center (to be completed by 1994-95)
20. International Plaza
21. Marina Mandarin Hotel
22. National Development Building
23. OCBC Center
24. Ocean Towers
25. Oriental Singapore Hotel
26. OUB Center
27. Pan Pacific Singapore Hotel
28. Parliament House
29. Pidemco Center
30. Raffles City Post Office
31. Raffles City Shopping Center
32. Raffles City Tower
33. Raffles Hotel
34. Robina House
35. SATA
36. Shaw Towers
37. Shell Tower
38. Shenton House
39. 78 Shenton Way
40. Shing Kwan House
41. SIA Building
42. Singapore Conference Hall
43. Singapore Cricket Club
44. Southpoint
45. Straits Trading Building
46. Supreme Court
47. TAS Building Robinson Road Post Office
48. Telok Ayer Market
49. Treasury Building
50. UIC Building
51. Victoria Theatre and Victoria Concert Hall
52. War Memorial Park
53. Westin Plaza Hotel
54. Westin Stamford Hotel

Retail District

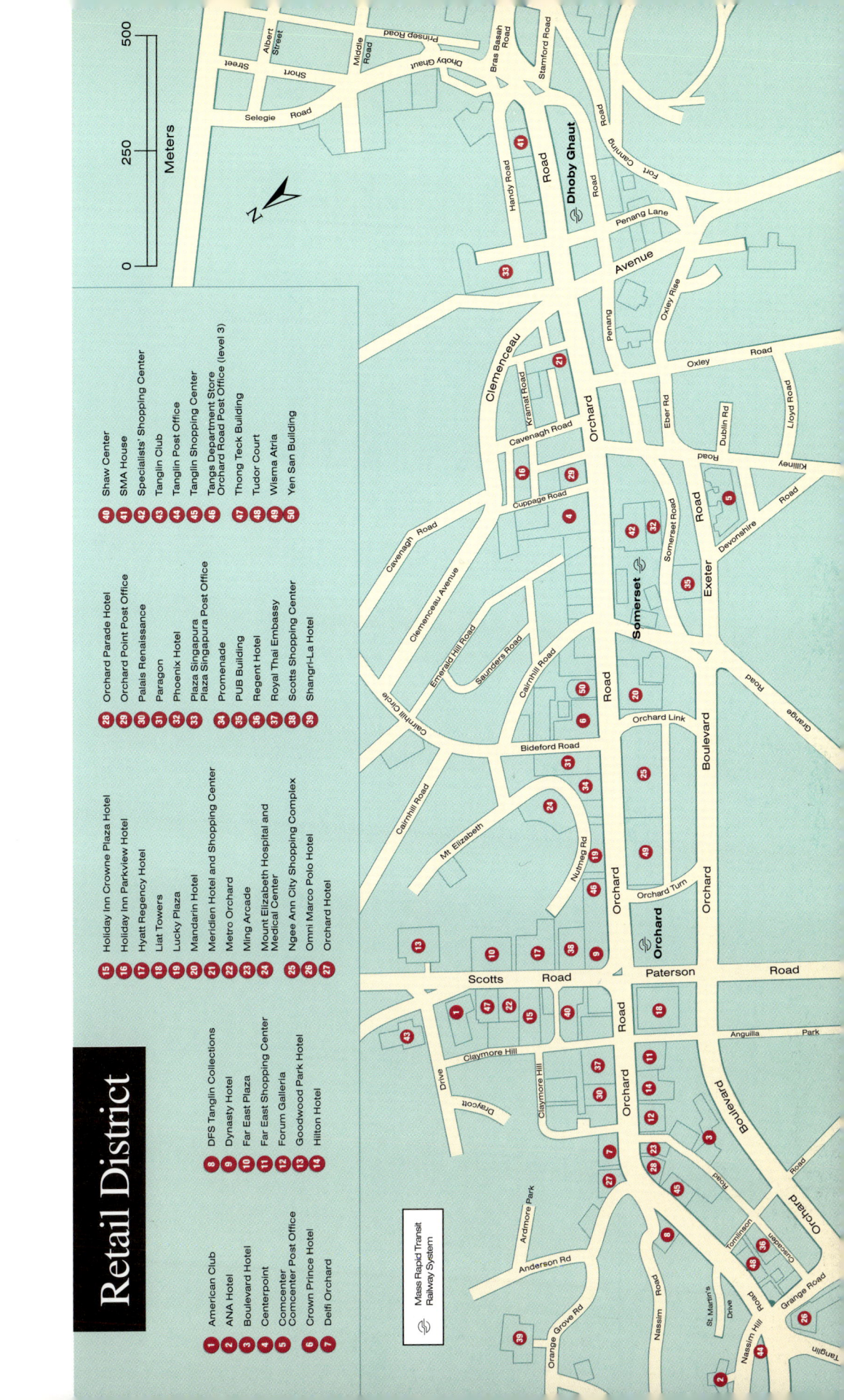

1. American Club
2. ANA Hotel
3. Boulevard Hotel
4. Centerpoint
5. Comcenter
   Comcenter Post Office
6. Crown Prince Hotel
7. Delfi Orchard
8. DFS Tanglin Collections
9. Dynasty Hotel
10. Far East Plaza
11. Far East Shopping Center
12. Forum Galleria
13. Goodwood Park Hotel
14. Hilton Hotel
15. Holiday Inn Crowne Plaza Hotel
16. Holiday Inn Parkview Hotel
17. Hyatt Regency Hotel
18. Liat Towers
19. Lucky Plaza
20. Mandarin Hotel
21. Meridien Hotel and Shopping Center
22. Metro Orchard
23. Ming Arcade
24. Mount Elizabeth Hospital and Medical Center
25. Ngee Ann City Shopping Complex
26. Omni Marco Polo Hotel
27. Orchard Hotel
28. Orchard Parade Hotel
29. Orchard Point Post Office
30. Palais Renaissance
31. Paragon
32. Phoenix Hotel
33. Plaza Singapura
    Plaza Singapura Post Office
34. Promenade
35. PUB Building
36. Regent Hotel
37. Royal Thai Embassy
38. Scotts Shopping Center
39. Shangri-La Hotel
40. Shaw Center
41. SMA House
42. Specialists' Shopping Center
43. Tanglin Club
44. Tanglin Post Office
45. Tanglin Shopping Center
46. Tangs Department Store
    Orchard Road Post Office (level 3)
47. Thong Teck Building
48. Tudor Court
49. Wisma Atria
50. Yen San Building

Mass Rapid Transit Railway System

Jurong

1. Raffles Country Club
2. Nanyang Technological Institute
3. Jurong Bird Park
4. Jurong Stadium
5. Chinese Garden
6. Japanese Garden
7. Omni Theater and Science Center
8. Jurong Country Club
9. Jurong Town Hall
10. Ngee Ann Polytechnic
11. West Coast Park
12. National University of Singapore

Reciprocity

After you have been entertained by your Singaporean Chinese associates, it is proper to return the favor unless time or other constraints make it impossible. A good time to have a return banquet is on the eve of your departure from Singapore or at the conclusion of the business at hand.

If possible, a third party should relay your invitations to the Chinese Singaporeans. If for some reason they must refuse the invitation, they will feel more comfortable telling the third party than speaking directly to you.

While Singapore is full of Western-style restaurants, it is advisable to make reservations at a Chinese restaurant where you are sure to get good service and food. Your Chinese guests are likely to appreciate it more than Western fare. Banquets are priced per person and cover all expenses except alcohol. There is no need to order specific dishes, although you may do so. Good restaurateurs know how to prepare adequately for a banquet.

THE MALAY CULTURE OF SINGAPORE

The second largest ethnic group in Singapore consists of the Malays, who comprise about 15 percent of the population. Their cultural habits are often very different from those of the Chinese. Although Malay life has in some ways changed drastically in recent years, they are a conservative people who have been surprisingly successful in maintaining traditional ways of life in the metropolis that is today's Singapore. The following sections discuss some of the major values that shape Malay culture.

Islam

The foundation of Malay life is Islam. As Muslims, they are expected to pray five times a day to Allah and to observe the teachings of the Koran. Malays are Sunni Muslims, not the more fundamentalist Shiite Muslims who predominate in Iran. Malays have been affected by the Islamic revival of the past two decades, but in contrast to more conservative Muslim societies, Malay women rarely wear veils, and Singapore's Malays in general seem able to function effectively in a modern environment without either touting or forsaking their religious creed. Their beliefs set forth very strict rules for behavior, including dietary laws and rules for social interaction between sexes.

Budi

The Malay people adhere to a moral system for behavior called *budi*. *Budi* concerns both outward social relations and internal personal ethics. The basic guidelines for *budi* are respect, courtesy, veneration of elders and parents, and harmonious relations within the family and among all members of society. Working for harmony and avoiding animosity are hallmarks of idealized Malay behavior.

Malays emphasize that responsibility and clean living are superior to the pursuit of material gain. Although diligence and hard work are important, individual greed should never be placed above group harmony. While many Malays have a standard of living below that of their Chinese counterparts, in recent years their economic conditions have improved.

The Family

As it is for Chinese, the family is the basic social unit for Malays. They are a collectivist people who place family welfare before individual needs. Children are dearly loved in the Malay household, and parents love to hold babies and toddlers. Fathers play a more active and important role in child rearing than they do in most other societies. Children honor their parent's wishes and never openly disagree with them or show anger. They are responsible for the welfare of their parents in old age.

Prohibitions

Like all other Muslims, Malays are forbidden to eat pork, drink alcohol or to eat meat and chicken not slaughtered by Muslims in accordance with prescribed methods. Forbidden foods are called *haram*, a word that means unclean. Properly prepared foods are called *halal*. When hosting Muslim associates, make perfectly sure that no *haram* food is prepared. Muslim cooks and servants should never be expected to handle such foods, which would be offensive.

Women in Malay Society and Business

As devout Muslims, Malay women have very definite societal roles that strictly differentiate them from men. However, many younger women are more open and cosmopolitan than their older counterparts. Many Malay women do work in businesses, although they often observe customs particular to their society.

Muslim women tend to keep most of their skin covered, exposing only the hands and head. In recent years, they have returned to wearing the traditional headdress, called the *tudung*. The *tudung* covers the head, neck, and hair. Only the face is exposed. In the workplace, many women wear Western-style attire adapted to cover their arms and legs.

At special feasts and at prayer, women traditionally sit apart from men. At home during festivals, Malays sit on the floor on colorful mats. In more conservative homes, women sit at one end of the living room and men at the other. When sitting, women tuck their legs to one side under their clothes and away from view. Feet should never be exposed for others to see. In the main prayer hall of a mosque, women are segregated from men.

Western Women and Malays

Because Malays live in Singapore's pluralistic society, they do not expect non-Malay women to observe their own strict codes. Malays are accustomed to seeing Western women engaging in sightseeing and business, and they are not likely to be chauvinistic in business dealings with them. Out of courtesy, though, Western women should dress modestly around Malays and behave with a certain formality. They should avoid touching men, even Western men, in front of Malays, and they should maintain a respectful distance when working with a Malay man.

Introductions and Greetings

The Salaam Among Malay people, Western-style handshaking is not the standard salutation. Instead, they usually *salaam*, which is done with two hands. Among Muslim peers, Malays outstretch both arms and lightly touch each other's hands. They then both bring their hands back to their chests and cover their hearts. Among persons of unequal age or status, the more senior person uses only one hand, and the junior person uses both hands as a sign of respect. Women *salaam* with each other, but they often first kneel, tucking their legs beneath their dress. After touching hands, they may bring their hands up close to the nose and mouth as if they were praying, and then cover the heart. Women and men rarely *salaam* but instead smile and nod slightly.

Handshaking When two close male friends meet, they usually forego the *salaam*. Instead, they will shake hands, using the right hand and a single firm shake. However, it is not uncommon for Malays to use the standard Western handshake when meeting non-Malays. This is especially common in a business environment. A foreigner who is aware of proper Malay greetings should *salaam*, because doing so shows respect for Malay culture. Some younger, non-traditional Malays commonly use handshakes with each other, including members of the opposite sex.

Handshaking may or may not be appropriate in interactions between men and women when one person is not a Malay. Foreigners, both women and men, should wait for the Malay to initiate handshaking. If no such gesture is made, the proper thing to do is smile and make nod slightly.

Spoken Greetings The many spoken greetings between Malays are usually related to the deeply spiritual nature of the Malay people. They often use the Arabic greeting that is traditional among Muslims everywhere, saying *Assalam alaikum* (Peace unto you) and responding with *Walaikum salam* (And also to you). This greeting is restricted to Muslims only. Non-Muslims can show their respect for tradition by saying the English words, but they should not use Arabic. Of course, a simple "How are you?" is also acceptable, as is asking friends about their family's or children's health.

Names

The Malay way of naming people is quite unusual. Although the people trace their lineage through the paternal line, Malays do not have family names as do Westerners and Chinese. Instead, male children add their father's personal name to their own personal name, placing the word *bin*, meaning "son of," between the two names.

For example, if a boy's name is Ali and his father's name is Mohammed, he is known as Ali bin Mohammed. If the boy grows up and has a son named Abdul, that child will be known as Abdul *bin* Ali. In this way, names survive no more than two generations. In English, Abdul *bin* Ali should be called *Mr. Abdul* (not *Mr. Ali*, for that is his father's name).

In much the same way, female Malays add their father's names to their own, but the linking word is *binti*, meaning "daughter of." A woman named Benazir who is the daughter of Abdul *bin* Ali will be known as *Benazir binti Abdul*. She does not take the name of her husband when she marries. Instead, she is called *Mrs. Benazir*. In the business world, however, Malay women sometimes will use their husband's last name in the Western style.

"Right" Etiquette

Among Malay and Indian people, it is taboo to give or receive some things with the left hand. The left hand is considered unclean, and for a very good reason. This is because Malays and Indians don't use toilet paper. Instead, after using the toilet, they use a scoop held in the right hand to pour water from a bucket over their backside, and they use the left hand to rinse.

Many Westerners cannot help but feel revolted by this mundane procedure employed by millions of people around the world. Think of it instead as a chance to drop your cultural baggage, if only temporarily. No matter how you feel about the implications, when you deal with Malays and Indians, you should remember to use only your right hand to pass money, gifts, and all else.

When dining, Malays and Indians do not traditionally use personal silverware or chopsticks but rely solely on their right hands. Spoons are used to serve food from the main dish to an individual plate or bowl.

The proper way of eating with the right hand is to hold the fingers very close together slightly cupped. Then scoop up a small amount of food with the fingers. Bend over the plate or bowl and push the food into your mouth with the thumb. Your tongue may stick out a little to receive the food. It is impolite to grab too much food at a time or to hold food in the palm. (It is also extremely messy!)

At the table, it is permissible to hold a glass or pass a plate with the left hand if the right hand is soiled with food. If the right hand is clean, then use it instead.

Body Language

Here is a short list of dos and don'ts to remember when interacting with Malay people:

Pointing It is impolite to point at a person with the forefinger. Instead, use the whole open *right* hand. You may also use the thumb of the right hand with the other fingers closed into a fist.

Dogs Muslims are forbidden to come in contact with dogs, and pictures of dogs are considered improper. Never give a Muslim child a stuffed toy dog or clothing printed with dog images.

Touching the head Never touch the head of a Malay child (or for that matter an adult). Malays consider the head to be the home of the soul and too sacred to touch.

Pig skin Never give a gift containing pig leather or skin of uncertain origin to a Muslim.

Removing shoes Remove your shoes when you enter a person's home.

Exposing feet Feet are considered unclean, and they should therefore not be casually displayed. Never point your feet at another person.

Sitting When men sit on the floor, they should sit cross-legged. Women should sit with their legs tucked under them out of sight.

The Koran Never touch a Koran in Arabic unless you are Muslim and you have first performed the proper ablutions. There is no taboo on touching translated Korans, but as a rule don't touch anyone's Koran except your own.

THE INDIAN COMMUNITY IN SINGAPORE

Although the Indians in Singapore comprise only about 7 percent of the total population, they play a significant role in its society. Many are shopkeepers, businessmen, politicians, or lawyers. Indeed, Indians have been in Singapore longer than the Chinese. The first Indians in Singapore were traders who arrived in about 100 AD. The name *Singapore* is derived from two Sanskrit words: *singha*, meaning lion, and *pura*, meaning port.

The Indian community of Singapore is very diverse. Ten native languages are spoken among Singaporean Indians, and religious affiliations are almost as diverse. Muslims, Hindus, Christians, Sikhs, Zoroastrians, and Jains can all be found in the community.

The largest percentage of Indians speak Tamil (about 64 percent). The Tamil language is one of Singapore's official languages, and Tamil culture remains strong. Tamils are mostly Hindus (about 57 percent of all Indians in Singapore are). Indian Muslims are also numerous, comprising about 22 percent of Singapore's Muslim population.

The Indians in Singapore form a spectrum ranging from extremely modern and Westernized to extremely conservative and traditional. The traditional beliefs and habits discussed in the following sections may not all be observed by many Indians today, but there are certainly those in Singapore to whom they apply.

Hindus

Hinduism is as much a cultural lifestyle as it is a religion, and for Hindus there is no difference between the two. Hinduism is characterized by a belief in social castes and the concept of *karma*. Devout Hindus aspire to liberation from a cycle of birth, death, and rebirth by living virtuous lives.

Castes Although the caste system is contrary to modern concepts of equality, is still very strong among Hindus. They believe that a person born into a certain caste should remain in that caste for his or her lifetime and seek to be a model member. If your life is virtuous, then you will have higher status in your next incarnation. If you are immoral or live improperly, then you will live your next life in a lower caste or even as an animal.

Karma The rule that dictates a person's position in life is called *karma*. *Karma* resembles the Western concept of fate, but it is not exactly the same. A person's *karma* is dictated by the sum total of the good or bad deeds that he or she has done in previous lifetimes. All a person's actions have a direct effect on his or her future lives.

Nirvana *Nirvana* is the ultimate goal of all devout Hindus. It is the final release of a person from temporal existence, when the individual soul merges with the one absolute Universal Soul. Only those who have attained the highest caste, called *Brahman* (the priestly class), are likely to attain *Nirvana*.

Religious Prohibitions Hindus believe that cows are sacred animals and that they should never be eaten. Alcohol and tobacco are also frowned upon, although many Indians smoke a traditional cigarette called a *beedie*. Tamil cuisine includes some meat dishes, such as chicken, but many Tamils are strict vegetarians.

Muslims

Many Indians are Muslims. They follow the same religious observances described earlier in this chapter for Malay Muslims.

Sikhs

Their turbans, long beards, and ceremonial knives worn at the waist make Sikhs easy to identify. They never cut their hair. Their religion is a mix of

Islam and Hinduism, although they do not recognize the caste system.

Names and Forms of Address

Most Indians in Singapore do not use a family name, and like the Malay place their father's name after their own. Between their name and their father's name they will place the characters s/o if they are male (meaning *son of*) or d/o if they are female (meaning *daughter of*). An Indian male may place his father's first initial at the head of his own name.

All male Sikhs adopt the last name of *Singh*, and females take the last name *Kaur*. These names do not signify clan allegiance but rather identify them as members of the Sikh faith. Like other Indians, they place their father's name after their own.

TRADE DELEGATIONS

There are a few important points to remember when you send a trade delegation to Singapore. First, keep in mind that Asians are group-oriented people and that they are more comfortable functioning as members of a group than as individuals. Generally, they assume that this is true of all people. They are confused when members of a visiting group speak as individuals and make statements that are contradictory or inconsistent with the stated views of the group as a whole. Individual opinions are not wanted. Therefore, every trade delegation should have a designated speaker, who is also its most senior member. Singaporeans will look to that member for all major communication and accept his words as the words of the entire organization.

Singaporeans are really concerned about the status that an individual holds in a company or organization. They will evaluate the seriousness of a trade delegation by the rank of its members, and a delegation is not likely to succeed if the Singaporeans know that its head is a junior executive. Likewise, they will wish to match your delegation with executives of similar status from their own organization. It is wise to send them a list of the delegates who will attend that gives their ranks in the company and to request that they do the same. If the Singaporean company sends someone to a meeting who is obviously of lower rank, the chances are that it is either not particularly interested in you or that it is unaware of the status of the members of your delegation.

A trade delegation should be led by older members of the company who have at least middle-level executive rank. They should be patient, genial, and persistent, and have extensive cross-cultural experience. Ideally, they have already had some experience in Singapore or Asia, and have enough rank to make decisions on the spot without fear of repercussions from the home office.

Little in the way of serious business will be accomplished on the first day of a delegation's visit to Singapore. This is a time for getting to know one another and for feeling out the personalities who will be involved in later negotiations. Use this time to get to know the Singaporeans, and try to determine the status of all the members and their likely relations with one other. The Singapore groups senior leader will be the only spokesman for the group on substantive issues, but other members will probably have some say in decision making. Singaporeans place great stock in consensus. Most likely they will debate their position on the business at hand among themselves, but never in front of the foreign delegation.

Negotiation Etiquette

When arranging for negotiations with Singaporean companies it is customary to give them as much detail on the issue to be discussed as is reasonable, plus notice of all delegation members who will be present. The team leader's name should be listed first. Other members should be listed in order of seniority or importance for the deal. The number of negotiation members can vary from two to 10, depending on the nature of the business. The Singaporean side will try to match their team members with the visiting team.

Negotiations are often held in meeting rooms at the place of business. A functionary escorts the members of the visiting delegation to the meeting room as soon as they arrive. The Singaporean team will already be there. The head of the visiting delegation should enter the meeting room first.

After a round of handshaking and smiles, the visitors are seated at the negotiation table. The table is usually rectangular, and teams sit opposite each other, with the heads of delegations sitting eye to eye. Other team members are arrayed next to delegation heads, often in descending order of importance. Most likely, the guest delegation will be seated facing the door, a common Asian courtesy. Tea or other drinks are provided.

Do not expect to jump into substantive negotiations right away. Some small talk is usually necessary in order to get the ball rolling, and this time can also be used to get a feel for those present. Singaporeans like to know with whom they are dealing. The subject of business usually comes up naturally after the participants feel comfortable enough to begin.

Entering Substantive Talks

After initial courtesies, the head of the host delegation usually delivers a short welcoming speech and then turns the floor over to the head of the guest delegation. Visitors customarily speak first in negotiations. In some ways, this can be to the host's advantage, but participants usually know enough about each other's positions through prior communication that there are few surprises. As noted earlier of trade delegations, Singaporeans look to the senior leader for all meaningful dialogue. Conflicting statements from different team members are to be avoided, and team members should speak only when they are asked to do so.

Singaporean businesspeople do not like surprises in negotiations, so it may be wise to lay out your basic position at this time. It can also be useful to distribute sheets stating your main points. Anything that you can do to clarify their understanding of your position is fine, but in the initial stage, your presentation may need to involve only the big picture. Details can be saved for later in the talks. However, in some forms of negotiations, the Singaporeans will expect a very serious and in-depth presentation covering all the major details and answering all foreseeable questions at the very outset of talks. A typical opening statement highlighting the major topics that need to be discussed can last between five and ten minutes.

After the visitor outlines his team's position, the host team leader takes the floor and answers point by point, remedying any perceived omissions. From this point on, the negotiation process runs with the rhythm of a formal debate, not an open-ended chat. The Asian approach is often first to gain a holistic view of the entire proposal, then to break it down into specific chunks, at which time concrete issues and problems can be discussed. Use your own judgment in the talks, and adopt methods that are naturally suitable for your particular subject.

TYPES OF NEGOTIATIONS

Foreigners may encounter a few unfamiliar negotiating scenarios when dealing with Singaporeans. As already noted, one of the easiest and quickest kinds of business that a foreigner can do with Singaporean is to import goods produced there into another country. Negotiations of this type aim to arrive at a straight purchasing agreement. Issues of quality control and time and method of payment and shipment will be discussed. Negotiations for such an agreement can be concluded within a matter of hours if both sides have prepared adequately before the meeting.

A very different type of business relationship is one between the Singaporean government and a foreign company. Good personal connections can be useful when dealing with the Singaporean government. However, the factor that has the most weight is the quality and reputation that your side can bring into a project. Foreign businesspeople will do well to develop an excellent presentation and make sure that they have a thorough knowledge of every detail of the project being negotiated.

In other negotiations, such as those for cooperative efforts between companies or for marketing to Singapore, personal relationships remain the preeminent factor. But Singaporeans play hardball in negotiations, even when there is a personal relationship. They may even try to use the relationship as leverage in the business negotiation.

Singaporean Negotiating Tactics

Singaporean negotiators are shrewd and use many tactics. This section reviews some of the most common ones.

- Controlling the schedule and location. Negotiations with Singaporean businesspeople are often held in Singapore. They are aware that foreigners must spend a good deal of time and money to come to Singapore, and that they do not want to go away empty-handed. The Singaporeans may appear at the negotiating table seemingly indifferent to the success or failure of the meeting, and then make excessive demands on the foreign side.
- Threatening to do business elsewhere. Singaporeans may tell you that they can easily do business with someone else for example, the Japanese or the Germans if their demands are not met.
- Using friendship as a way of gaining concessions. Singaporeans who have established relations with foreigners may remind them that true friends would aim to out an agreement of maximum mutual benefit. Be sure that the benefits in your agreement are not all one-way.
- Showing anger. Although the display of anger is not acceptable under Confucian morality, Singaporean negotiators may show calculated anger to put pressure on the opposite side, which may be afraid of losing the contract.
- Sensing the foreigner's fear of failure. If the Singapore side knows that you are committed 100 percent to procuring a contract and that you are fearful of not succeeding, they are likely to increase their demands for concessions.
- Knowing when you need to leave. If the Singaporean side knows the date of your departure, they may delay substantive negotiations until the day before you plan to leave in order to pressure you into a hasty agreement. If possible, make departure reservations for several different dates, and be willing to stay longer than anticipated if there is a real chance for success.
- Attrition. Singaporean negotiators are patient and can stretch out the negotiations in order to wear you down. Jet lag and excessive entertaining in the evening can also take the edge off a foreign negotiator's attentiveness.

- Using your own words and looking for inconsistencies. Singaporeans take careful notes at discussions and they have been known to quote a foreigner's own words in order to refute his current position.
- Playing off competitors. Singaporeans may invite several competing companies to negotiate at the same time, and they will tell you about it to apply pressure.
- Inflating prices and hiding the real bottom line. The Singaporean side may appear to give in to your demand for lower prices, but their original stated price may have been abnormally high.

Tips for Foreign Negotiators

A number of tactics may be helpful for foreign negotiators working in Singapore.

- Be absolutely prepared. The effective negotiator has a thorough knowledge of every aspect of the business deal. At least one member of your negotiating team should have an in-depth technical knowledge of your area and be able to display it to the Singaporeans. Be prepared to give a lengthy and detailed presentation on your side of the deal.
- Play off competitors. If the going gets tough, you may let the Singaporeans know that they are not the only game in town. Competition is cutthroat among Asian producers, and you can probably find other sources for what your counterpart has to offer. Also, if price is the problem, you may be able to strike a cheaper deal in China or Southeast Asia. If quality is the concern, Japanese companies may be able to outperform Singaporean producers.
- Be willing to cut your losses and go home. Let the Singaporeans know that failure to agree is an acceptable alternative to a bad deal.
- Cover every detail of the contract before you sign it. Talk over the entire contract with the Singaporean side. Be sure that everyone understands his duties and obligations.
- Take copious and careful notes. Review what the Singaporean side has said, and ask for clarification on any possible ambiguities.
- Pad your price. Do as the Singaporeans do. Start out high, and be willing to give a little from there.
- Remain calm and impersonal during negotiations. Don't show your agitation, lest the Singaporeans know your sensitive areas. Even if you were good buddies the night before, a standoffish personal attitude in negotiations lets the Singaporeans know that your first priority is good business.
- Be patient. Asians believe that Westerners are always in a hurry, and they may try to get you to sign an agreement before you have adequate time to review the details.
- State your commitment to work toward a fair deal. Tell the Singaporeans that your relationship can only be strengthened by a mutually beneficial arrangement.
- Be willing to compromise, but don't give anything away easily.

FURTHER READING

The preceding discussion of business culture and etiquette in Singapore is by no means complete. The sources described in this section can give the reader some additional insight. Because business in Singapore is dominated by Chinese, works that study Chinese behavior are useful even when they do not directly pertain to Singapore.

Culture Shock! Singapore, by JoAnn Meriwether Craig. Portland, Ore.: Graphic Arts Center Publishing Company, 1993. ISBN 1-55868-108-6. $10.95. A readable overview of the major aspects of Singaporean culture that includes tips for business travelers and a long section of advice for expatriates, the book stresses learning to function in an alien environment and overcoming the inevitable culture shock of living in a foreign land.

Dealing With the Chinese, by Scott D. Seligman. New York: Warner Books, 1989. ISBN 0-446-38994-3. $12.95. A detailed examination of business relationships between Chinese and foreigners, whose author has extensive experience in mainland China and uses personal stories and examples to illustrate Chinese behavior and etiquette.

Chinese Etiquette and Ethics in Business, by Boye De Mente. Lincolnwood, Ill.: NTC Business Books, 1989. ISBN 0-8442-8525-0. $14.95. A broad cultural survey of Chinese morals and values related to business interaction, mostly in mainland China.

Do's and Taboos Around the World, edited by Roger Axtell. New York: John Wiley and Sons, 1990. ISBN 0-471-52119-1. $10.95. A humorous and insightful bestseller compiled by the Parker Pen Company on social customs around the world includes a specific section on Singapore and relevant information about Malaysia, India, and various Chinese societies.

Gestures: The Do's and Taboos of Body Language Around the World, by Roger Axtell. New York: John Wiley and Sons, 1991. ISBN 0-471-53672-5. A follow-up to the preceding book focused on body languages in different cultures.

Demographics

AT A GLANCE

The figures given here are the best available, but sources vary in comprehensiveness, in definition of categories, and in reliability. Sources include the United Nations, the World Bank, the International Monetary Fund, and the Singapore government. The value of demographics lies not just in raw numbers but in trends, and the trends illustrated here are accurate.

POPULATION GROWTH RATE AND PROJECTIONS

Average annual growth rate (percent)

1970-80	1980-91	1991-2000
2.0%	1.7%	1.5%

Age structure of population (percent)

	1991	2025
Under 15 years old	22.9	18.3
15 - 64 years old	70.7	63.5
64 years and over	6.4	18.2

POPULATION BY AGE & SEX (1990)

Age	Total	Male	Female
All ages	2,690,100	1,360,500	1,329,600
birth - 4	223,000	115,500	107,500
5 - 9	205,100	106,400	98,600
10 - 14	197,600	102,200	95,400
15 - 19	220,400	113,800	106,600
20 - 24	229,800	116,800	113,000
25 - 29	280,700	141,500	139,200
30 - 34	290,500	147,200	143,300
35 - 39	251,400	127,500	123,900
40 - 44	203,300	102,800	100,500
45 - 49	127,200	64,000	63,100
50 - 54	117,100	59,000	58,200
55 - 59	99,400	49,600	49,800
60 - 64	82,500	40,800	41,600
65 - 69	59,300	29,200	30,200
70 - 74	44,200	20,300	24,100
75 - 79	31,800	13,700	18,100
80 +	26,600	10,000	16,600

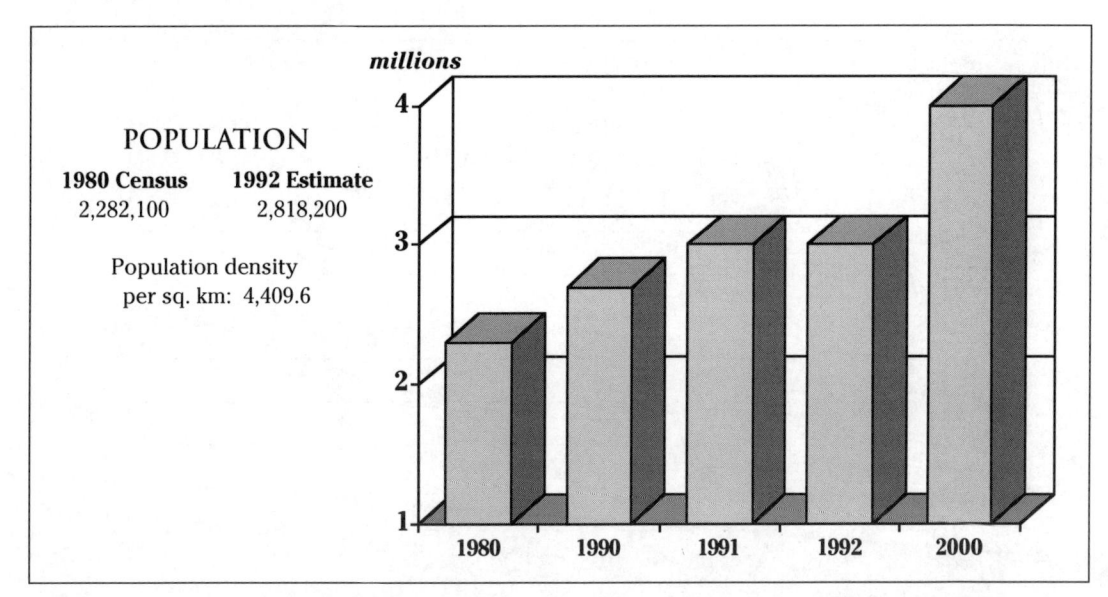

POPULATION

1980 Census	1992 Estimate
2,282,100	2,818,200

Population density per sq. km: 4,409.6

VITAL STATISTICS

Live births (1991):	49,114	(18 per 1,000)
Deaths (1991):	13,876	(5 per 1,000)
Life expectancy at birth	**1960**	65
	1990	74
Fertility rate	**1970**	3.1
	1991	1.8
	2000 (proj)	1.9
Women of childbearing age	**1965**	45%
(percentage of all women)	**1991**	60%

ETHNIC GROUPS
(1992 estimate)

Group	Population
Chinese	2,187,200
Malay	399,400
Indian	199,600
Others	32,000
TOTAL	2,818,200

RELIGIOUS AFFILIATION

Buddhism	29 percent
Christianity	19 percent
Islam	16 percent
Taoism	13 percent
Hinduism	6 percent
Other or none	17 percent

EDUCATION-1991

Category	Institutions	Teachers	Enrollment
Primary	202	9,843	260,286
Secondary	186	9,200	185,713
Technical and vocational	23	1,586	28,871
Universities, colleges	6	4,959	60,373
TOTAL	**417**	**25,588**	**535,243**

Consumer Price Increases
(percent)

Average 1975-84	1985	1986	1987	1988	1989	1990	1991	1992
3.7	0.5	-1.4	0.5	1.5	2.4	3.4	3.4	2.5

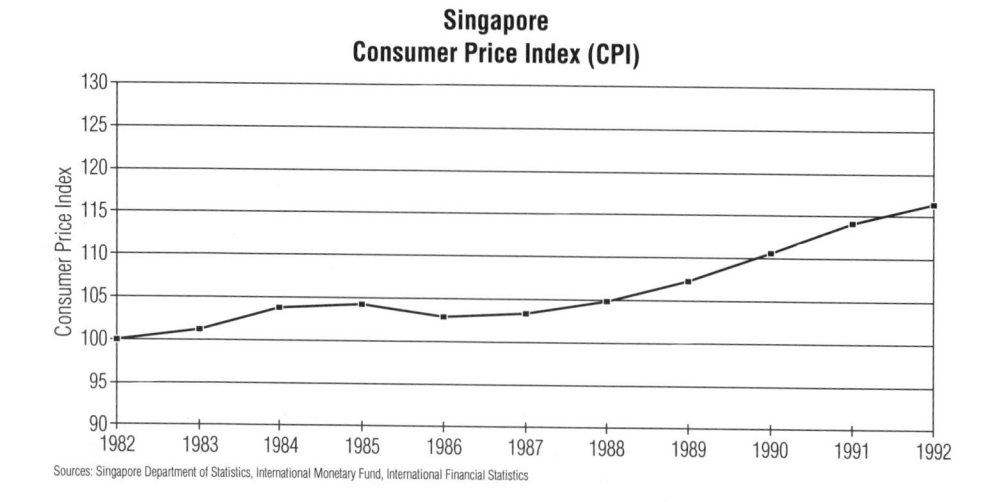

Singapore Consumer Price Index (CPI)

Sources: Singapore Department of Statistics, International Monetary Fund, International Financial Statistics

MANUFACTURING WAGES
Monthly, in Singapore dollars

GNP per capita (1992):
S$23,569 (US$14,731)

Average annual growth rate
1980-91: 5.3 percent

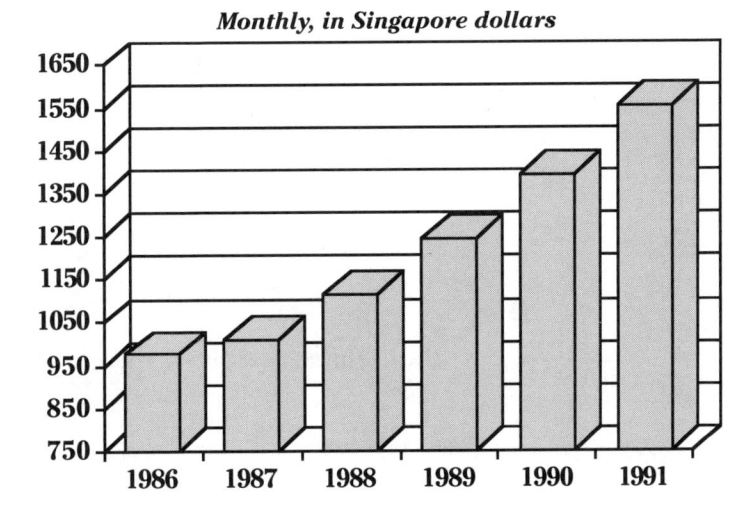

Year	Wage
1986	~970
1987	~990
1988	~1100
1989	~1250
1990	~1400
1991	~1550

CONSUMPTION PATTERNS
(percentage of total households possessing or using goods and services)

CATEGORY	Percent
Refrigerator	97.8
Television	97.5
Telephone line	95.3
Washing machine	80.5
Videocassette recorder	75.1
Vacuum cleaner	53.6
Life insurance	50.2
Air conditioner	35.3
Car	31.1
Credit card	26.0
Microwave oven	21.9
Personal computer	20.2
Laser/compact disc player	19.6
Stocks and bonds	15.8
Piano/organ	11.8
Motorcycle/scooter	10.6
Health & medical insurance	9.7
Portable phone	6.5
Country club membership	2.4
Second apartment/condominium	1.9
Second house	1.3

PRICE INDEX BY CATEGORY
(Sept 1987-August 1988 = 100)

Category	1989	1990	1991
Food	101.7	102.5	104.0
Housing	101.6	107.5	109.5
Clothing	102.7	103.7	105.5
Transportation & communication	105.5	113.7	121.6
Miscellaneous	103.9	108.2	114.4
All items	102.8	106.3	110.0

Average Annual Rate of Inflation
(percent)

1970-80	1980-91
5.9	1.9

ALLOCATION OF HOUSEHOLD CONSUMPTION
(percent of total)

Food	Clothing	Rent, fuel, utilities	Medical care	Transport, communication	Education	Other
19	8	11	7	13	12	30

NUTRITION

	Calories			% of Calorie Requirement			Protein (grams)	
1980	1986	1990	1967/70	1980	1984/6	1980	1986	1990
2,723	3,080	3,114	114	136	124	76	87	87

HEALTH

Health expenditures (**1990**): US$658 million
Per capita: US$219.00

Health expenditures as a percentage of GDP (**1990**)
Total: 1.9 Public: 1.1 Private: 0.8

Population per doctor
1970	1990
1,370	820

Tobacco consumption per year (kilograms per capita adult)
1974-76	1990	2000
6.3	3.4	3.2

ENERGY CONSUMPTION
(kilograms per capita of coal equivalent)

1980	1988	1990
4,937	4,872	5,607

MOTOR VEHICLES IN USE
(in thousands)

	1982	1985	1988	1990
Passenger	175	236	251	287
Commercial	95	118	117	127

COMMUNICATION CHANNELS

Daily newspapers

Circulation (000)			Number			Per 1,000 Persons		
1979	1986	1988	1979	1986	1988	1979	1986	1988
588	924	763	11	10	8	246	357	289

Televisions and radios (1990)

Radios: 1,750,000
Televisions: 1,025,000

Telephones

Number (000)			Per 100 Persons		
1980	1987	1990	1980	1987	1990
702	1,164	1,040	29	44	39

Marketing

HOW TO ENTER THE SINGAPORE MARKET

The question is not "How do I enter the Singapore market?" but rather "How can I afford not enter it?" If you're in business anywhere in the world, chances are very good that something in your business came from or through Singapore at some point in its existence. Singapore is an open market and a free port—by some measures, the largest port in the world. The city-state is located at the crossroads of international trade routes and enjoys easy access to various parts of the world and Southeast Asia in particular. More than 700 shipping lines link Singapore to more than 300 ports, while more than 50 airlines link Changi Airport—ranked 13th in the world in cargo handling—to more than 100 cities in 54 countries. Above all, Singapore is a gateway to the markets of Southeast Asia. The business climate is as warm and inviting as Singapore's tropical climate, especially when compared with the often cold and hostile business climate that prevails in certain other East Asian nations:

- More than 96 percent of imports enter duty free, making them competitive with local products.
- Aside from a very few product areas, Singapore has no restrictive trade or investment policies of any consequence.
- There are no taxes on capital gains, development, or imports, although the government will impose a 3 percent sales tax on goods and services beginning April 1, 1994.
- There is almost no red tape. The government doesn't interfere to stifle business. In fact, it encourages foreign firms to export to Singapore. And through its Economic Development Board (EDB), the government offers a wide range of incentive programs to stimulate foreign investment.
- Foreign investors don't have to take on local joint venture partners, enter into licensing agreements with Singaporean firms, or cede management control to local interests. No laws insert the government into the relationships between foreign firms and Singaporean agents and distributors.
- Foreign firms have ready access to credit and modern banking and financial services.
- The language of business is English.
- Singapore has an excellent physical and business infrastructure.

Singapore seems to be telling business around the world, "If what you're doing is legal, come do it here." Whatever it is you have to sell, the cash-rich Singaporeans will either buy it or find someone, somewhere, who will.

SEVEN WAYS TO ENTER THE SINGAPORE MARKET

1. Establish a representative office

Preferred for products needing heavy after-sales service and cultivation of close relationships with clients—for example, software, computers, appliances, and sophisticated or large-scale equipment.

Advantages Allows you to retain a competitive edge in prompt service, customer commitment, and consulting aspects of a sale; suggests to your buyers that you have an interest and presence in domestic markets, promoting the image of stability and long-term availability; enables use of Singapore as a regional marketing and sales support base; takes advantage of Singapore tax incentives.

Disadvantages Cost of office, plus added costs of specializing to serve immediate customers' needs.

2. Exhibit at trade fairs

Preferred for product promotion; available only if trade fair includes your product.

Advantages Provides contacts with major and smaller buyers and foreign and local industry representatives; hands-on demonstration techniques in-

Five Ways to Build a Good Overseas Relationship

1. Be careful in choosing overseas distributors.

This is crucial. Whether you choose to go with a subsidiary, agent, export trading company, export management company, dealer, distributor, or your own setup, you must investigate the potential and pitfalls of each. Pay personal visits to potential partners to assure yourself of their long-term commitment to you and your product, and their experience, ability, reputation, and financial stability. Rather than relying on bank or credit sources for information on a prospective distributor's financial stability and resources, hire an independent expert to advise you. The keys to the McDonald's Corporation's foreign success, says spokesman Brad Trask, is a meticulous search for partners that focuses on "shared philosophies, past business conduct, and dedication. After all, we're asking a businessman to give up two years to be absorbed into the McDonald's way of business. We want to be sure we're right for each other."

2. Treat your overseas distributors as equals of their domestic counterparts.

Your overseas distributors aren't some poor family relations entitled only to crumbs and handouts. They are critical to your company's future success, a division equal to any domestic division. Offer them advertising campaigns, discount programs, sales incentives, special credit terms, warranty deals, and service programs that are equivalent to those you offer your domestic distributors and tailored to meet the special needs of their country and market.

Also take into account the fact that distributors of export goods need to act more independently of manufacturers and marketers than do domestic distributors because of the differences in trade laws and practices, and the vagaries of international communications and transportation.

3. Learn the do's and taboos.

Each country does business in its own way, a process developed over years to match the history, culture, and precepts of the people. Ignore these practices and you lose. "McDonald's system has enough leeway in it to allow the local businessmen to do what they have to do to succeed," Trask says. Thus every new McDonald's in Thailand holds a "staff night" just before the grand opening. The families of the youthful employees descend en masse to be served McDonald's meals in an atmosphere that they can see for themselves is clean and wholesome. (Refer to "Business Culture" chapter for a detailed discussion.)

4. Be flexible in forming partnerships.

US companies in particular are notoriously obsessed with gaining a majority share of a joint venture, the type of partnership most favored by East Asian governments. One reason is accounting: Revenue can show up on the books at home only when the stake is more than 50 percent. Another reason is the US Foreign Corrupt Practices Act, which makes US citizens and companies liable for the conduct of their overseas partners; the idea, presumptuous at best, is that majority control translates into effective control of the minority partner.

Here, Japanese practices are illuminating. Ownership is yet another area where the Japanese have succeeded; they see a two-sided relationship where US investors see themselves as the superior partner in knowledge, finances, technology, and culture—in other words, know-it-alls. Westerners have a lot to learn about flexibility in business relationships.

Finally, keep in mind that there is more than one way to do business overseas and that changing laws or market conditions will often force you to consider other options. Where a distributorship may be best at first, a joint venture or a licensing agreement may be the way to go later.

5. Concentrate on the relationship.

We cannot overemphasize this point. The Confucian culture of East Asia emphasizes personal relationships above all else. Building a good relationship takes time, patience, courtesy, reliability, dignity, honorable conduct, and farsightedness. A poorly developed relationship dooms even your best marketing efforts to failure from the start. One US computer maker made a great mistake when it fired its Asian distributor after a falling-out. The dismissal, handled in a typically abrupt Western way, caused the man to lose face, and ruined all the relationships the company had built through this man.

> ### Five Ways (cont'd.)
>
> For three years afterward, company executives couldn't find another distributor because no one would talk to them. Not only did the company lose untold millions of dollars in sales, but it took $40 million in advertising to create enough consumer-driven demand for local distributors to even consider meeting with the firm.
>
> So do your very best to build a sound, trusting, and profitable relationship with your overseas partners. They are putting themselves on the line for you, spending time, money, and energy in hopes of future rewards and a solid, long-term relationship.
>
> And by all means, don't expect your foreign distributors to jump through hoops on a moment's notice. For example, they need price protection so they don't lose money on your price changes. If they buy your product for S$100 and a month later you cut your price to S$90, you have to give them credit so they don't get stuck with inventory at the higher price. If you raise your price, you have to honor your prior commitment while you give ample notice of the increase.
>
> With their focus on long-term personal relationships and on mutual respect and trust, East Asians, in particular, make honorable partners once you have gained their confidence by showing them they have yours.

crease product awareness; very high attendance is typical in Singapore.

Disadvantages Market limited to attendees; intense competition with other products targeted for same industry; hectic atmosphere often prevents deep and serious contacts.

3. Get a trading firm to act as distributor or agent

Usual route for imports, at least initially; preferred for consumer products with well-established competitors or for nonconsumer products (business and vertical market applications software, industrial machinery, electronic parts) aimed at government or commercial institutions; good for products requiring before- and after-sales service, and supervision of agents and distributors; agents often act as distributors and vice versa; usually responsible for advertising and promotion.

Advantages Removes need to create your own marketing and distribution structure; knows local needs and customs; often aware of opportunities before bids are announced; knows ins and outs of bidding; monitors and promotes smaller sales, which can add up over time; maintains spare parts inventories; provides after-sales services; markets to both wholesalers and retailers.

Disadvantages Middleman fees raise cost of product in local market; marketing limited by agent's preferences and biases; need to monitor practice of Singapore agents or distributors appointing subagents or distributors in neighboring markets; large size and complexity of trading firms can limit their attention to your product.

4. Do direct marketing

Preferred for consumer products (auto parts and accessories, small appliances, consumer electronic products) but also common among industries (including factories) that want to avoid middlemen costs.

Contact local importers who market through warehouse stores or other retail outlets.

Advantages Direct access to large consumer market.

Disadvantages Extreme competition from other producers, because such stores market a large number of products.

Advertise in industry-specific trade journals or magazines.

Advantages Economical and effective means to increase product awareness among large numbers of consumers; good for testing the market.

Disadvantages Limited time; costs of extended advertising can be high.

5. Open your own distributorship or retail stores

Preferred for companies with a large array of products to offer (for example, auto accessories).

Advantages Direct market access, which allows you to keep prices low and competitive (by eliminating middlemen), control sales environment (type of building, training of sales personnel), and improve quality and service (by reducing the gap between you and the end user).

Disadvantages High cost to establish, maintain, and staff; need to overcome bias of consumers towards already established local merchants; language and cultural barriers.

6. Negotiate a joint venture with a local company

Preferred for consumer products and food; also high-tech products that must be modified for sale in local market, that are in growing international demand, and that are protected by copyright, patent, or similar intellectual property laws (for example, high-tech software).

Advantages Gives ability to create specialized

products aimed at particular needs of consumers in domestic market; allows use of local company's marketing and other contacts.

Disadvantages Allows for technology transfer, potential infringement on design and technology rights, and resulting enforcement problems (Singapore is notorious as a home for software pirates and weak on enforcement of software copyrights, although tough on violators when it catches them).

7. Enter a bid on projects

Primarily for sales to public organizations; best done through local representative or joint venture.

Advantages Successful bid can further your product's reputation in domestic markets.

Disadvantages Price concessions may be needed for success; occasional obstructionist, obfuscatory regulations for foreign firms.

ADVERTISING

Advertising is widely used in marketing in Singapore, which has several advertising and public relations firms that can provide you or your agent with the complete range of public relations services. Product identification can be developed to associate the merchandise with either the manufacturer or the local distributor, depending on your preferences.

Some companies highlight the origins of the products; others choose to either localize or internationalize the image. Several large trading companies do their own advertising. This practice results in the product or service's being associated with the distributor rather than the foreign supplier.

Foreign suppliers can purchase space in local newspapers. When an export company buys local advertising space, it often uses a format developed by a local advertising firm to ensure that its marketing campaign will conform to local customs and preferences. The most effective media advertising is done in the daily newspapers (circulation 700,925 in the four official languages—English, Chinese, Malay, and Tamil). Most of the Singaporean business and consumer public can be reached in the English-language press.

Rising family incomes make radio and television a very popular medium for advertising. More than 70 percent of the population listens daily to Singapore Broadcasting Corporation (SBC) radio broadcasts. SBC's television network operates 155 hours weekly on three separate channels, offering a full range of programs in color. Both radio and television offer substantial commercial time and program sponsorships. Advertising in more popular foreign publications also reaches the Singapore market. Foreign consumer, news, and trade magazines have a wide circulation, particularly among the more affluent international business and professional customers, many of whom are headquartered in Singapore.

Most forms of advertising familiar to the exporter are available in Singapore. The type of promotion depends, of course, on the product and the target market. Flyers, billboards, store displays, radio, newspapers, and television are all effective marketing techniques in Singapore.

HELPING YOUR COMPANY LEARN TO LOVE EXPORT MARKETING

Five In-House Rules

1. Eliminate as much guesswork as you can

Expert export consultation is usually time and money well spent. You need a well thought out marketing plan. You cannot get into successful exporting by accident. It's not a simple matter of saying, "Let's sell our product in Singapore." You need to know that your product will, in fact, sell and how you're going to sell it. First, do you need to do anything obvious to your product? Who is your buyer? How are you going to find him? How is he going to find you? Do you need to advertise? Exhibit at a trade fair? How much can you expect to sell? Can you sell more than one product? A plan may be the only way you can begin to uncover hidden traps and costs before you get overly involved in a fiasco. While you may be able to see an opportunity, knowing how to exploit it isn't necessarily a simple matter. You must plot and plan and prepare.

2. Just go for it

We're not suggesting you throw caution to the winds, but sometimes your "plan" may be to use a shotgun approach—rather than the more tightly targeted rifle approach—and just blast away to see if you hit anything. You can narrow things down later. If your product is new to the market, there may be precious little marketing information, and you may have essentially no other choice. Two scenarios illustrate these points: Two companies decide to begin selling similar products in East Asia, which has never seen such products before. Company A hires a market research firm, which spends six months and US$50,000 to come up with a detailed plan. Company Z sends its president to a trade fair—not to exhibit but just to look around and meet people. He follows that trip up with two others. On the last one his new associates present him with his first order. Company Z also spent six months and US$50,000 investigating doing export business, but it has an order to show for it, while Company A only has an unproven plan.

3. Get your bosses to back you up and stick with the program

Whether your company consists of 10, 50, 500, or 5,000 people—or just you—and whether you're the head of the company, the chief financial officer, or the person leading the exporting charge, there must be an explicit commitment to sustain the initial setbacks and financial requirements of export marketing. You must be sure that the firm is committed to the long-term: Don't waste money by abandoning the project too early.

International marketing consultants report that because results don't show up in the first few months, the international marketing and advertising budget is *invariably* the first to be cut in any company that doesn't have money to burn. Such short-sighted budgetary decisions are responsible for innumerable premature failures in exporting.

The hard fact is that exports don't bring in money as quickly as domestic sales. It takes time and persistence for an international marketing effort to succeed. There are many hurdles to overcome—personal, political, cultural, and legal, among others. It will be at least six to nine months before you and your overseas associates can even begin to expect to see glimmers of success. And it may be even longer. Be patient, keep a close but not a suffocating watch on your international marketing efforts, and give the venture a chance to develop.

4. Avoid an internal tug-of-war

Consultants report that one of the biggest obstacles to successful export marketing in larger companies is internal conflict between divisions within a company. Domestic marketing battles international marketing while each is also warring with engineering, and everybody fights with the bean counters. All the complex strategies, relationship building, and legal and cultural accommodations that export marketing requires mean that support and teamwork are crucial to the success of the venture.

5. Stick with export marketing even when business booms at home

Exporting isn't something to fall back on when your domestic market falters. Nor is it something to put on the back burner when business is booming at home. It is difficult to ease your way into exporting. All the complex strategies, relationship-building, legal and cultural accommodations, and financial and management investment, and blood, sweat, and tears that export marketing requires means that a clear commitment is necessary from the beginning. Any other attitude as good as dooms the venture from

---

Five Ways to Help Your Local Agent

1. Make frequent visits to Singapore to support your agent's efforts. This helps to build the relationship, without which no amount of effort can succeed in Singapore. Keep in mind that your competitors are also paying personal visits to their agents and customers. And invite your agent to your country to reciprocate his hospitality and familiarize him with your country and your company.

2. Hold many demonstrations and exhibits of your products. For suppliers to Singaporean manufacturers, the value of sales presentations at factories cannot be overemphasized. Factory engineers and managers are directly responsible for the equipment and machinery to be purchased, and they have a great deal of influence over the decision to buy.

3. Increase the distribution of promotional brochures and technical data to potential buyers, libraries, and industry associations. When your agent makes personal sales calls, your potential customers won't be completely in the dark.

4. Improve follow-up on initial sales leads. Let your agent know you're backing him up with whatever it takes to pursue the lead. "All of our foreign partners know that they have the support of a large system behind them," McDonald's spokesman Brad Trask says. "The support system is available on request." Make it easy to request help.

5. Deliver on time. If you don't, you can believe that someone else will. Failure to deliver on time not only makes your agent lose face and thereby undermines your relationship, but it jeopardizes your sales. There's not much you can do to make ships go faster or airlines schedule more flights, but you can stockpile your products in Singapore to ensure that your agent has a steady supply.

the start, and you may as well forget it. We can't overstress this aspect: Take the long-range view or don't play at all. Decide that you're going to export and that you're in it for the long haul as a viable money-making full-fledged division within your company.

McDonald's Corporation spokesperson Brad Trask, commenting on his company's overwhelming international success, notes, "We're a very long-term focused company. We do things with patience; we're very deliberate. We're there to stay, not to take the money and run." And Texas Instruments, which has suffered recent losses in its semiconductor business, has made a considered move into long-term joint ventures in East Asia, banking that these investments will provide a big payoff five years down the road.

SEVEN RULES FOR SELLING YOUR PRODUCT

1. Respect the individuality of each market

The profit motive generally operates cross-culturally and the nationals of most countries, especially within a given region, will have much in common with one another. However, there will also be substantial differences, enough to cause a generic marketing program to fall flat on its face and even build ill-will in the process. You may have some success with this sort of one-size-fits-all approach, but you won't be able to build a solid operation or maximize profits this way. "Japan proves this point phenomenally," says Steve Provost, KFC's vice president of International Public Affairs. "Our first three restaurants in Tokyo were modeled after our American restaurants, and all three failed within six months. Then we listened to our Japanese partner, who suggested we open smaller restaurants. We've never looked back." What works in Japan doesn't necessarily work elsewhere. Singaporean tastes may be more similar to US tastes than to Japanese, or may differ in other ways.

2. Adapt your product to the foreign market

Markets are individual, and you may well need to tailor your products to suit individual needs. As the United States' Big Three automakers have yet to learn, it's hard to sell a left-hand-drive car in a right-hand-drive country. White may be a popular color in your country, but may also be seen as the color of death in your foreign market. Dress, styles, and designs considered fashionably tasteful at home can cause offense abroad. One major US computer manufacturer endured years of costly marketing miscalculations before it realized that the US is only one-third of its market, and that the other two-thirds required somewhat different products as well as different approaches.

You can avoid this company's multi-million dollar mistakes by avoiding lazy and culturally-biased thinking. A foreign country has official regulations and cultural preferences that differ from those of your own. Learn about these differences, respect them, and adapt your product accordingly. Often it won't even take that much thought, money, or effort. Kentucky Fried Chicken offers a salmon sandwich in Japan, fried plantains in Mexico, and tabouleh in the Middle East— and 450 other locally specific menu items worldwide. And even the highly standardized McDonald's serves pineapple pie in Thailand, teriyaki burgers and tatsuda sandwiches (chicken with ginger and soy) in Japan, spicy sauces with burgers in Malaysia (prepared according to Muslim guidelines), and a seasonal durian fruit shake in Singapore.

3. Don't get greedy

Price your product to match the market you're entering. Don't try to take maximum profits in the first year. Take the long-term view. It's what your competitors are doing, and they're in it for the long haul. Singaporeans can be very price-conscious. When you're pricing your product, include in your calculations the demand for spare parts, components, and auxiliary equipment. Add-on profits from these sources can help keep the primary product price down and therefore more competitive.

4. Demand quality

A poor-quality product can ambush the best-laid marketing plans. Singaporeans may look at price first, but they also want value and won't buy junk no matter how cheap. And there's just too much competition to make it worth your while to put this adage to the test. Whatever market you gain initially will rapidly fall apart if you have a casual attitude towards quality. And it is hard to come back from an initial quality-based flop. On then other hand, a product with a justified reputation for high quality and good value creates its own potential for market and price expansions.

5. Back up your sales with service

Some products demand more work than others— more sales effort, more after-sales service, more hand-holding of the distributor, and more contact with the end user. The channel you select is crucial here. Paradoxically in this age of ubiquitous and lightning-fast communications and saturation advertising, people rely more than ever on word of mouth to sort out the truth from hyperbole. Nothing will sink your product faster than a reputation for poor or nonexistent service and after-sales support. US firms in particular need to do some serious reputation building for such after-sales service. Although Singaporeans often see US products as superior in quality and performance, they rate Japanese after-sales service as vastly better. And guess whose products they buy.

Consider setting up your own service facility. If you're looking for a Singaporean agent to handle your product, look for one who has qualified maintenance people already familiar with your type of product or who can handle your service needs with a little judicious training. And make sure that this partner understands how important service and support are to you and to your future relationship with him.

6. Notice that foreigners speak a different language

Your sales, service, and warranty information may contain a wealth of information but if it's not in their language, you leave the foreign distributors, sales and service personnel, and consumers out in the cold. English is an official language in Singapore, but you will get much broader exposure if you can operate in Chinese or Malaysian. It's expensive to translate everything into another language, but it's necessary.

7. Focus on specific geographic areas and markets

To avoid wasteful spending, focus your marketing efforts. A lack of focus means that you're wasting your money, time, and energies. A lack of specificity means that your foreign operations may get too big too fast. Not only does this cost more than the local business can justify or support, it also can translate into an impersonal attitude towards sales and service and the relationships you've working so hard to build. Instead concentrate your time, money, and efforts on a specific market niche and work on building the all-important business relationships that will carry you over the many obstacles to successful export marketing.

Business Entities & Formation

FORMS OF BUSINESS ORGANIZATION

Singapore offers Singaporeans and foreign nationals a wide range of options for establishing a business. Alternatives include setting up a company, partnership, sole proprietorship, or foreign branch office or representative office. An entity can be incorporated and registered under the provisions of the Companies Act or established as a registered business according the Business Registration Act. The specific type of business entity selected will be determined by the objectives, circumstances, degree of control desired, and anticipated duration of the investment as well as by available land and labor requirements and tax or other investment incentives. The range of likely solutions to most business needs is fairly narrow.

Companies

A company is an entity that has been organized and registered for profit-seeking purposes. Under Singapore's Companies Act, two or more persons may form an incorporated company by completing a memorandum of association to do so and complying with registration requirements. In fact, with a few exceptions, any form of entity organized for profit-making purposes must register as a company once it exceeds 20 members.

Singapore recognizes four types of companies: companies limited by shares, companies limited by guarantee, companies limited by share guarantee, and unlimited companies. Foreigners are allowed to do business under any of these formats, but most foreign and local businesses organize as a company limited by shares. The legal concepts and regulatory frameworks governing Singaporean companies limited by shares derive from British common law and are similar to those familiar in Western legal and business practice.

Companies Limited by Shares The company limited by shares is not only the corporate structure most often used in Singapore by both Singaporean and foreign firms, but it also is the form most familiar to and preferred by the authorities. Such a company can be 100 percent foreign owned, and most local subsidiaries of foreign companies are organized in this fashion. In practice, the authorities may insist that there be at least some local equity participation for the company to be eligible for incentives in some areas.

The company's initial capital as well as all subsequent earnings can be freely repatriated, although earnings are subject to tax withholding. This withholding can be offset to some extent by corporate income taxes paid to avoid double taxation. In theory, investment incentives are available to any business regardless of its structure. However, in practice companies limited by shares are seen as representing more serious business commitments, and they are thus more likely to receive more and larger concessions from the Economic Development Board (EDB), which grants specific incentives from the range prescribed by statute.

A company limited by shares must have at least two members whose liability is limited to the amount of paid-in capital plus any amount subscribed and committed to but as yet unpaid. There is no upper limit on the number of shareholders.

The other three forms of companies represent special cases primarily of interest to trade and professional organizations that prefer a corporate structure to a partnership structure.

Companies Limited by Guarantee A company limited by guarantee must have at least two members whose liability is limited to the actual amounts that each has contributed to company assets. This form of business entity is usually limited to specialized organizations seeking corporate status, and it is not generally of interest to foreign investors.

Companies Limited by Shares and Guarantee A company limited by shares and guarantee must have at least two members whose liability is limited to the sum of the shares that they have purchased and any additional amounts that they have contributed. This

> **GLOSSARY**
>
> **Economic Development Board (EDB)** The Economic Development Board, a statutory board that operates under the Ministry of Trade and Industry, is the government agency that has responsibility for promoting foreign investment in Singapore. It allocates incentives to new investments. Currently its focus is on the promotion of capital-intensive investments in manufacturing and services, where Singapore has a comparative advantage.
>
> **Ministry of Finance (MOF)** The Ministry of Finance is the primary regulatory agency with administration oversight according to the Business Registration Act and the Companies Act. This includes responsibility for the Registry of Companies and Businesses.
>
> **Ministry of Trade and Industry (MTI)** The Ministry of Trade and Industry controls the manufacture of certain goods in accordance with the Control of Manufactures Act. Its responsibilities include the approval of import licenses on specific goods.
>
> **Monetary Authority of Singapore (MAS)** The Monetary Authority of Singapore functions primarily as the government's central bank. In addition to having oversight for financial services and regulatory operations, the MAS is responsible for monetary and exchange rate policies.
>
> **Registry of Companies and Businesses (RCB)** Those who wish to set up a business entity in Singapore must register with the Registry of Companies and Businesses. The RCB is also a center of information on existing business entities.
>
> **Singapore Trade Development Board (STDB)** The Singapore Trade Development Board was set up to develop and expand Singapore's international trade by promoting exports of goods and services and encouraging foreign companies to use Singapore as their global trading hub.

type of arrangement is of interest to specialized organizations requiring secured financial backing. It is not generally of interest to foreign investors.

Unlimited Companies An unlimited company requires a minimum of two shareholders each of whom bears both unlimited liability and joint and several liability for the company's obligations. This form of business entity is usually limited to specialized organizations seeking the benefits of incorporation. As such, it is not generally of interest to foreign investors, even if they were not deterred by the prospect of unlimited liability.

Public Versus Private Companies Limited companies can be either private or public entities. Most companies that foreign interests organize in Singapore are public companies. A company's status is indicated by its name, which must end with the English words "Private" or "Limited" or the Malay words "Berhad," (for limited), or "Sendirian Berhad," (for private). These terms are abbreviated Ltd., Pte Ltd., Bhd., and Sdn Bhd., respectively.

A private company must meet all the following criteria: it must have no more than 50 shareholders (it can get around this requirement by not counting current and former employees who have shares in the company), it must restrict the right to transfer shares to outsiders, it must not offer any of its shares or debt instruments to the public, and it must not accept monetary deposits from the public. If it fails to meet any of these criteria, it must register as a public company or, if already registered as a private company, it must convert to a public company. Conversely, a public company that meets all the standards for a private company may convert to private status.

Unlike a private company, a public company must issue a prospectus after it has been organized, and it can register to offer shares and debt to the public as part of its funding. A private company can begin operations immediately without the added delay and expense occasioned by this aspect of the registration procedure required of a public company.

In general, a private company faces fewer regulatory requirements than a public company. It is quicker, easier, and less expensive to set up. A private company could provide a suitable structure for a foreign company that does not intend to sell shares or borrow funds in Singapore.

A private company that has no more than 20 shareholders—or that is wholly owned by the government and been declared exempt by the Ministry of Finance (MOF)—can register as an exempt private company. Exempt private status is desirable because it relieves the company from the necessity of filing financial statements with the Registrar of Companies where these statements are available for public inspection. An exempt private company can instead file a certificate of solvency prepared by a certified public accountant, attesting to its financial strength and indicating that an audit has been performed. Such certificates do not disclose financial data, which remains private. Exempt status also allows self-dealing in the form of company loans to directors or to companies connected with its directors.

A corporation is prohibited from holding shares in an exempt private company, and none of a private company's shares can represent beneficial holdings, either directly or indirectly through a nominee, of a corporation. This regulation prevents the Singaporean subsidiary of a foreign company from being organized as an exempt private company. However, there is technically nothing to prevent a private foreign individual from forming an exempt private company.

Capital Requirements Singaporean law sets no minimum capital or equity participation requirements, nor does it require that any portion of the debt or equity capital of a company be held by nationals. Authorized capital must be equal to or greater than the issued capital, and any increase in either authorized or issued capital must be voted by the shareholders. Capital cannot be reduced without a court order.

A minimum of two capital shares is necessary to organize a company. However, a company can subsequently shrink to a single share held by a single shareholder if its bylaws allow it and the shareholders so vote. Companies involved in financial services, such as banking and insurance, are regulated by different authorities and must have additional shareholders and minimum capital in order to incorporate and operate.

Companies can issue common, preferred, or redeemable preferred stock as well as convertible debt. The common shares of public companies must carry voting rights, but preferred shares can be restricted with respect to voting, payment of dividends, and return of capital. All shares must have a par value and be issued in registered form. Shares can be paid for in full or issued upon partial payment, in which case the outstanding balance is a legal obligation of the subscriber. However, the Singapore stock exchange usually will not list issues that are not fully paid up, so this is not an option if the entity is interested in being traded on the local exchange.

The company can issue shares at a premium or at a discount to stated value with shareholder and court approval. It cannot use its share capital as collateral when it borrows. A company cannot buy its own shares, nor can it offer financial assistance or incentives to anyone to buy its shares, except in a limited way to help employees purchase shares.

Shareholders, Directors, Officers, and Corporate Governance Unless otherwise provided for by law, every company may define the specific rights and responsibilities of its shareholders, directors, and officers in its bylaws.

Every company must have at least two directors, one of whom is a legal resident of Singapore but not necessarily a Singaporean national. Every company must also have at least one secretary who is a Singaporean national. There are no other limitations on or prescriptions regarding the nationality of officers. Directors and officers must be natural persons: a corporation cannot be elected to fill such a position. These persons must be of legal age, but they cannot be over 70 years of age unless they have been voted in by a three-fourths majority of the shareholders.

Directors are chosen by the shareholders for a term, the duration of which is specified in the bylaws. Directors are not required to be shareholders unless the bylaws so specify. Directors may appoint management officers or, if allowed by the bylaws, elect one of their number to act as executive officer. Directors of public companies are prohibited from self-dealing with their company except in very narrowly defined areas, and they must disclose all holdings and benefits received.

Public companies must hold a statutory shareholders meeting within one to three months after the entity is declared eligible to begin operations. Every company must hold an annual general shareholders meeting no more than 18 months after incorporation and at least once a year thereafter. Annual meetings must be held within six months of the close of the fiscal year, and the main order of business at such meetings is the disclosure of earnings and the company's financial condition. Special meetings can be called at the behest of either the directors or the shareholders. Ordinary issues are resolved by a simple majority vote of shareholders in attendance—there is no specific quorum requirement—while special issues require a three-fourths majority.

Companies are not required to pay dividends, and they can pay dividends only out of profits, although profits are broadly defined. Dividends can be paid in cash or stock, and they are usually paid during the year, with a special final dividend being declared at the annual shareholders meeting after all the results are in.

Within one month of the annual meeting, the company must file a report that includes data about the meeting, the company's capital structure, its outstanding debt, its shareholders and their holdings, and its directors and officers. The report also includes the company's financial statements, all of which must be filed with the Registry of Companies and Businesses.

A qualified independent auditor must be appointed no later than 90 days after the company is incorporated. This official is charged with examining the company's business and financial condition, and he has authority and responsibility to inspect corporate books, records, and other documents. Auditors report to the shareholders at the annual meeting. They are elected for a term of one year, but they can be reelected, and there is no limit on the number of consecutive terms that they can serve.

Companies must maintain a physical office in Singapore where records are kept. Foreign companies must maintain a registered office that is open to the public for a minimum of five hours a day during regular business hours on all legal business days. A foreign company must appoint a local manager or agent who is officially responsible for all necessary actions and filings required of the company under the Companies Act. This individual is also personally liable for any penalties levied on the company for breaches of company or other law as it applies to the entity.

Dissolution Under the Companies Act, a company may dissolve voluntarily or as mandated by court order. Voluntary liquidation can be initiated either by the company or by its shareholders. If caused by insolvency, the dissolution is known as a *creditors' winding-up;* if other reasons are the cause and the company is not insolvent, it is called a *members' winding-up.* In either case, the company will generally manage its own dissolution. A creditor or the shareholders can initiate a *court winding-up,* in which case a receiver is appointed to liquidate the company.

Branch Offices

A branch office is any office that is registered and maintained in Singapore for business purposes for which the company's principal office holds ultimate responsibility. A branch office can carry on virtually any type of business activity, can earn income, and can repatriate earnings and capital. It is subject to local taxes on earnings and to withholding on remittances. Credits are given for income taxes paid to offset double taxation. Branch earnings are taxed at the same rate as company earnings.

A branch office is relatively easy to set up, and it costs less to start and maintain than an incorporated subsidiary company. However, a branch usually bears greater potential liability than a limited company, because a company incorporated in Singapore is considered a separate entity, and its liability is limited to its local capital, while a branch is considered to be dependent on its parent firm, which is implicitly liable for it to the full extent of the parent's total capital. This fact could mean that a branch format would be advantageous for a bank or financial services firm, because it could operate a branch in Singapore based on its total capital, not merely on the capital that it committed to form a Singaporean company.

Both the local business community and the government tend to view a branch office as less serious than a company and as demonstrating a relatively low level of commitment to do business in Singapore. For this reason, a branch office is unlikely to be granted the incentives that a company could expect to obtain from authorities.

Operating Requirements There are no minimum capital requirements for the establishment of a branch office, nor is there any local participation requirement. A foreign parent must appoint at least two resident agents to manage a branch. These individuals can be foreign nationals, but they must reside in Singapore, and they must hold an employment pass allowing them to work in the country. These agents are answerable for all company actions carried out by their office under the terms of the Companies Act, and they are personally liable for any penalties imposed on the company for breach of such requirements.

A branch office must maintain a physical location where it keeps its licenses, books, and shareholder registry. All documents must be maintained in English or translated into English and certified. Required financial documents include balance sheets and accounting data for the parent company and similar detailed information for branch operations. Branch offices must undergo an annual statutory audit by an accountant licensed to do business in Singapore.

There are no restrictions on the dissolution of a branch. However, the Registrar of Companies must be notified within one week of the cessation of business operations in Singapore or within one month if the cessation is the result of a change in the status of the parent company in its home jurisdiction.

Representative Offices

A representative office is any office operating in Singapore under the authorization of the company's principal office that engages in no direct profit-making activities. In practice, representative offices are involved primarily in overseeing local sales agents. Representative offices can be authorized in the areas of manufacturing, trade, and shipping. Other sectors and industries do not use representative offices, and authorities are reluctant to authorize them.

Representative offices are allowed to deal with the sales agents that the parent company has employed to sell its goods locally; handle customer inquiries; provide customer service to the extent of advising on product use, although they are not allowed to service or repair products; solicit orders; and assist in the planning and execution of local marketing campaigns.

They are expressly prohibited from engaging in business; signing contracts; consulting for a fee; taking responsibility for the transshipment of goods; conducting business in locales other than Singapore for which Singapore is an intermediate point; or using letters of credit to operate. Although representative offices can negotiate deals, they cannot close such deals or commit their parent to any binding agreements.

A foreign representative office is generally staffed with foreign managers and Singaporean support staff. No registered capital is required to set up a representative office. Because the office cannot earn income, it must be supported financially by its parent organization, which bears implicit liability for its activities. There are no special accounting requirements on funds transfers. As nonearning entities, representative offices are not liable for income taxes.

A representative office has no specific legal standing. However, it must obtain approval to operate from the Singapore Trade Development Board (STDB), which regulates its operations. There are no prescribed procedures for closing a representative office.

A representative office is a useful vehicle for a foreign company developing business interests in Singapore that wants to investigate the local business climate at first hand. Later it can upgrade its presence to that of a branch office or company. However, the functions of a representative office are sufficiently specialized—dealing primarily with the overseeing of existing sales conducted by contractual agents—that a firm without a preexisting local sales presence that is interested in gathering market intelligence would generally do better to hire outside market researchers than set up such an office.

Liaison Offices

Liaison offices are not recognized or used in Singapore. Functions that a separate liaison office would undertake are usually performed by a representative office.

Agents and Distributorships

Agents are individuals or firms that provide local sales representation for a foreign business in the buying of local goods or the selling of the foreign firm's goods in Singapore. Foreign entities interested in doing business in Singapore are not required to use local agents, and there is no specific regulation of agent relationships aside from the civil codes governing contractual relationships.

Because of its small geographic size, a single local agent can usually handle all a foreign firm's business in Singapore unless the firm is involved in a wide range of industries requiring a variety of specialized expertise. Goods imported into Singapore are usually handled through established trading firms, which often contract to provide after-sales support, which is considered an important factor in doing business in Singapore.

Information on sales and purchases in Singapore can be obtained from the STDB and from the Singapore International Chamber of Commerce (SICC).

Licensing and Technical Assistance Agreements

Licensing and technical assistance agreements are contractual agreements by foreign nationals or firms to license or sell specific technologies to Singaporean firms or individuals. In exchange, the local entity agrees to pay fees or royalties.

Such transfers and payments do not need approval and the terms and conditions of such agreements do not have to be reviewed. Licensing and technical assistance agreements do not need to be registered with any official agency, and there is no restriction on repatriation of proceeds from such agreements. Standard provisions of the civil code on contracts are the only regulatory framework for such arrangements.

Qualifying technology transfers in the manufacturing sector are eligible for full or partial exemption from withholding tax. These incentives are administered by the EDB, and they must be negotiated before the agreement is concluded.

Partnerships

A partnership is almost any unincorporated profit-making business in which contributions and assets are jointly agreed upon and held by two or more partners. Partnerships are registered under and regulated by the Business Registration Act. Because a partnership is not considered to be a separate legal person with permanent status, it must renew its registration every year.

Foreign individuals and companies may form partnerships, and this form of entity can be advantageous for specific foreign investment activities, especially those having a limited time frame. Although a partnership lacks standing as a separate legal person and it is therefore open to a substantial degree of potential liability, members of the local business community are familiar with such entities and generally view them as serious business formats for certain types of operations. Partnerships are legally eligible for incentives. However, except in certain limited areas, they generally do not receive many of the incentives routinely granted to companies.

Partnership Structure and Operation Partnerships can be either general or limited. In a general partnership, all partners bear unlimited liability for any debts and obligations incurred by the partnership. A limited partnership requires at least one partner to have unlimited liability, while the liability of one or more other partners is limited to the amount of their capital contribution. General partnerships are relatively common in Singapore, but limited partnerships are rare.

There are no legal minimum capital requirements for a partnership. Profits are taxable at the standard corporate rate. Under some circumstances, they may be taxed at a lower rate. All earnings are eligible for repatriation, subject to tax withholding. Credits are given to offset taxes already paid to avoid double taxation. A partnership is not required to file an annual financial statement, although an outside audit is usually required to satisfy tax authorities if the partnership has foreign members, is involved in complicated business dealings, or has a particularly high cash flow or profitability.

Partnerships are limited to 20 individuals. However, professional partnerships in such areas as law, accounting, and medicine, which are regulated by other agencies, may have more than 20 partners. Partners may be either foreign or domestic individuals or corporations. Foreign investors can form new partnerships or join existing partnerships. Partnerships can be composed of local and foreign investors or exclusively of foreign members, including those of different nationalities.

Foreign companies must assign a Singaporean national or a permanent resident to act as local manager. A foreign national with a valid employment pass good for a period of more than one year will generally be confirmed. The local manager is held accountable for the business obligations of the partnership.

Disputes among partners and between the partnership and third parties are generally handled under civil and criminal codes as though the claimants were individuals.

A partnership may dissolve under any of the following circumstances: the stated duration of the partnership has expired; all partners agree to dissolve the partnership; or the partnership has achieved, or failed to achieve, its stated objectives. There are no specific requirements for closing a partnership. Usually, it simply does not renew its registration after having concluded its activities.

Sole Proprietorships

Proprietorships represent a single individual engaged in a profit-seeking enterprise. No prior approvals are required to form a proprietorship, but it must register with the Registrar of Companies and Businesses. The proprietorship has no separate legal status, and it therefore leaves the proprietor open to unlimited liability. Both authorized foreign residents and Singaporean nationals may form sole proprietorships.

There are no minimum capital requirements. Earnings are subject to taxation at the standard corporate rate, and they are eligible for repatriation subject to withholding. There are no filing requirements, but the proprietor has to satisfy the tax authorities, which may require documentation or an audit. A sole proprietorship can dissolve at will, but the proprietor's personal liability is continuing. Although sole proprietorships are legal, they are not recognized as an appropriate vehicle for a serious foreign investor, and they cannot expect to benefit from incentives.

Joint Ventures

As in much of the rest of Asia, a joint venture in Singapore is a vague description that refers to a wide range of mutual agreements between contracting parties, often—but not always—of different nationalities. It is not a specific type of business structure or independent entity with legal standing, as it is in much Western law. Singapore recognizes joint ventures on a practical level, but there is no specific legal authorization for them. In practice, joint ventures in Singapore are formed to complete a specific project, not to establish an ongoing business relationship.

A joint venture can be established as a partnership or incorporated company. Partnerships can be preferable when business operations are expected to be concluded within a limited period of time. The laws governing a particular joint venture depend on whether it has been organized as a partnership or a company. Joint ventures can be made up of individual or corporate members, and they can consist completely of foreign participants or of foreigners and locals.

REGISTERING A BUSINESS

Singapore generally does not require a foreign business to obtain approval in order to invest or operate in the country. However, every entity that conducts business in Singapore must register with the appropriate authorities. This section outlines the procedures and documentation required to establish and register a business in Singapore. Entities are formed under the Business Registration Act or the Companies Act as administered by the Ministry of Finance (MOF) through the Registrar of Companies and Businesses (RCB).

A foreign company is not considered to be carrying on business in Singapore—and is therefore not required to establish or register a local business entity—if it limits its activities to the following:

- Maintaining a local bank account
- Making sales through an independent contractor
- Taking orders for goods or services to be delivered outside the country
- Taking part in an isolated transaction that is concluded in less than one month
- Investing its own funds locally
- Holding assets locally
- Borrowing or lending locally
- Collecting debts owed to it or securing collateral for debts owed to it.

Such activities must be incidental to the foreign company's general business operations, not its primary business, and they must not recur either frequently or on a regular basis. For instance, a company must register and comply with all local requirements if its primary business is investing, lending, or trading financial assets and it intends to operate in those lines of business in Singapore on an ongoing basis. However, it may conduct transactions in those activities without being classed as doing business in Singapore if they are only sporadic and incidental to its main business. Foreign investors involved in such activities should seek official confirmation that their specific operations are not considered to constitute local business activity requiring registration.

Although business formation and registration procedures in Singapore are relatively straightforward and simple compared to those in some other Asian countries, it is still recommended that those interested in setting up in Singapore seek local professional assistance because of the sophistication of the laws governing the establishment of such entities. Unincorporated businesses—proprietorships and partnerships that register under the Business Registration Act—can technically be registered without outside help, but only licensed accountants and lawyers are legally allowed to complete the formation and registration procedures for entities covered by the Companies Act, which governs the format chosen by most foreign entities.

The complex nature of the regulations governing such procedures makes it highly advisable that in addition to conferring with government authorities, individuals or firms wishing to do business in Singapore obtain legal and accounting assistance early on in the decision process, no matter what level of presence they are contemplating. Such assistance will help to ensure that individual or corporate enterprises comply with regulatory requirements and procedures and help to prevent problems. (Refer to "Important Addresses" chapter for a partial listing of government agencies, legal offices, and accounting firms.)

Licensing Entities doing business in Singapore are required to register with the RCB. Incorporated companies require a certificate of incorporation. Unincorporated businesses require a certificate of registration. Some entities also need a license from the separate authorities that regulate special activities.

Special Registration for Businesses Some areas of business require special authorization or licensing. Such businesses are allowed to register only after they have obtained the requisite authorizations. For financial industries, such as insurance, banking, securities brokerage and trading, and related financial services, approval is obtained through the Monetary Authority of Singapore (MAS). Approval and licensing for specified manufactures included in the First and Second Schedules of goods listed under the Control of Manufactures Act—motor vehicles, detergents, some tobacco products, matches, some alcoholic beverages, fireworks, certain iron and steel products, airconditioners, and refrigerators—are obtained from the Registrar of Manufactures of the Economic Development Board. Consult local authorities for specifics regarding these areas and the requirements necessary to obtain special licenses.

Restrictions Private enterprises are restricted from operating in areas deemed vital to the national interest and from those in which the public sector operates on a monopoly basis. Restricted areas include weapons and munitions manufacture, public transportation, public utilities, telecommunications, and the media, including broadcasting and newspaper, magazine, and book publishing. The bylaws of some public, semipublic, and private companies in Singapore limit foreign participation. Examples include the national airline and certain shipping firms. Foreigners may own no more than 40 percent of the equity in locally incorporated banks. Additional restrictions are placed on foreign ownership of real estate. Consult local authorities regarding specific projects.

Fees and Expenses Company registration fees are figured on a sliding scale based on the company's authorized capital. The minimum registration fee is S$1,200—about US$750. The maximum fee is S$35,000—about US$22,000. A company without share capital pays a fee of S$600—about US$375. The fee for registering a branch office is the same as the fee for incorporating a company. Fees for registering unincorporated entities can be characterized as nominal processing fees.

Establishing a company requires professional assistance from certified public accountants and attorneys. Professional fees and associated costs for establishing any form of company are generally less than S$3,500—about US$2,200. However, the complexity of the business involved can cause total costs to rise substantially.

Basic Authorizations Needed and Applications Procedures

Specific provisions exist for the formation and registration of each type of organizational entity. However, all entities must complete the following basic steps: first, an entity must obtain clearance for the use of a specific business name. Second, it must submit a copy of its memorandum of association and company information to the RCB.

Foreign companies do not need to receive prior investment approval, but they must ascertain whether the project requires any specific authorizations and they must negotiate incentives with the EDB before they incorporate and register. They must also submit documentation on their parent company to the RCB.

A certified English-language translation of all documents in a foreign language must be submitted in all registration procedures.

Summary of Registration Procedures by Type of Entity

Companies in General A company must file for name clearance. It must then submit its memorandum of association and supplementary information. Once it has received approval for its name and paid its fees, it receives a certificate of incorporation from the RCB allowing it to begin operations. A public company must file a prospectus before beginning business if it is going to sell shares to the public. If it does not intend to sell shares, it files a statement in lieu of prospectus. A private company does not need to issue a prospectus.

Foreign Companies In addition to the procedures just outlined for companies in general, a foreign company must also submit documentation on its parent company and documentation on its own proposed structure and operations in Singapore.

Branch Offices A branch follows the same procedure for registering as those followed by a foreign company. It is issued a certificate of registration after its name has been approved and it has filed documentation with the RCB.

Partnerships and Sole Proprietorships Partnerships and sole proprietorships must register with the RCB's Business Registry. Registrations must be renewed annually.

Representative Offices A representative office of a foreign company must obtain approval from the Singapore Trade Development Board (STDB) in order to open. Approval is generally granted upon presentation of documentation on the parent company and an operations plan for the proposed office.

Licensing and Technical Assistance Agreements No approval, review, or registration is required for licensing and technical agreements. Participants should negotiate eligibility for incentives with the EDB before signing any agreement.

Joint Ventures A joint venture requires no advanced approval. The resulting entity must be registered with the RCB either as a company or as a partnership, following the procedures just outlined. Investors should negotiate eligibility for incentives with the EDB before initiating the agreement.

Basic Procedures

Company Registration Requirements Every company must submit an application for name approval to the RCB. The RCB's Directory Service must be checked to see that the proposed name is available for use. If the name is available, the applicant can reserve it for use for a period of two months while the necessary registration procedures are completed. If the business has not been registered under the chosen name at the end of that period, the name becomes available for use by another business.

The company must then submit the following materials to the RCB:

- Memorandum of association, the equivalent of articles of incorporation; these must include the company name, the objectives for which it is being formed, the authorized share capital, and other such information as the relevant portions of the Companies Act require;
- Articles of association, or bylaws, that stipulate procedures for governing the entity; if no articles are submitted or if those submitted do not cover all relevant items, the Companies Law position will be used as the default;
- Notarized certificate of identity listing the officers and subscribers, a statutory declaration of compliance with legal requirements made by the accountant or lawyer handling the incorporation, and a statutory authorization for the directors and secretary to act in governing the company and their acceptance; and
- Required fees.

When all these items have been received and approved, the RCB issues a certificate of incorporation. The directors then adopt a company seal, allocate shares, and register the company's official location with the RCB. Such registration completes the process.

Foreign Company Specifics Foreign companies must also provide the following information to the RCB before establishing a place of business:

- Certified copy of its certificate of incorporation or equivalent document from its place of origin;
- Certified copy of its home country memorandum and articles of association or equivalent documents;
- Roster of directors with full particulars;
- Memorandum specifying powers accorded to any local directors;
- Power of attorney executed by the parent company appointing its local representatives;
- Notice of the location of local offices; and
- Statutory declaration made by the agents of the company.

Branch Office Requirements Procedures for registering a branch office are the same as those for registering a foreign company.

Partnership and Sole Proprietorship Requirements Partnerships and sole proprietorships, which by definition are not incorporated, must register with the Registrar of Businesses. Applicants must submit an application for name approval to the RCB. The RCB's Directory Service must be checked to see that the proposed name is available for use. If it is, the applicant can reserve it for use for a period of two months while the necessary registration procedures are completed. If the business has not been registered under the chosen name at the end of that period, the name becomes available for use by another business.

Partnerships should be prepared to submit copies of their partnership agreements. Registration for such entities must be renewed annually.

Representative Office Procedures Application to open a representative office is made to the STDB. The application must include annual reports of the parent company for the preceding three years and a copy of the certificate of incorporation of the parent company.

TEN REMINDERS, RECOMMENDATIONS, AND RULES

1. Sole proprietorships and partnerships are registered under the terms of the Business Registration Act with the Registrar of Businesses. Incorporated companies and foreign companies, including their branches, are registered under the terms of the Companies Act with the Registrar of Companies. Both are operated by the Registrar of Companies and Businesses, the office of the administrator who oversees the operations of the Registry of Companies and Businesses (RCB).
2. Businesses involving banking, insurance, securities, and other financial services are regulated under separate statutes administered by the Monetary Authority of Singapore (MAS). Manufacturing businesses are regulated under separate statutes administered by the Economic Development Board (EDB). Before November 1993, shipping was regulated separately by the Singapore Trade Development Board (STDB). Since then, the industry has not required special approval, and shipping operations are now registered using standard procedures directly with the RCB.
3. A company limited by shares is the most common vehicle for foreigners doing business in Singapore. A branch office is the next most common format. Partnerships are also relatively common vehicles for business in Singapore, especially among professional groups and partners uniting to undertake projects of a limited duration.
4. No prior approvals are required for most types of foreign investment in Singapore. However, investors must negotiate with the Economic Development Board (EDB) to obtain incentives before making their investment. Incentives are not granted automatically, and some businesses and forms of business—particularly companies limited by shares—have a better chance than others of securing favorable incentive treatment.
5. In general, there are no restrictions on foreign ownership in any recognized type of entity. However, there are some restrictions on foreign participation in some specific activities and sectors of the economy.
6. Companies limited by shares are viewed favorably by the government and the local business community as demonstrating a high level of commitment to doing business in Singapore. Branch offices are not viewed as reflecting a high level of commitment, and they are therefore less prestigious. Partnerships may be preferable to branch offices from the standpoint of local recognition, particularly in cases where the project has a limited duration.
7. Foreign investors considering establishing a business in Singapore can contact the Economic Development Board (EDB), the Singapore Trade Development Board (STDB), the Monetary Authority of Singapore (MAS), and the Singapore International Chamber of Commerce (SICC) for information and assistance.
8. There are no exchange controls and no restrictions are imposed on the movement of capital into or out of Singapore, including the repatriation of profits derived from foreign investments. Foreign businesses thus do not have to register their flows of funds in order to obtain foreign exchange. They do have to account for such flows for tax purposes.
9. Approval to register a business entity is generally rapid. Filing the appropriate documentation with the RCB usually takes between one and two weeks.
10. Foreign businesses must appoint a Singaporean national or permanent resident to manage local business operations. Such local representatives are legally and personally responsible for the activities of the entity that they manage and for compliance with all relevant requirements.

USEFUL ADDRESSES

In addition to the government agencies listed, individuals or firms should contact chambers of commerce, embassies, banks and financial service centers, local consultants, lawyers, and resident foreign businesses for assistance and information. (Refer to "Important Addresses" chapter for a more complete listing.)

Economic Development Board (EDB)
250 North Bridge Road #24-00
Raffles City Tower
Singapore 0617
Tel: [65] 3362288 Fax: [65] 3396077
Tlx: 26233 SINEDB

Monetary Authority of Singapore (MAS)
10 Shenton Way
MAS Building
Singapore 0207
Tel: [65] 2255577 Fax: [65] 2299491
Tlx: 28174 ORCHID

Registry of Companies and Businesses (RCB)
10 Anson Rd. #05-01/15
International Plaza
Singapore 0207
Tel: [65] 2278551 Fax: [65] 2251676
Tel: [65] 2283731 (general information)
Tel: [65] 2283715 (registration of companies)
Tel: [65] 2283720 (registration of businesses)
Tlx: 28395 RCB

Singapore International Chamber of Commerce (SICC)
6 Raffles Quay #05-00, Denmark House
Singapore 0104
Tel: [65] 2241255 Fax: [65] 2242785
Tlx: 25235 INTCHAM

Singapore Trade Development Board (STDB)
1 Maritime Square #10-40 (Lobby D)
World Trade Centre, Telok Blangah Road
Singapore 0409
Tel: [65] 2719388 Fax: [65] 2740770, 2782518
Tlx: 28617 TRADEV

FURTHER READING

The preceding discussion is provided as a basic guide for individuals interested in doing business in Singapore. The resources described here provide further information on company law, investment, taxation and accounting procedures, and procedural requirements.

Doing Business in Singapore, Ernst & Young. New York: Ernst & Young International, 1991. Available in the United States from Ernst & Young, 787 Seventh Avenue, New York, NY 10019, USA; Tel: [1] (212) 773-3000. Available in Singapore from Ernst & Young, 10 Collyer Quay, #21-01 Ocean Building, Singapore 0104; Tel: [65] 5357777. Provides an overview of the investment environment, taxation, business organizational structures, business practices, and accounting requirements in Singapore.

Doing Business in Singapore, Price Waterhouse. Los Angeles: Price Waterhouse World Firm Limited, 1993. Available in the United States from Price Waterhouse, 400 South Hope Street, Los Angeles, CA 90071-2889, USA; Tel: [1] (213) 236-3000. Available in Singapore from Price Waterhouse, 6 Battery Road #32-00, Singapore 0104; Tel: [65] 225-6066. Covers the investment and business environment in Singapore and its audit, accounting, and taxation requirements.

Marketing in Singapore, Washington, D.C.: US-D.O.C. Overseas Business Reports, International Trade Administration, US Department of Commerce, 1989. Available from the United States Government Printing Office, Washington, DC 20402, USA. Further information: US-D.O.C. Office of the Pacific Basin; Tel: [1] (202) 482-3877/2522. An annual overseas business report with general information and brief overviews of foreign trade, economic and industry trends, marketing, transportation, trade regulations, and investment in Singapore.

Labor

THE LABOR ECONOMY

For a small city-state with few natural resources, Singapore has achieved tremendous economic growth in a short time. One need only compare the current standard of living of Singaporeans with that of two decades ago. Per capita GNP in 1992 was S$23,569.60 (US$14,731), almost six times the level of 1970 at constant prices. The source of this success is manifold: a trade system that encourages foreign investment, consistent economic policies, and a stable political environment, to name only three. But the factor that may be most important to Singapore's success is its labor economy—the productivity of its labor force, its policy on human resource development, labor-employer relations, and the wages and benefits that workers receive. Singapore has excelled in this respect, and it continues to do so. Nevertheless, wealth and an increasingly high-tech economy have brought new challenges, which Singapore must overcome if it is to continue along its path of impressive economic development.

One of the most critical labor issues that Singapore faces is how to ensure an adequate supply of workers. Because its population is limited, Singapore must either demand more of its local labor force or bring in additional workers from outside. However, Singapore already has a significant population of foreign workers; nonresidents comprised nearly 20 percent of the work force in 1992. And government officials, in the face of growing public sentiment against opening the country's doors any wider, are reluctant to admit or recruit more foreign personnel. Therefore, Singapore's policymakers are looking for ways of enhancing Singapore's domestic labor resources. Efforts in this direction include attracting women to the work force, extending the retirement age, improving education and upgrading skills, and limiting wage hikes.

Population

Singapore conducts so much international trade and commerce that it is easy to forget how small its population is. If we include nonresidents, the city-state's population in of 1992 was 2.82 million. In that year, population grew at an annual rate of 2 percent—almost half the rate in 1989.

The country's major ethnic groups are Chinese (77.6 percent), Malay (14.2 percent) and Indian (7.1 percent). The languages spoken in Singapore are English, Mandarin, Malay, Tamil, and various Chinese dialects. English is the lingua franca of government, education, and business.

Labor Force and Distribution by Sector

Singapore's labor force totaled 1.6 million in 1992, and it was growing at an annual rate of 2.7 percent. While total population and labor force growth rates declined throughout the 1980s, growth of the labor force has accelerated slightly each year since 1990. Growth of the labor force is due primarily to increased numbers of foreign workers and increasing participation by married women and older workers.

In 1991 64.4 percent of Singapore's work force was employed in services, 34.8 percent in industry, and just 0.8 percent in agriculture. The number of people in the service sector grew by almost 5 percent over the preceding year and by almost 2.4 percent in industry. However, the agricultural sector shrank by 22.4 percent from its 1990 level.

Labor Availability

Despite a heavy influx of foreign workers, Singapore's rapid growth and declining birth rate have put increasing pressure on labor resources in recent years. One of the most worrisome trends for employers is an increase in job turnover among workers in certain occupations and industries. The high turnover rate relates directly to the labor shortage. A recent survey of Singapore manufacturers found that, for the first time, job turnover was one of their top concerns.

Production, technical, accounting, and clerical workers are the most likely to change jobs. In 1992 the hardest-hit industry was the trading industry, in which turnover rose to 25 percent for production and semiskilled technical workers and to 11 percent for clerical workers. In manufacturing, job-hopping reached 18 percent for both production and clerical workers. In contrast, managers and executives showed a decline to 9 percent from 13 percent recorded in 1991.

Requirements for skilled technical and professional personnel can usually be met by the local work force. Nevertheless, the government is working actively to increase the overall skill level of workers. It aims not only to meet the current needs of employers but also to move toward increasingly high-value-added, knowledge-intensive industries.

Foreign Workers

Singapore's small work force of 1.6 million limits the labor supply at all skill levels. Shortages have become particularly acute due to the recent strong economic growth. To make up for the lack of sufficient production and assembly line workers, the government selectively allows firms to bring in foreign workers, mainly from neighboring Southeast Asian countries.

In 1990, the last year for which complete figures are available, Singapore had 311,300 nonresident foreigners, roughly two-thirds of whom were employed. Although the shortages continue, government officials are now concerned that the country is beginning to show signs of strain, and efforts have been made to cap the growth of imported labor.

Since April 1990 firms have had to pay a per capita fee to use foreign production and assembly workers. The fee varies with skill level and industry. For workers in the construction industry, employers pay a monthly levy of S$250 (about US$155) for skilled workers and S$400 (about US$250) for unskilled workers. Employers of foreign workers in the maritime industry pay a monthly fee of S$250 for skilled workers and S$350 for unskilled. The permit fee for a foreign worker in the manufacturing industry is S$300 (about US$187) for the first 35 percent of a company's work force and S$450 (about US$280) for each additional worker after that. Individuals who employ a foreign maid are required to pay a monthly fee of S$300. Such fees can amount to as much as 30 percent of base salary.

All foreign workers entering Singapore to take up or continue employment must obtain a work permit from the Controller of Work Permits. Generally, only workers earning S$1,500 (about US$935) per month or less must have work permits.

For further information contact:

Work Permit and Employment Department
18 Havelock Road, #03-01
Singapore
Tel: [65] 5383033

To compensate for the shortage in certain fields of professionals and skilled technicians (often defined as those earning more than S$1,500 per month), the government encourages foreign professionals to work in Singapore. There is no limit on the number of skilled professionals who may enter the country provided employers receive approval from the government's Economic Development Board (EDB). Employment passes or a professional visit pass are issued on a case by case or on a specific-need basis. There is usually little difficulty in obtaining such a pass when applicants are upper-level managers of foreign corporations, well-qualified specialists, or personnel required for the start up of a new industrial, financial, or service venture. Employers should contact either the government's Economic Development Board or the Controller of Immigration.

Comparative Unemployment 1990-1991

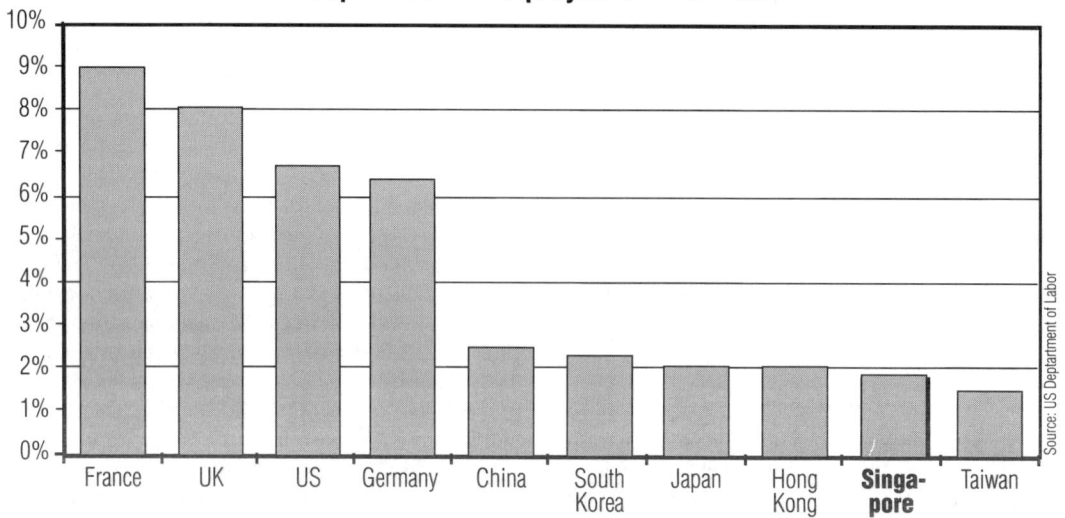

Country	Unemployment
France	9.0%
UK	8.0%
US	6.7%
Germany	6.4%
China	2.4%
South Korea	2.2%
Japan	2.0%
Hong Kong	2.0%
Singapore	1.8%
Taiwan	1.4%

Source: US Department of Labor

EDB, Immigration Facilitation Unit
International Manpower Division
250 North Bridge Road, #20-00
Raffles City Tower
Singapore 0617
Tel: [65] 3362288 Fax: [65] 3396077
Tlx: 26233 SINEDB

Immigration Department
Employment Pass Section
95 South Bridge Road
#08-26 Pidemco Centre
Singapore 0105
Tel: [65] 5301866 Fax: [65] 5301840

Singapore is also trying to promote the country as a destination for Hong Kong emigrants planning to leave the territory before sovereignty is transferred to China in 1997. About 33,000 Hong Kong families have been given rights to 10-year residences in Singapore. But despite promotional efforts, only a few thousand Hong Kong families have relocated. The majority of Hong Kong emigrants have preferred such countries as Australia, Canada, and the United States.

Unemployment Trends

Singapore enjoyed low unemployment levels throughout the 1980s and early 1990s. This trend is likely to continue for the foreseeable future. Singapore's unemployment rate was 14 percent in 1960. By 1988 unemployment had reached 3.3 percent, falling to 2.2 percent in 1989, and remaining at or below 2 percent since then. For all intents and purposes, Singapore has no structural unemployment.

HUMAN RESOURCES

With virtually no natural resources and restricted to a small geographical space, Singapore relies heavily for its economic success on the talents and skills of its people. Education and training are therefore critical to continued development. To maximize its human resources, Singapore tries to follow three basic strategies: educate each individual to his or her maximum potential, develop skills that meet the needs of industry and business, and promote continuous training and retraining.

Workers in most fields and at most skill levels are known for their competence and familiarity with the latest technologies. International firms consistently rank the Singaporean worker among the highest in the world for industriousness, efficiency, and adaptability. More than anything, such a ranking reflects Singapore's investment in education and training, which amounts to about 20 percent of the government's spending every year. Nevertheless, there are simply not enough Singaporeans to go around.

Education and Attitudes Toward Learning

Primary education is free, and more than 94 percent of the children between the ages of 6 and 17 are enrolled in school. The adult literacy rate in 1991 was 90.7 percent. Most Singaporeans speak English, which is the language of business and government. For the vast majority, however, English is a second or even third language after Mandarin and other Asian languages. In keeping with their multilingual heritage, many young people are also learning French, German, and Japanese, among other widely spoken languages.

In addition to public primary and secondary schools, there are some 25 international schools in Singapore. Designed to meet the needs of foreign and local residents alike, these schools range from preschool to college preparatory. They offer instruction in various languages, and they are based on the British, American, French, German, Dutch, Italian, and Japanese systems, among others.

The National University of Singapore, the Nanyang Technological University, and several technical and vocational institutes currently produce about 22,000 graduates annually in a variety of fields. There are 29 research scientists and engineers per 10,000 in the work force, and the ratio is expected to increase to 40 per 10,000 by 1995.

Training

Singapore's commitment to worker training is just as important as its commitment to formal education. The government has asked companies to spend 4 percent of their payroll on worker training, but the request is nonbinding. In addition, Singapore's policymakers are promoting the creation of an entire training system involving businesses, government agencies, labor groups, and various training institutes. Currently, special government training programs operate in cooperation with small, medium, and large corporations to make use of their expertise and knowledge of technological applications and trends for producing a more skilled workforce.

For example, many international manufacturing corporations have cooperated with the EDB, Singapore's leading government agency for industrial planning, as part of the Continual Upgrading Training Program. The program promotes the application by engineers and technicians of new technologies, such as artificial intelligence, in various industries.

To encourage firms to retrain workers on a continuous basis, the Singaporean government recently established the Skills Development Fund. The fund is financed by a levy on employers, who are required to pay 1 percent of the base pay of any worker earning less than S$1,200 (about US$750) per month. Firms may then qualify to use money from the fund to finance the training programs that they undertake. Interested firms should contact the EDB for more information.

In addition to company-based training programs, the Institute of Technical Education offers a wide range of programs for those just finishing school and for workers at virtually every stage in their careers. The New Apprenticeship System provides on- and off-the-job training to recent school graduates. Workers can take part in regular continuing education and training programs or the Modular Skills Training Scheme, which is offered on a part-time basis so as not to disrupt a person's job. Other options include the Basic Education for Skills Training program, the Secondary Education program, and the Training Initiative for Mature Employees for workers 40 years of age and older.

Institute of Technical Education
PSA Building, 460 Alexandra Road #28-00
Singapore 0511
Tel: [65] 7757800 Fax: [65] 2731372

Women in the Work Force

As one way of minimizing its dependency on foreign workers, Singapore has tentatively begun to encourage women to enter the work force. Attracted by increased opportunity and propelled by better qualifications than ever before, women have joined the ranks of the employed in record numbers. Nevertheless, roughly one-half of Singaporean women of working age remain economically inactive. The participation rate for the female labor force rose from 48.4 percent in 1990 to 50.5 percent in 1991. This increase can be compared with a rise in the participation rate for the male labor force from 78.3 percent to 79.8 percent in the same period.

One dilemma facing policymakers is how to reconcile the need for more working women with the government's goal of encouraging marriage, childbearing, and traditional family values. One solution is to rely more on foreign maids, many of whom are Filipinas, to manage the domestic sphere, although some feel that such reliance would still undermine family life. Another possibility, which is gaining popularity, is for women to take part-time positions. Firms are generally open to hiring women, but women can expect to run into barriers when attempting to move up too aggressively in a traditionally male-dominated environment. There are no equal opportunity regulations in Singapore.

CONDITIONS OF EMPLOYMENT

Singapore's standards for conditions of employment are similar to those of industrialized countries in North America and Western Europe. In fact, Singapore's standards for such things as working hours and vacations are more in line with the prevailing standards of countries from advanced Western nations than they are with those of Singapore's developing Asian neighbors, such as Korea, and Taiwan. Singapore's Employment Act sets minimum standards for most aspects of employment conditions. Its provisions include guidelines for working hours, overtime, and vacation time as well as sick leave and maternity leave. It also covers bonus payments.

Working Hours, Overtime, and Vacations

The Employment Act limits normal working hours to 44 hours per week. Including overtime, the average number of hours worked per week in 1991 was 46.7. Typically, clerical staff work an average of 40 to 42 hours per week, excluding overtime.

Pay for work in excess of 44 hours per week is one-and-a-half times the basic hourly rate. Pay for time worked on holidays and normal days off is two-and-a-half times the normal rate. Overtime may not exceed 72 hours per month.

According to the Employment Act, all employees are entitled to seven days of paid annual vacation leave during their first year of service and to an additional day of paid leave for every year after that to a maximum of 14 extra days or 21 days in total.

Yearly vacation time for executives and upper-level managers is not covered by legislation. It depends on the individual employment contract. However, top local managers and executives usually take three to four weeks of paid vacation a year. Top foreign personnel typically have four to six weeks.

Special Leave

The Employment Act provides guidelines for employee sick leave and maternity leave for women. After one consecutive year of service with the same employer, a worker is entitled to 14 days of paid sick leave per year and up to 60 days if hospitalization is required. A female employee is entitled to maternity leave of four weeks before and four weeks after childbirth with full pay provided that she has been at the same company for at least 180 days before taking leave. This policy applies only to a woman's first two children. The employer pays for medical examinations incurred during sick leave or maternity leave. Some companies extend this benefit to the members of an employee's immediate family.

Termination of Employment

The Employment Act stipulates that employers must observe the following guidelines when they terminate an employment contract: they must give one day's notice if the individual has been employed for less than 26 weeks; one week's notice if the individual has been employed for 26 weeks but less than two years; two weeks' notice if the individual has

Comparative Average Weekly Wages - 1991

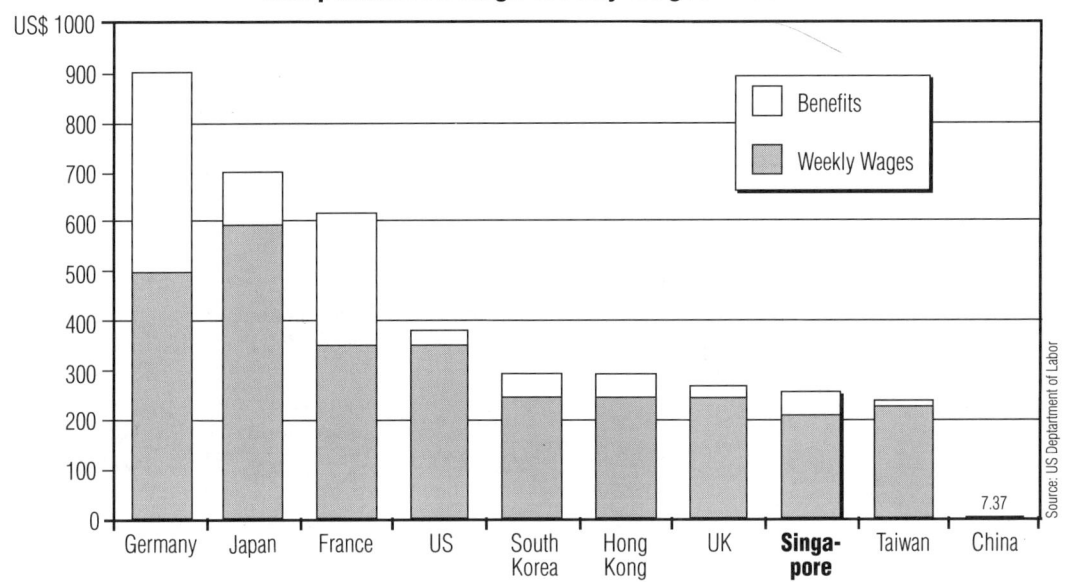

Chart showing weekly wages and benefits (US$) for: Germany (~900), Japan (~700), France (~620), US (~380), South Korea (~295), Hong Kong (~295), UK (~270), Singapore (~255), Taiwan (~235), China (7.37). Source: US Department of Labor

been employed for at least two but less than five years; and four weeks' notice if the individual has been employed for five years or more.

In the event that an employer does not observe these minimum guidelines, he can be required to pay compensation to a worker. The amount of compensation usually depends on the individual circumstance.

WAGES AND BENEFITS

Wages and salaries in Singapore are comparatively higher than they are in most other developing countries in Asia but lower than wages in other industrialized and newly-industrialized countries in the region, including Japan, Korea, and Hong Kong. It is customary for employers to pay workers an annual bonus. Health insurance and retirement are covered in a comprehensive social security scheme called the Central Provident Fund (CPF), to which both employers and employees contribute. All employers are required to be insured against work-related accidents. Employers can expect to pay supplemental benefits and bonuses amounting to around 25 percent of a workers' average earnings.

Wages, Salaries, and Bonuses

The Singaporean government exerts a great deal of control over wage policy and other matters affecting labor conditions. It operates through a National Wages Council (NWC), a tripartite body with representatives from government, business, and labor. With the government's approval, the NWC sets guidelines for annual wage negotiations between employers and workers. However, government dictates have been relatively ineffective in this area, as the labor shortage has caused employers to bid up wages in competition for scarce employees. The government has been uncharacteristically tentative in forcing this issue because it fears interfering with market forces.

In the mid-1980s the government introduced the flexi-wage concept, under which annual wage increases lag behind growth in productivity growth. Firms performing well reward employees with higher bonuses instead of wage increases. The measure seeks to relieve companies of pressure to constantly raise wages, and to motivate workers to take a direct interest in the company's profitability.

A 1992 address by Prime Minister Goh Chok Tong summarized the government's view on labor and wage policy. Goh noted that the 7.6 percent real increase in wages in 1991 greatly exceeded growth of productivity—1.5 percent—a trend that stretched back to 1988. Mr. Goh stressed the importance of linking productivity and wage growth to ensure international competitiveness. In fact, this has been one of the government's ongoing themes since the early 1970s, when it first established a national productivity board. By closely monitoring trends in productivity growth from year to year, the government determines an appropriate wage policy and then tries to influence labor-employer negotiations through the National Wages Council.

The salaries of managers, executives, and professionals vary considerably with ability, experience, and type of business. In addition, annual bonuses and other benefits, such as a company car, club membership, and full retirement fund coverage, can add significantly to basic salaries.

Bonuses are common at all levels of employment, and they, too, vary widely. Usually, employers pay an annual bonus of at least one month's wages or salary after a full year of service, but bonuses can be as much as two to four months' pay.

Minimum Wage

There is no minimum wage in Singapore. However, the relatively small size of the country's labor market and the existence of labor shortages means that few workers are willing to accept wages below the average prevailing levels.

Average Weekly Earnings by Industry - 1991

Industry	US$*	% Change from 1990
All Industry	204.60	5.6%
Manufacturing	190.30	7.5%
Construction	174.80	8.1%
Commerce	174.00	5.1%
Transportation	205.70	0.7%
Finance	264.60	4.5%
Other Services	226.10	6.2%

Source: US Department of Labor
* US$1.00 = S$1.656

Average Weekly Earnings by Occupation - 1991

Occupation	Base (US$*)	Total (US$*)
Manager	256.71	661.30
Professional	159.90	441.90
Technician	105.36	271.60
Clerical	61.57	158.60
Service	61.30	157.90
Production	71.27	183.60
Machine Operator	56.76	146.20

Source: US Department of Labor, Singapore Department of Statistics
* US$1.00 = S$1.61

Accident Insurance

The Workmen's Compensation Act requires employers to carry accident insurance. Workers who are injured or afflicted with occupational diseases in the course of work are entitled to compensation under the act. A worker who sustains temporary injuries receives normal pay during his or her recovery.

When death results from work-related causes, the maximum compensation is S$78,000 (about US$48,600), and the minimum is S$26,000 (about US$16,200). For permanent disabilities, the maximum compensation is S$105,000, while the minimum is S$35,000 (about US$65,400 to US$21,800).

Social Security and the Central Provident Fund

In addition to basic pay, employers are required by law to contribute 17.5 percent of an employee's base pay to the Central Provident Fund (CPF), a government-managed social security program. Employees pay 22.5 percent of their base pay to the CPF. The government's long-term goal is for employers and employees each to pay 20 percent.

Most foreign employers and workers are also required to contribute to the CPF, although foreign nationals with employment or professional visit passes (as opposed to work permits) may apply for an exemption.

Monthly contributions are subject to a cap of S$1,050 for the employer and S$1,350 for the employee (about US$654 and US$841, respectively). These figures are based on a pay ceiling of S$6,000 (about US$3,738) per month. To encourage the continued employment of older workers, rates of contribution start to decline at age 55. For workers between 55 and 59, employer and employee each pay 12.5 percent. The rate drops to 7.5 percent each for workers between 60 and 64, and for workers aged 65 and above, the rate is 5 percent each.

Workers covered under the CPF have three separate accounts: an Ordinary Account, a Medisave Account, and a Special Account. Thirty percent of a worker's contribution is retained in the Ordinary Account, 6 percent in the Medisave Account, and the remaining 64 percent in the Special Account. CPF members receive a market-regulated interest rate on their savings. In the first half of 1992, the interest rate was 4.59 percent.

Employees may draw from their CPF savings when they reach the voluntary retirement age of 55. (In view of the worsening labor shortage the retirement age may soon change to 60.) They may also withdraw their savings if they leave Singapore or Peninsular Malaysia permanently, become permanently disabled, or lose their sanity. CPF members who reach retirement age must have set aside a minimum of S$32,700 (about US$20,375). Those with insufficient savings when they retire can have their accounts supplemented by contributions from children or other family members.

Aside from its principal function of providing for retirement, the CPF may also be used for any of the following purposes: to purchase government-built apartment or private residential and nonresidential properties; to purchase CPF life insurance coverage; to pay hospitalization expenses incurred by CPF members or their immediate family; to purchase catastrophic illness insurance for CPF members or their immediate family; to invest in approved shares, loan stocks, unit trusts, and gold; and to finance the further education of CPF members or the education of their children. The government has recently begun a program to encourage workers to invest their CPF accounts in listed financial investments both to boost national ownership and returns as well as to provide a floor under its financial markets.

LABOR RELATIONS

International country-risk analysts and business executives routinely cite good labor-employer relations as one reason for Singapore's strong competitiveness and attractiveness as a commercial center. Singapore has seen very little labor unrest. Apart from one work stoppage in 1986, there have been no strikes since the late 1970s. The government discourages labor unrest—making it difficult for unions to strike—and generally supports management.

Generally, organized labor has a high regard for national, as opposed to industry-specific, interests, and this gives employer organizations and the government a comfortable framework for negotiation and the implementation of a national economic agenda. Thus, during a recession in 1985, labor agreed to a policy of wage restraint. Now, the government is again calling for slow wage increases as growth remains sluggish.

Unions and the Labor Movement

In 1991 there were 86 registered employee trade unions in Singapore, with 217,086 members. The unionization rate has dropped steadily since the 1970s. In 1991 14 percent of all workers belonged to a union compared to 16.3 percent in 1989 and 21.9 percent in 1980. Workers are not required to join a union, but there is usually worker representation at large companies, especially in the manufacturing industry.

In early 1992 the National Trade Union Congress (NTUC), which represents almost 99 percent of Singapore's unionized workers and all but nine of its unions, launched a massive unionization drive Hurt by the growing number of free riders, who do not pay dues but still receive union-negotiated wages, the NTUC has concentrated on improving the benefits available specifically to union members.

Currently, the NTUC owns a taxi company, a radio station, an insurance cooperative, a holiday resort, and a large grocery chain. It has been actively promoting one of its best known perks, a 10 percent rebate on total yearly expenditures at NTUC Fair Price grocery stores. Recently, the NTUC announced that it was considering taking over management of a government owned hospital, a move that many feel would boost membership.

National Trades Union Congress (NTUC)
Trade Union House, Shenton Way
Singapore 0106
Tel: [65] 2226555 Fax: [65] 2205110

Employer Organizations

Five employer unions have registered under the Trade Unions Act. The leader is the Singapore National Employers Federation (SNEF). Together with the National Trades Union Congress and a government representative, the SNEF conducts wage and other labor-related negotiations at the National Wages Council.

Singapore National Employees Federation (SNEF)
Tanglin Shopping Centre
10-01/04, 19 Tanglin Road
Singapore 1024
Tel: [65] 2358911 Fax: [65] 2353904

Industrial Arbitration

The Industrial Relations Act provides the framework for collective bargaining, negotiation, and conciliation. Under the act, the Industrial Arbitration Court, which has the status of a high court, may make awards or give directions to the parties in a dispute Depending on the nature of the dispute, the court may consist of the president alone or of the president and a representative each from the employers' and employees' panels.

Besides arbitrating disputes, the Industrial Arbitration Court certifies collective agreements made by employers and trade unions. Through its registrar, the court also provides mediation services to disputing parties.

Industrial Arbitration Court
2nd Storey, City Hall
St. Andrew's Road
Singapore 0617
Tel: [65] 3378191 Fax: [65] 3307215

Business Law

INTRODUCTION

Singapore is reforming its business laws, particularly as they relate to the activities of foreign businesses. Changes are frequent, and the goal is to improve the climate for foreign investment. However, many rules and regulations that significantly affect foreign businesses are found in unpublished government advisories and internal policy statements, rather than in the statutes. You should carefully investigate the status of the legal requirements that may affect your particular business activities. The information in this chapter is intended to emphasize the important issues in commercial law, but it should not replace legal advice or council. You should be certain to review your business activities with an attorney familiar with international transactions, the laws of Singapore, and the laws of your own country. Refer to "Important Addresses" chapter for a list of legal firms in Singapore.

BASIS OF SINGAPORE'S LEGAL SYSTEM

Singapore is a common law country, which means that its courts rely on judicial precedent, as set by court decisions in earlier cases, in determining disputes currently under litigation.

Singapore came under British control in the early 1800s, and English law was adopted there in 1826 with only necessary changes to ensure that the local inhabitants were not unduly burdened. The country was administered by England until 1959, with the exception of three years of Japanese occupation. In 1959 Singapore was granted self-government and subsequently joined the Federation of Malaya. The independent Republic of Singapore was established in 1965 when Singapore seceded from the Federation.

Singapore's laws, particularly its commercial laws, are largely based on English statutes. There is some influence also from India, Australia, and Malaysia. The Singaporean courts refer extensively to cases in all of these countries for legal guidelines in the absence of, or for the interpretation of, local law.

STRUCTURE OF SINGAPORE'S GOVERNMENT AND LAWS

Singapore's President is the head of state, but executive powers are in fact exercised by the Prime Minister and a cabinet. All cabinet members are also elected members of Parliament, which holds legislative powers. The court system is independent of the other government branches. Singapore's Supreme Court consists of three divisions: the High Court (first appellate level, hearing both criminal and civil matters), the Court of Appeal (hearing civil appeals from the High Court), and the Court of Criminal Appeals (hearing criminal appeals from the High Court). From the Supreme Court, parties may further appeal to Singapore's highest court, the Judicial Committee of the Privy Council, sitting in London. (Judicial Committee Act, Cap. 8.) Note: Cap. refers to the chapter headings of the Singapore Statutes.

Singapore legislation consists of the Acts of Parliament, collected in the *Singapore Statutes,* plus subsidiary legislation, which include all rules and regulations made under the authority of an Act of Parliament and usually by a minister or other authorized person.

LAWS GOVERNING BUSINESS IN SINGAPORE

Under the Civil Law Act (Cap. 30, sec. 5), Singapore applies the same law as would be applied in England at the time with respect to questions or issues related to the law of partnerships; corporations; banks; banking; principals; agents; carriers by air, land, and sea; marine, average, life, and fire insurance; and mercantile law generally. Exceptions to this Act are made for all matters for which Singapore has

BUSINESS LAW
TABLE OF CONTENTS

Foreign Corrupt Practices Act	191
Legal Glossary	192
International Sales Contract Provisions	194
Charges Over Bank Deposits	197
Law Digest	197
Agency	197
Aliens	197
Assignments	197
Commercial Register	197
Contracts	198
Copyrights and Registered Designs	199
Foreign Exchange	200
Frauds, Statute of	200
Intellectual Property Rights	200
Patents	200
Power of Attorney	200
Principal and Agent	200
Sales	200
Statute of Frauds	204
Trademarks	204

GEOGRAPHICAL SCOPE OF SINGAPORE LAWS

Laws digested in this section are in force throughout the Republic of Singapore.

PRACTICAL APPLICATION OF SINGAPORE LAWS

Contracts Singapore businesses often use contracts with terms that resemble those in British agreements. These contracts usually emphasize guidelines for continuing relationships. Singaporean business owners respect contract obligations. If they sign a contract, they intend to perform and are aware of the consequences of breach.

Dispute Resolution Singapore is not an overly litigious society. Informal negotiations are preferred for the resolution of contract or other disagreements. If negotiations fail, firms will mediate before resorting to litigation. Practical solutions that will work for all parties are preferred to strict interpretations of contract rights and damage awards for breach.

Role of Attorneys The role of attorneys in commercial matters is expanding in Singapore. Parties commonly have their transactions reviewed by attorneys, and often negotiate with the assistance of counsel. Many large foreign law firms have opened offices in Singapore.

Intellectual Property and Trademark Rights The protection for patents in Singapore is minimal. However, Singapore has made recent amendments to its copyright and trademark laws to improve legal protection and enforcement. The trademark laws extend to service marks.

RELATED SECTIONS

Refer to "Corporate Taxation" and "Personal Taxation" chapters for a discussion of tax issues, and "Business Entities & Formation" chapter for an explanation of the types of entities that may be created in Singapore.

its own laws, for the holding or transferring of interests in immovable property, for laws giving effect to treaties or international agreements not joined by Singapore, and for laws that impose registration, licensing, other controlling procedures, or penalties for purposes of regulating business activities.

The registration, formation, and operation of entities that carry on business in Singapore are regulated by the Companies Act. (*See* Commercial Register.) Contract rights are governed generally by common law principles and by specific legislation, such as the Companies Act and the Frustrated Contract Act. (*See* Contracts.) Contracts for the sale of goods are regulated by the British Sale of Goods Act and Unfair Contract Terms Act, as applied through the Civil Law Act (Cap. 30, sec. 5). (*See* Sales.)

Intellectual property and trademark rights are protected under the Copyright Act and the Trade Mark Act. (*See* Copyright and Registered Designs; and Trademarks.) Patents are not protected by Singapore law, although patents registered in Britain are recognized in Singapore. (*See* Patents.)

Foreign Corrupt Practices Act

United States business owners are subject to the Foreign Corrupt Practices Act (FCPA). The FCPA makes it unlawful for any United States citizen or firm (or any person who acts on behalf of a US citizen or firm) to use a means of US interstate commerce (examples: mail, telephone, telegram, or electronic mail) to offer, pay, transfer, promise to pay or transfer, or authorize a payment, transfer, or promise of money or anything of value to any foreign appointed or elected government official, foreign political party, or candidate for a foreign political office for a corrupt purpose (that is, to influence a discretionary act or decision of the official), and for the purpose of obtaining or retaining business.

It is also unlawful for a US business owner to make such an offer, promise, payment, or transfer to any person if the US business owner knows, or has reason to know, that the person will offer, give, or promise directly or indirectly all or any part of the payment to a foreign government official, political party, or candidate. For purposes of the FCPA, the term *knowledge* means *actual knowledge*—the business owner in fact knew that the offer, payment, or transfer was included in the transaction—and *implied knowledge*—the business owner should have known from the facts and circumstances of a transaction that the agent paid a bribe but failed to carry out a reasonable investigation into the transaction. A business owner should make a reasonable investigation into a transaction if, for example, the sales representative requests a higher commission on a particular sale for no apparent reason, the buyer is a foreign government, the product has a military use, or the buyer's country is one in which bribes are considered customary in business relationships.

The FCPA also contains provisions applicable to US publicly held companies concerning financial record keeping and internal accounting controls.

Legal Payments

The provisions of the FCPA do not prohibit payments made to *facilitate* a routine government action. A facilitating payment is one made in connection with an action that a foreign official must perform as part of the job. In comparison, a corrupt payment is made to influence an official's discretionary decision. For example, payments are not generally considered corrupt if made to cover an official's overtime required to expedite the processing of export documentation for a legal shipment of merchandise or to cover the expense of additional crew to handle a shipment.

A person charged with violating FCPA provisions may assert as a defense that the payment was lawful under the written laws and regulations of the foreign country and therefore was not for a corrupt purpose. Alternatively, a person may contend that the payment was associated with demonstrating a product or performing a preexisting contractual obligation and therefore was not for obtaining or retaining business.

Enforcing Agencies and Penalties

Criminal Proceedings The Department of Justice prosecutes criminal proceedings for FCPA violations. Firms are subject to a fine of up to US$2 million. Officers, directors, employees, agents, and stockholders are subject to fines of up to US$100,000, imprisonment for up to five years, or both.

A US business owner may also be charged under other federal criminal laws, and on conviction may be liable for fines of up to US$250,000 or up to twice the amount of the gross gain or gross loss, provided the defendant derived pecuniary gain from the offense or caused pecuniary loss to another person.

Civil Proceedings Two agencies are responsible for enforcing civil provisions of the FCPA: the Department of Justice handles actions against domestic concerns, and the Securities and Exchange Commission (SEC) files actions against issuers. Civil fines of up to US$100,000 may be imposed on a firm; any officer, director, employee, or agent of a firm; or any stockholder acting for a firm. In addition, the appropriate government agency may seek an injunction against a person or firm that has violated or is about to violate FCPA provisions.

Conduct that constitutes a violation of FCPA provisions may also give rise to a cause of action under the federal Racketeer-Influenced and Corrupt Organizations Act (RICO), as well as under a similar state statute if enacted in the state with jurisdiction over the US business owner.

Administrative Penalties A person or firm that is held to have violated any FCPA provisions may be barred from doing business with the US government. Indictment alone may result in suspension of the right to do business with the government.

Department of Justice Opinion Procedure

Any person may request the Department of Justice to issue a statement of opinion on whether specific proposed business conduct would be considered a violation of the FCPA. The opinion procedure is detailed in 28 C.F.R. Part 77. If the Department of Justice issues an opinion stating that certain conduct conforms with current enforcement policy, conduct in accordance with that opinion is presumed to comply with FCPA provisions.

Legal Glossary

Acceptance An unconditional assent to an offer or an assent to an offer conditioned on only minor changes that do not affect any material terms of the offer. *See also* Counteroffer, Offer.

Accord and satisfaction A means of discharging a contract or cause of action by which the parties agree (the accord) to alter their obligations and then perform (the satisfaction) the new obligations. Example: a seller who cannot obtain red fabric dye according to contract specifications and threatens to breach the contract may enter into an accord and satisfaction with the buyer to provide blue-dyed fabric for a slightly lower price.

Agency A relationship between one individual or legal entity (the agent) who represents, acts on behalf of, and binds another individual or legal entity (the principal) in accordance with the principal's request or instruction. *See also* Agent, Power of Attorney, Principal.

Agent A person authorized to act on behalf of another person (the principal). Example: a sales representative is an agent of the seller. *See also* Agency, Power of Attorney, Principal.

Authentication The act of conferring legal authenticity on a written document, typically made by a notary public who attests and certifies that the document is in proper legal form and that it is executed by a person identified as having authority to do so.

Bailment A delivery of goods or personal property by one person (the bailor) to another (the bailee) on an express or implied contract and for a particular purpose related to the goods while in possession of the bailee, who has a duty to redeliver them to the bailor or otherwise dispose of them in accordance with the bailor's instructions once the purpose has been accomplished. Example: a bailment arises when a seller delivers goods to a shipping company with instructions to transport them to a buyer at a certain destination.

Bill of exchange A written instrument signed by a person (the drawer) and addressed to another person (the drawee), typically a bank, ordering the drawee to pay unconditionally a stated sum of money to yet another person (the payee) on demand or at a future time.

Bona fide In or with good faith, honesty, and sincerity. Example: a bona fide purchaser is one who buys goods for value and without knowledge of fraud or unfair dealing in the transaction. Knowledge of fraud or unfair dealing may be implied if the purchaser should have reasonably known that the transaction involved deceit, as when goods susceptible to piracy are provided without documentation of their origin.

Capacity to contract Legal competency to make a contract. A party has capacity to contract if he or she has attained the age required by law and has the mental ability to understand the nature of contract obligations.

Carrier An individual or legal entity that is in the business of transporting passengers or goods for hire.

Chattel An item of personal property.

Chose in action *See* Thing in action.

Counteroffer A reply to an offer that adds to, limits, or modifies materially the terms of the offer. A seller who accepts a buyer's offer, but informs the buyer that the goods will be of a different color, has made a counteroffer. *See also* Acceptance, Offer.

Equitable assignment An assignment that does not meet statutory requirements but that the courts will nevertheless recognize and enforce in equity, that is, to do justice between the parties.

Ex parte By one party or side only. Example: an application ex parte is a request that is made by only one of the parties involved in an action.

Negotiable instrument A written document that can be transferred merely by endorsement or delivery. Example: a check or bill of exchange is a negotiable instrument.

Legal Glossary (cont'd.)

Nexus A party's connection with, or presence in, a place that is sufficient to subject in fairness that party to the jurisdiction of the court or government located there.

Offer A proposal that is made to a certain individual or legal entity to enter into a contract, that is definite in its terms, and that indicates the party making the offer's intent to be bound by an acceptance. Example: an order delivered to a seller to buy a product on certain terms is an offer, but an advertisement sent to many potential buyers is not. *See also* Acceptance, Counteroffer.

Power of attorney A written document by which one person (the principal) authorizes another person (the agent) to perform stated acts on the principal's behalf. Example: a principal may execute a special power of attorney authorizing an agent to sign a specific contract or a general power of attorney authorizing the agent to sign all contracts for the principal.

Prima facie Presumption of fact as true unless contradicted by other evidence. Example: in the absence of contrary evidence, a party to a bill of exchange is prima facie deemed to be a holder in due course, that is, a party who took the bill in good faith and for value and who may thus enforce payment of the bill.

Principal A person who authorizes another party (the agent) to act on the principal's behalf.

Quantum meruit The amount earned. Example: damages for failure to complete a contract may be awarded on a quantum meruit basis, which means the value of the work actually completed.

Rescind A contracting party's right to cancel the contract. Example: a contract may give one party a right to rescind if the other party fails to perform within a reasonable time. A court may grant rescission of a contract if performance has become impossible.

Rescission *See* Rescind.

Seal A mark or sign that is used to witness and authenticate the signing of an instrument, contract, or other document. Example: a corporation uses a seal to authenticate its contracts and records of its corporate acts.

Statute of Frauds A law that requires designated documents to be in writing in order to be enforced by a court. Example: contracting parties may agree orally to convey land, but a court may not enforce that contract, and may not award damages for breach, unless the contract was written.

Thing in action A right to bring an action to recover personal property, money, damages, or a debt. Example: a seller who has a right to recover payment for goods and who is not in possession of the buyer's payment has a thing in action, that is, a right to procure payment by lawsuit.

Ultra vires An act performed without the authority to do so. Example: a company is permitted to undertake transactions only within the objects stated in its charter, and a person who acts on behalf of the company but outside the scope of the company's charter commits an ultra vires act.

International Sales Contract Provisions

When dealing internationally, you must consider the business practices and legal requirements of the country where the buyer or seller is located. For a small, one-time sale, an invoice may be commonly accepted. For a more involved business transaction, a formal written contract may be preferable to define clearly the rights, responsibilities, and remedies of all parties. The laws of your country or the foreign country may require a written contract and may even specify all or some of the contract terms. *See also* Contracts and Sales for specific laws on contracts and the sale of goods.

Parties generally have freedom to agree to any contract terms that they desire. Whether a contract term is valid in a particular country is of concern only if you have to seek enforcement. Thus, you have fairly broad flexibility in negotiating contract terms. However, you should always be certain to come to a definite understanding on four issues: the goods (quantity, type, and quality); the time of delivery; the price; and the time and means of payment.

You need to consider the following clauses when you negotiate an international sales contract.

Contract date

Specify the date when the contract is signed. This date is particularly important if payment or delivery times are fixed in reference to it—for example, "shipment within 30 days of the contract date."

Identification of parties

Designate the names of the parties, and describe their relation to each other.

Goods

Description Describe the type and quality of the goods. You may simply indicate a model number, or you may have to attach detailed lists, plans, drawings, or other specifications. This clause should be clear enough that both parties fully understand the specifications and have no discretion in interpreting them.

Quantity Specify the number of units, or other measure of quantity, of the goods. If the goods are measured by weight, you should specify net weight, dry weight, or drained weight. If the goods are prepackaged and are subject to weight restrictions in the end market, you may want to ensure that the seller will provide the goods that will comply with those restrictions.

Price Indicate the price per unit or other measure, such as per pound or ton, and the extended price.

Packaging arrangements

Set forth packaging specifications, especially for goods that can be damaged in transit. At a minimum, this provision should require the seller to package the goods in such a way as to withstand transportation. If special packaging requirements are necessary to meet consumer and product liability standards in the end market, you should specify them also.

Transportation arrangements

Carrier Name a preferred carrier for transporting the goods. You should designate a particular carrier if, for example, a carrier offers you special pricing or is better able than others to transport the product.

Storage Specify any particular requirements for storage of the goods before or during shipment, such as security arrangements, special climate demands, and weather protection needs.

Notice provisions Require the seller to notify the buyer when the goods are ready for delivery or pickup, particularly if the goods are perishable or fluctuate in value. If your transaction is time-sensitive, you could even provide for several notices to allow the buyer to track the goods and take steps to minimize damages if delivery is delayed.

Shipping time State the exact date for shipping or provide for shipment within a reasonable time from the contract date. If this clause is included and the seller fails to ship on time, the buyer may claim a right to cancel the contract, even if the goods have been shipped, provided that the buyer has not yet accepted delivery.

Costs and charges

Specify which party is to pay the additional costs and charges related to the sale.

Duties and taxes Designate the party that will be

responsible for import, export, and other fees and taxes and for obtaining all required licenses. For example, a party may be made responsible for paying the duties, taxes, and charges imposed by that party's own country, since that party is best situated to know the legal requirements of that country.

Insurance costs Identify the party that will pay costs of insuring the goods in transit. This is a critical provision because the party responsible bears the risk if the goods are lost during transit. A seller is typically responsible for insurance until title to the goods passes to the buyer, at which time the buyer becomes responsible for insurance or becomes the named beneficiary under the seller's insurance policy.

Handling and transport Specify the party that will pay shipping, handling, packaging, security, and any other costs related to transportation, which should be specified.

Terms defined Explain the meaning of all abbreviations—for example, FAS (free alongside ship), FOB (free on board), CIF (cost, insurance, and freight)—used in your contract to assign responsibility and costs for goods, transportation, and insurance. If you define your own terms, you can make the definitions specific to your own circumstances and needs. As an alternative, you may agree to adopt a particular standard, such as the Revised American Foreign Trade Definitions or Incoterms 1990. In either case, this clause should be clear enough that both parties understand when each is responsible for insuring the goods.

Insurance or risk of loss protection

Specify the insurance required, the beneficiary of the policy, the party who will obtain the insurance, and the date by which it will have been obtained.

Payment provisions

Provisions for payment vary with such factors as the length of the relationship between the contracting parties, the extent of trust between them, and the availability of certain forms of payment within a particular country. A seller will typically seek the most secure form of payment before committing to shipment, while a buyer wants the goods cleared through customs and delivered in satisfactory condition before remitting full payment.

Method of payment State the means by which payment will be tendered—for example, prepayment in cash, traveler's checks, or bank check; delivery of a documentary letter of credit or documents against payment; credit card, credit on open account, or credit for a specified number of days.

Medium of exchange Designate the currency to be used—for example, US currency, currency of the country of origin, or currency of a third country.

Exchange rate Specify a fixed exchange rate for the price stated in the contract. You may use this clause to lock in a specific price and ensure against fluctuating currency values.

Import documentation

Require that the seller be responsible for presenting to customs all required documentation for the shipment.

Inspection rights

Provide that the buyer has a right to inspect goods before taking delivery to determine whether the goods meet the contract specifications. This clause should specify the person who will do the inspection—for example, the buyer, a third party, a licensed inspector; the location where the inspection will occur—for example at the seller's plant, the buyer's warehouse, a receiving dock; the time at which the inspection will occur; the need for a certified document of inspection; and any requirements related to the return of nonconforming goods, such as payment of return freight by the seller.

Warranty provisions

Limit or extend any implied warranties, and define any express warranties on property fitness and quality. The contract may, for example, state that the seller warrants that the goods are of merchantable quality, are fit for any purpose for which they would ordinarily be used, or are fit for a particular purpose requested by the buyer.

International Sales Contract Provisions (cont'd.)

The seller may also warrant that the goods will be of the same quality as any sample or model that the seller has furnished as representative of the goods. Finally, the seller may warrant that the goods will be packaged in a specific way or in a way that will adequately preserve and protect the goods.

Indemnity

Agree that one party will hold the other harmless from damages that arise from specific causes, such as the design or manufacture of a product.

Enforcement and Remedies

Time is of the essence Stipulate that timely performance of the contract is essential. The inclusion of this clause allows a party to claim breach merely because the other party fails to perform within the time prescribed in the contract. Common in United States contracts, a clause of this type is considered less important in other countries.

Modification Require the parties to make all changes to the contract in advance and in a signed written modification.

Cancellation State the reasons for which either party may cancel the contract and the notice required for cancellation.

Contingencies Specify any events that must occur before a party is obligated to perform the contract. For example, you may agree that the seller has no duty to ship goods until the buyer forwards documents that secure the payment for the goods.

Governing law Choose the law of a specific jurisdiction to control any interpretation of the contract terms. The law that you choose will usually affect where you can sue or enforce a judgment and what rules and procedures will be applied.

Choice of forum Identify the place where a dispute may be settled—for example, the country of origin of the goods, the country of destination, or a third country that is convenient to both parties.

Arbitration provisions Agree to arbitration as an alternative to litigation for the resolution of any disputes that arise. You should agree to arbitrate only if you seriously intend to settle disputes in this way. If you agree to arbitrate but later file suit, the court is likely to uphold the arbitration clause and force you to settle your dispute as you agreed under the contract.

An arbitration clause should specify whether arbitration is binding or nonbinding on the parties; the place where arbitration will be conducted (which should be a country that has adopted a convention for enforcing arbitration awards, such as the United Nations Convention on Recognition and Enforcement of Foreign Awards); the procedure by which an arbitration award may be enforced; the rules governing the arbitration, such as the United Nations Commission on International Trade Law Model Rules; the institute that will administer the arbitration, such as the International Chamber of Commerce (Paris), the American Arbitration Association (New York), the Japan Commercial Arbitration Association, the United Nations Economic and Social Commission for Asia and the Pacific, the London Court of Arbitration, or the United Nations Commission International Trade Law; the law that will govern procedural issues or the merits of the dispute; any limitations on the selection of arbitrators (for example, a national of a disputing party may be excluded from being an arbitrator); the qualifications or expertise of the arbitrators; the language in which the arbitration will be conducted; and the availability of translations and translators if needed.

Severability Provide that individual clauses can be removed from the contract without affecting the validity of the contract as a whole. This clause is important because it provides that, if one clause is declared invalid and unenforceable for any reason, the rest of the contract remains in force.

LAW DIGEST

References to "Cap." mean the Chapter of the Singapore Statutes.

AGENCY

See Principal and Agent.

ALIENS

Any individual who is not a Singaporean citizen must have an entry permit or a valid pass to enter Singapore. To work in Singapore, a foreigner must have an employment pass, which is issued by Singaporean Immigration authorities. (Immigration Act, Cap. 8.)

ASSIGNMENTS

A person is entitled to assign any debt or legal chose in action. (Civil Law Act, Cap. 30, sec. 4(6).) The statute requires the assignment to be absolute, written, and signed by the assignor. The assignor must give express written notice of the assignment to the debtor, trustee, or other person from whom the assignor would have been entitled to receive or claim the debt or legal chose in action. An assignment transfers only the interests held by the assignor; that is, the assignee takes the assignment subject to any other interests that have priority over the assignor's interest. (Civil Law Act, Cap. 30, sec. 4(6).)

The courts will generally enforce an equitable assignment if the assignor seeks enforcement. An equitable assignment is one that does not comply with the statute. Thus, a conditional assignment is equitable.

No assignment is enforceable if it is contrary to public policy.

COMMERCIAL REGISTER

Prior to carrying on business in Singapore, a foreign business must first file specific documents—such as a certificate of incorporation, charter, articles, and list of directors—with the Registrar of Companies. (Companies Act, Cap. 50, sec. 368.) "Carrying on business" means establishing a share transfer or registration office in Singapore or administering, managing, or dealing with property located in Singapore. (Companies Act, Cap. 50, sec. 366(1).) This phrase does not include an isolated transaction completed within 31 days and not repeated from time to time, nor does it include an order solicited by an independent contractor and accepted outside Singapore. (Companies Act, Cap. 50, sec. 366(2).) A business that stops carrying on business in Singapore must notify the Regis-

---

Charges Over Bank Deposits

A recent amendment to the Civil Law Act permits a bank to take a charge over its own customers' deposits. This reverses a previous English High Court decision and will undoubtedly relieve the banking community's concern over such charges.

The English case of Re Charge Card Services Limited has now been reversed in Singapore by the Civil Law (Amendment) Act 1993. In that case, the English High Court held that where a customer has deposited money with a bank, it was not possible for the customer to charge the deposit to the bank as security for credit facilities. The reasoning of the court was that the bank deposit is a debt due from the bank to the depositor. The property being charged is therefore a right by the customer to sue the bank for the amount of the deposit. A charge of property amounts in law to an assignment of the property to the chargee, subject to a right of reassignment when the secured liabilities have been discharged. A charge of the bank deposit will therefore amount to an assignment of the right to sue the bank back to the bank. The court held that this was logically impossible as the bank cannot sue itself, and therefore cannot accept an assignment of a right to sue itself. The purported assignment would therefore operate as a set-off, where possible.

This decision has created difficulties for banks and financial institutions in England as well as other legal systems which derive their principles from English law. The decision has therefore been heavily criticised, both by academics and practitioners.

This decision has now been reversed in Singapore. The new legislation declares that a person can create, in favour of another person, a legal or equitable charge or mortgage over the first person's interest in a debt enforceable against the second person. The provision is retroactive in that the declaration states that the person is able, and has always been able, to create such a charge. It therefore validates all such charges created both before and after the coming into force of this Act, whether or not the charge was created before or after the decision in Re Charge Card Services Limited.

Reprinted from Asia Pacific Legal Developments Bulletin, vol. 8, no. 3, Baker & McKenzie, Sept. 1993, with permission of the author, Edmund H.M. Leow, Singapore, and the law firm of Baker & McKenzie, Singapore.

trar within seven days from the time it stops. (Companies Act, Cap. 50, sec. 377(1).)

At the time a business registers, it must appoint an agent resident in Singapore to act as its agents to accept service of process and notices for the company and to ensure compliance with the Companies Act. (Companies Act, Cap. 50, sec. 366(1).) The appointment must be made by a written memorandum or power of attorney that is verified and that is executed or under seal. (Companies Act, Cap. 50, sec. 366(1).) If an appointed agent stops acting on behalf of a company in Singapore, the company must appoint a new agent and must file the required documents with the Registrar within 21 days. (Companies Act, Cap. 50, sec. 370(5).)

CONTRACTS

Common Law Governs Singapore's laws governing contracts are similar to those of England. Parties have freedom to agree to any terms they desire, with a few exceptions created by the Unfair Contract Terms Act for sales contracts. (*See also* Sales.)

A contract is enforceable if it constitutes an agreement between two or more parties for one (the promisor) to do something in return for something (mutual consideration) from the other (the promisee). In this context, the term "agreement" means an offer by one party that has been accepted by the other party, but only if both parties intend to be bound.

Capacity to Contract If at the time of entering into a contract a party is a minor, that party is entitled to have a court declare the contract void, unless the contract is for goods necessary to maintain health and welfare. Thus, a contracting party may raise minority age as a defense against a party who seeks to enforce the contract. A contract is for such necessary goods, and thus is not voidable on grounds of minority, if at the time the goods are sold and delivered they are suitable to the minor's condition in life and actual requirements. (British Sale of Goods Act, sec. 3(1), 3(3).)

With respect to commercial contracts, the age of majority is unclear. Singapore applies the common-law rule—21 years. However, in mercantile matters, English law is applied in Singapore (Civil Law Act, Cap. 30, sec. 5), and the age of majority has been reduced in England to 18 years. (British Family Law Reform Act.) The issue is whether, in the context of a commercial contract, majority age is a question of mercantile law, in which event English law would apply. Until this issue is resolved, parties should assume that the older age—21 years—is the age of majority in Singapore.

A person who makes a contract while mentally incapacitated or drunk is bound by the contract. However, a court may declare the contract void if the person can prove that, at the time of making the contract, he or she did not know what was happening and the other party did not know of the incapacity. If the contract was for necessary goods, such a person will be required to pay a reasonable price for them. (British Sale of Goods Act, sec. 3(1), 3(3).)

An ultra vires agreement with a business entity is generally valid and enforceable against that entity. However, before the agreement has been performed, a party may apply to a court to restrain performance, and the court may do so on whatever terms it deems just. (Companies Act, Cap. 50.)

Unfair Terms The courts generally dislike contract terms that exempt one party from liability under the contract. However, a court will uphold such a term if it is clear and precise, it does not violate public policy, and it is not prohibited by statute. (*See also* Sales.)

Excuses for Nonperformance Frustration is an excuse for nonperformance—that is, a party may be relieved from the contract obligation if performance becomes impossible. Performance is considered frustrated if any one of the following occurs and makes performance impossible: (1) A certain event fails to happen, provided the contract was based on an assumption that the event would take place; (2) an item is accidentally destroyed, provided performance of the contract depended on the continued existence of that item; (3) the circumstances change so fundamentally that the result of any performance would be inconsistent with the original intent of the parties; (4) the law is changed such that the transaction is rendered illegal; or (5) if the contract is for personal services, the performer becomes unable to perform because of illness or death that is not self-induced. (Frustrated Contract Act, Cap. 31.)

If a contract becomes frustrated before either party has performed, the parties may simply cancel the contract, relieving both parties of their obligations. With certain exceptions, the law provides the parties with specific relief if they have partially or completely performed. Thus, a party who has remitted an advance payment may generally recover that payment when performance is frustrated, and the party is excused from paying any sum that comes due thereafter. However, a party (such as a manufacturer) who has incurred expenses in at least partially performing the contract before the frustrating event occurs is entitled to retain an advance payment, or to recover funds from the other party, up to the amount that will cover those expenses. If one party has received some benefit under the contract (such as a buyer who accepts a partial shipment of goods) before frustration occurs, a court may order that party to pay an amount deemed to equal the benefit received (such as the part of the total contract price that is proportionate to the value of the goods received). (Frustrated Contract Act, Cap. 31.)

The Frustrated Contract Act is inapplicable to a contract that expressly sets forth remedies in the event that performance is frustrated. It is also inapplicable to contracts for the hire of a ship (charter party), carriage of goods by sea, insurance, sale of goods covered under the British Sale of Goods Act, or sale of specific goods that have perished. (Frustrated Contract Act, Cap. 31.) (*See also* Sales.)

Releases, Reformations, and Novations Before either party has performed, both may agree to waive all rights and discharge all obligations under the contract, resulting in a complete release. A party who can perform may release or partially release the other from fulfilling the contract—for example, a seller may agree to a reduced price if the buyer is having financial difficulties and has already disposed of the goods. If contracting parties completely discharge their contract obligations and negotiate a new contract, the replacement agreement is a novation. Parties to a contract may negotiate an accord and satisfaction, in which they agree to different obligations but keep the same terms as in their initial agreement. An accord and satisfaction may in fact be a novation if the parties replace their initial contract entirely.

Remedies for Breach of Contract A party who seeks judicial relief for a breach of contract may request the remedy of rescission, under which the court will restore the parties to their original positions before the contract was made. For example, a court may order rescission if a buyer made an advance payment but never received the goods, in which event the seller would be required to refund the advance payment, placing the parties in the same position as if the contract had never been made.

A contracting party who has suffered a loss because of the other party's breach may sue for damages. Damages are not awarded to penalize the breaching party, but to place the injured party in the same position as he or she would have been if the contract had been performed. For example, if a seller fails to perform and the buyer is forced to obtain goods from another supplier at a higher market price, the buyer may sue for damages equal to the difference between the market price and the contract price. An injured party has a duty to minimize (mitigate) the damages if possible.

If one party has partially performed a contract at the time the other party breaches, the performing party may seek quantum meruit damages. That is, the party may request the court to award an amount of money based on the value of the partial performance.

Contracting parties may set forth an amount of damages to be paid on breach—known as liquidated damages. A court will uphold such a contract clause only to the extent that the amount approximates the actual loss arising from the breach. If the court finds that the amount of liquidated damages is extravagant or unconscionable, it will refuse to award that amount on the ground that the amount operates as a penalty.

The judicial remedy of specific performance is a court order that requires a party to undertake a particular act, such as the completion of a transaction. This remedy is unavailable if damages are adequate, the court cannot easily supervise the act, the contract is for personal services, the party is a minor, or the contract is for a loan of money. In a majority of business contracts, damages are adequate, and therefore specific performance is unavailable. However, a party may seek specific performance, for example, if a contract involves unique goods that cannot be obtained from another source.

COPYRIGHTS AND REGISTERED DESIGNS

Works Connected With Singapore The Singapore Copyright Act 1988 (Cap. 63) secures protection only for copyrights on works connected with Singapore—that is, the work must have been published first in Singapore or the author must be a Singapore citizen or resident, or both. (Copyright Act 1988, Cap. 63, sec. 27.)

The Act protects against the following without approval from the copyright owner or licensee: reproduction, publication, performance, broadcast, retransmission, or adaptation of original artistic, dramatic, musical, and literary works; reproduction, broadcasting, or public showing of cinematograph films; reproduction or broadcasting of sound recordings, including those associated with films; reproduction or rebroadcasting of television and sound broadcasts and cable programs; or reproduction of the typographical arrangement of published editions of works.

A creator of a work is not required to register the copyright in Singapore. A copyright is effective for 50 years from the date of the first publication.

International Works Copyrights on works not connected with Singapore may be protected in Singapore by bilateral agreement with the country where the copyright is recognized. Thus, copyrights registered in the United States and England are protected in Singapore by bilateral agreements. (Copyright International Protection Regulations 1987; Copyright Act 1988, Cap. 63, secs. 184, 202.)

Registered Designs Any designs registered under the United Kingdom Design Protection Act are protected in Singapore. (United Kingdom Design Protection Act, Cap. 339.) The term *design* refers to any original aspect of the internal or external shape or configuration of all or part of an article.

A design does not include any method or principle

of construction, nor any features of shape or configuration that allow the article to be connected to or placed in, around, or against another article so that either article may function. If the creator of an article intends that it will form an integral part of another article, the design is not registrable if any features depend on the appearance of the other article, unless the features constitute a special original design.

The designer owns the design rights, except that the employer owns the rights if the design is created by an employee in the course of employment or under commission. The owner has the exclusive right to reproduce the design for commercial purposes and may seek legal remedies to protect that right against persons who infringe on it—that is, against persons who copy the design or article or who knowingly deal with designs or articles that infringe on the rights. A registered design is afforded 25 years of protection from the earliest of the year in which the design was first recorded in the design document or the year in which the article was first made from the design. However, this period is reduced if the article is exploited commercially in the first five years of that period.

See also Patents; and Trademarks.

FOREIGN EXCHANGE

Singapore laws do not restrict foreign exchange. Transfers of funds in and out of Singapore are made freely.

FRAUDS, STATUTE OF

See Statute of Frauds.

INTELLECTUAL PROPERTY RIGHTS

See Copyright and Registered Designs; Patents; and Trademarks.

PATENTS

Singapore law does not provide for the protection of patents. A patent registered under the United Kingdom Patents Act is protected in Singapore. A grant of a patent under the United Kingdom Patents Act is effective for 20 years.

POWER OF ATTORNEY

No specific legislation governing powers of attorney exists in Singapore. A power of attorney is recognized in Singapore whether executed in or out of the country.

PRINCIPAL AND AGENT

An agency relationship may usually be created orally or by the conduct of the parties. A written document—power of attorney—is not needed except for agents who dispose of immovable property in Singapore or who are appointed by a firm carrying on business in Singapore. (*See* Commercial Register; and Power of Attorney.)

A principal is generally liable for the acts of an agent, provided the agent acts with the express or apparent authority of the principal. A principal may also ratify unauthorized acts of an agent, therefore becoming liable for those acts.

Under the statute that makes English mercantile law applicable in Singapore (Civil Law Act, Cap. 30, sec. 5), mercantile agents in Singapore are subject to the British Factors Act. A mercantile agent is one who, in the customary course of business as an agent, has authority with or without the owner's consent to sell goods, consign goods for sale, buy goods, or raise money using the goods as security. (Factors Act, sec. 1(1), 2(1).) A mercantile agent has a general lien on the goods in his or her possession and on the proceeds from any sale of the goods for the balance of the account with the principal. (Factors Act, sec. 3.)

SALES

Applicable Law Under Singapore's Civil Law Act, the British Sale of Goods Act of 1979 (BSGA) applies in Singapore. (Civil Law Act, Cap. 30, sec. 5.) The BSGA governs formation of contracts for the sale of goods, express and implied contract terms, rights of contracting parties and third parties, performance of the contract, and remedies for nonperformance.

Contract for Sale of Goods A contract for the sale of goods is a contract by which a seller transfers or agrees to transfer the ownership of goods—that is, personal chattels (BSGA, sec. 6(1))—to a buyer for a price that must be a sum of money. (BSGA, sec. 2(1).) A contract for the sale of goods is formed only if one party makes an offer and, before the offer is withdrawn, the other party accepts it unconditionally. A conditional acceptance is considered a counteroffer, in which event a contract arises only if the counteroffer is unconditionally accepted before if is withdrawn.

A contract of sale may be written, oral, partly written and partly oral, or implied from the conduct of the parties. (BSGA, sec. 4(1).) However, contracting parties should be cautious of agreeing orally to change or contradict any written terms of their contract. In a judicial action to enforce a contract, a party's evidence of oral terms is usually inadmissible for purposes of proving that the written terms are incorrect or were modified. (Evidence Act, Cap. 5, sec. 94.)

A contract for the sale of goods may be absolute—that is, unconditional—or conditional—that is, subject to the satisfaction of a condition. If the contract is conditional, the obligations arise only if the condition is met. (BSGA, sec. 4(2).) For example, a contract may be made conditional on the confirmation of a buyer's credit, and if the buyer's credit cannot be confirmed, neither party is obligated to perform the contract, which in effect is canceled.

Implied Terms and Warranties With some exceptions, the BSGA implies certain provisions into contracts for the sale of goods.

The BSGA implies that the term *month* as used in a contract for the sale of goods means a calendar month, unless otherwise expressly defined in the contract. (BSGA, sec. 10(3).)

If a contract for the sale of goods fails to state a price, or a means of fixing a price (see BSGA, secs. 8(1), 9(1)), for the goods, the contract is voidable—that is, a party has the option of requesting a court to declare the contract invalid. Alternatively, the seller may seek legal relief based on the duty of the buyer to pay for any goods accepted, and the court is authorized to fix a reasonable price for the goods in accordance with the circumstances existing at the time of the sale. (BSGA, sec. 8(3).)

With respect to contracts that provide for the transfer of all ownership rights to the goods, the BSGA implies all of the following (BSGA, sec. 12(1), (2)): The seller has a legal interest in the goods that entitles the seller to sell them or, at the time the goods are to be transferred to the buyer, the seller will have such an interest. The seller warrants that the goods are free from any charge or encumbrance undisclosed or unknown to the buyer before the contract was made, and they will remain free from such interests until ownership passes from the seller.

The buyer will enjoy quiet possession to the goods, subject only to disturbance from the owner or person entitled to the benefit of any charge or encumbrance that the seller has disclosed or that is known to the buyer.

A seller may agree to transfer only those interests that the seller, or a third person, might have at the time of sale. In such event, the BSGA implied terms for a transfer of all ownership rights are inapplicable. Instead, the BSGA implies into the contract for sale both of the following (BSGA, sec. 12(4), (5)):

The seller warrants that all charges or encumbrances known to the seller and not to the buyer have been disclosed to the buyer before making the contract.

The seller warrants that the buyer's quiet possession of the goods will not be disturbed by (1) the seller; (2) the third person, if the transfer is of only a third person's ownership interests in the goods; and (3) any person that claims through or under the seller or the third person, except in relation to a charge or encumbrance that was disclosed to or known by the buyer before making the contract.

If a seller describes the goods by words, the BSGA implies into the contract for sale a warranty by the seller that the goods will correspond with the description. If the goods do not so correspond and the buyer relied on the seller's description in entering into the contract, the buyer may reject the goods and cancel the contract. (BSGA, sec. 13(1).)

If a buyer is provided with samples of the goods and the parties intend those samples to serve as a description of the goods, the BSGA also implies certain terms into the contract. The intent of the parties that the samples will be descriptive of the goods may be expressly stated in the contract or may be implied from common dealings between them or customary usage in the trade. The implied terms are as follows: (1) the goods will correspond with the sample in quality; (2) the buyer will have a reasonable opportunity to compare the goods with the sample; and (3) the goods will be free from defects that would render them unmerchantable and that would not be apparent on reasonable examination of the sample. A buyer who is not given a reasonable opportunity to compare the goods with the sample is entitled to reject them. (BSGA, sec. 15(2).)

For any goods sold in the course of business to consumers, the BSGA implies the following warranties related to the fitness of the product: (1) At the time of the sale, the goods are of merchantable quality—that is, they are reasonably fit for the purpose for which goods of that kind are commonly bought under similar circumstances—except for any defects that were specifically drawn to the buyer's attention or, if the buyer examined the goods before making the contract, any defects that should have been revealed during that inspection; and (2) if at the time of sale the buyer informed the seller of a particular purpose for which the goods would be used and the buyer reasonably relied on the seller's skill and judgment, the goods are fit for that purpose. (BSGA, sec. 14.) These warranties are not implied into nonconsumer contracts—that is, contracts between merchants. However, the warranties will apply if the parties annex them to the contract or if the warranties are customarily used in the trade. Subject to the Unfair Contract Terms Act, merchants may exclude or restrict these implied warranties in nonconsumer contracts, but not in consumer contracts. (BSGA, sec. 14(4).)

Exceptions to Implied Terms and Warranties The terms implied by the BSGA may be varied or eliminated by express agreement of the parties, a course of dealing between the parties, or customary usage in a particular trade. However, modification or elimination of the terms implied by the BSGA is limited by the British Unfair Contract Terms Act.

Unfair Contract Terms A party's right to change or eliminate terms implied under the BSGA is subject to the British Unfair Contract Terms Act (BUCTA). There is some question, however, as to whether BUCTA applies in Singapore. The general provision of Singapore's Civil Law Act that applies the law of England to mercantile matters in Singapore (Cap. 30, sec. 5) would seem to include BUCTA. However, BUCTA also regulates hire-purchase agreements. The Singapore Civil Law Act expressly excludes application of any English law that governs hire-purchase agreements (Cap. 30, sec. 5), for which Singapore has its own law (Hire-Purchase Act, Cap. 192.).

Assuming that BUCTA has at least limited application in Singapore, parties to any contract for the sale of goods, regardless of whether the contract is with a consumer, are not permitted to restrict any of the warranties of title implied by section 12(1) and (2) of the BSGA into a contract that transfers the full ownership interest in the goods. (BUCTA, sec. 6.) Nor can parties to a consumer contract restrict or exclude any of the BSGA implied terms for contracts for the sale of goods by description (words or samples) or for the sale of goods in the course of business. If the contract is not with a consumer, the terms in these contracts may be limited, but only if the restriction is reasonable. For purposes of BUCTA, a *consumer contract* is one that is made between a seller in the course of business and a buyer not in the course of business for goods that are of a type ordinarily supplied for private use or consumption. (BUCTA, sec. 12). *See* Implied Terms above.

Terms of Essence A contract term of essence is one that is of such vital importance that a party's failure to comply exactly constitutes a breach of the contract. For terms that are not of essence, substantial performance is adequate to avoid breach.

Parties to a contract for sale of goods may expressly stipulate that specific terms are of essence. Unless so stipulated, the time of payment is not generally of essence. (BSGA, sec. 10(1).) The time for delivery of goods is usually of essence, meaning that failure to deliver on time entitles the buyer to reject the goods. (See BSGA sec. 10(2).)

Acceptance of Goods Under the BSGA, a buyer has a duty to accept and pay for the goods described in the contract for sale. (BSGA, sec. 27(1).) To determine whether the goods conform to the contract, the buyer has a right to examine and inspect them, and the seller must allow the buyer a reasonable opportunity to do so. (BSGA, sec. 34.) The buyer is assumed to have accepted the goods under any one of three situations: (1) the buyer expressly indicates acceptance to the seller; (2) the buyer acts in relation to the goods in a manner that is inconsistent with the seller's ownership of the goods; or (3) the buyer retains the goods for a reasonable time without rejecting them. (BSGA, sec. 27(1).)

Risk of Loss The BSGA places the risk of loss from damage or destruction of the goods subject to a contract for sale on the seller until ownership passes to the buyer. When ownership passes, the risk is transferred to the buyer. The parties may modify these risk provisions by express or implied agreement. If delivery of the goods is delayed because of one party's fault, that party bears the risk for damage or loss to the goods that might not have occurred but for the delay. (BSGA, sec. 20.)

Buyer's Remedies If goods received do not conform to the contract or if a seller has otherwise breached a condition of the contract, the BSGA provides several alternative remedies for the buyer.

A buyer may reject the goods. Goods may be rejected if the quantity received is less than the quantity required by the contract, but only if the difference is more than a minimal amount—that is, a difference in quantity that would be unacceptable to a reasonable buyer under similar circumstances. (BSGA, sec. 30(1).) If the quantity of goods received is greater than the contract amount, the buyer may reject the additional goods. (BSGA, sec. 30(2).) If the goods are mixed, such that some comply with the contract and others do not, the buyer may reject the entire shipment. (BSGA, sec. 30(4).)

Even if goods conform to the contract, a buyer may still reject them if the seller has breached a condition of the contract—for example, if delivery is unreasonably delayed beyond the date fixed in the contract. However, if the buyer has already accepted part of the goods, whether the buyer is then entitled to reject the rest of the goods based on the seller's breach of a condition depends on whether the contract is nonseverable. A nonseverable contract is one in which two or more obligations of a party cannot be separately performed, and therefore failure to perform any one obligation constitutes breach of the entire contract. If the contract is nonseverable and the buyer has not yet accepted any goods, the buyer may reject all the goods based on the seller's breach of a condition; if the buyer has already accepted some goods despite the seller's breach, the buyer cannot then reject the remaining goods based on that same breach. For a severable contract—that is, one in which two or more obligations may be performed independently—the seller's breach of a condition related to one portion of the contract will allow the buyer to reject the goods for that part only. (BSGA, sec. 11(4).) For example, assume that the parties to a contract for sale of goods agree to two separate deliveries over a six-month period and further agree that the contract is nonseverable. If the first delivery is unreasonably late, the buyer may reject or accept it because the seller has breached the contract. The buyer's rejection of the first delivery will cancel the entire contract, in which event the buyer may also reject the second delivery. However, the buyer's acceptance of the first delivery in effect waives the seller's breach, and the buyer cannot later use that breach as a reason to reject the second delivery. On the other hand, if the parties had agreed that the contract would be severable, the buyer's acceptance or rejection of the first delivery would have no effect on the obligations of each party related to the second delivery.

Unless the parties agree otherwise in their contract, a buyer is not required to return rejected goods to the seller. The seller is required to provide for the return shipment, and risk of loss or damage is on the seller. However, the buyer is required to act reasonably with respect to the goods so as to avoid or mitigate damages that the seller may incur. The buyer should follow, for example, reasonable instructions from the seller to deliver the goods to the seller's shipper or to make them available for pick up by the seller's agent.

A buyer may accept the goods and sue for damages incurred because the goods are nonconforming or because the seller breached another condition of the contract. However, the buyer has a duty to pay the contract price of any goods accepted (BSGA, secs. 27(1), 30.), and therefore the buyer should expressly inform the seller of any problems and attempt to resolve the difficulty before accepting any of the goods. To reduce problems with the application of the BSGA provisions, the parties should include express terms in their contract to govern the provisional acceptance of goods when a dispute arises over a breach of the contract. For example, the contract may require the buyer to place payment for the goods into a secured fund until the dispute is concluded.

A buyer who never receives the goods may sue for damages for nondelivery. Damages for nondelivery are measured as the difference between the contract price and the market or current price at the time fixed for delivery by the contract. (BSGA, sec. 51.) Alternatively, the buyer may seek specific performance of the contract if damages are inadequate relief, such as when the buyer cannot obtain comparable goods from another supplier. (BSGA, sec. 52.)

If a seller breaches a warranty express or implied in the contract, the buyer may set off the damages resulting from that breach against the price payable under the contract. Alternatively, the buyer may sue the seller for damages. The damages recoverable are those that directly and naturally result in the ordinary course of events from the breach. If the goods do not meet the quality warranted, the measure of damages is the difference between the value of the goods at the time of delivery and the value that they would have been had the warranty been met. (BSGA, sec. 53.)

Seller's Remedies If a buyer breaches a contract for the sale of goods, the seller has a number of alternative remedies. The availability of these remedies generally depends on whether ownership or possession of the goods has passed to the buyer at the time of the breach.

If a buyer fails to pay for goods as of the date required by the contract, the seller may sue to recover the price, regardless of whether ownership to the goods has passed. (BSGA, sec. 49.)

If a buyer fails to pay for goods as of the date required by the contract, and if ownership but not possession of the goods has passed to the buyer, the seller may withhold delivery and impose a lien on the goods. The seller has a right to retain possession of the goods until payment is made. If some of the goods have already been delivered to the buyer, the seller may retain and impose a lien against the remainder. The seller loses the right to claim a lien by transferring the goods to a carrier or other bailee for purposes of delivering them to the buyer, unless the seller reserves a right of disposal. (BSGA, sec. 43(1).) By reserving a right of disposal in goods that are shipped, the seller retains ownership in the goods until the buyer has met a certain condition, such as delivery to the seller of documents of credit or payment. (BSGA, sec. 19.)

If a buyer fails to pay for goods as of the date required by the contract, and ownership of the goods has not yet been transferred to the buyer, the seller may withhold delivery until payment is made (BSGA, secs. 39(2), 44) or may resell the goods to another buyer (BSGA, sec. 39(1)(c)). If the seller has already shipped the goods, the seller may stop the goods in transit and direct the shipper to return the goods to the seller. (BSGA, sec. 39(1)(b).)

If a buyer wrongfully neglects or refuses to accept and pay for the goods, the seller may sue for the price of the goods as set forth in the contract. Alternatively, the seller may sue to recover the damages that directly and naturally resulted in the ordinary course of events from the buyer's breach. Such damages are measured as the difference between the contract price and the market or current price at the time fixed in the contract for acceptance of the goods or, if no such time is fixed, at the time the buyer neglected or refused to accept them.

If a buyer breaches a contract for the sale of goods before ownership of the goods passes to the buyer, the seller may also seek specific performance of the contract. (BSGA, sec. 52.) A court will order specific performance if an award of damages is an inadequate remedy. For example, a seller may be awarded specific performance if the contract is for identified goods that are not commonly marketable.

Frustration of Contract Under the BSGA, the parties may avoid their contract obligations, and thus cancel their contract, if performance becomes impossible because (1) the goods were destroyed at the time the contract was made but without the seller's knowledge (BSGA, sec. 6); or (2) the contract was for the sale of identified goods that were destroyed after the contract was made, without the seller's fault, and before ownership (and therefore risk of loss) passed to the buyer (BSGA, sec. 7).

STATUTE OF FRAUDS

A statute of frauds is a law that requires a contract to be written in order to be enforceable in a court. In Singapore, the following contracts must be written to be enforced: bills of exchange and promissory notes (Bills of Exchange Act, Cap. 280.); leases of real property in excess of three years. (Land Titles Act.); an agreement to transfer the shares of a registered company (Company Act, Cap. 50.); and hire-purchase agreements. (Hire-Purchase Act, Cap. 192.)

TRADEMARKS

Governing Laws Trademarks are governed by the Trade Marks Act (Cap. 332) and the Trade Mark Rules, both of which are based on British trademark law. The Trade Mark Rules provide forms for applications and other procedures required by the Act and contain a schedule that sets forth all fees for trademark registration and other procedures.

Effect of Registration A valid registration of a trademark gives the registrant the exclusive right to use the mark in relation to the goods or services for which the mark is registered. (Trade Marks Act, Cap. 332, sec. 45.) The mark is treated as personal property of the registrant, and the registrant may therefore assign the mark in the same manner as personal property. (Trade Marks Act, Cap. 332, sec. 44.) The registrant may also protect the mark and recover damages caused by another person's infringement of the registrant's property right in the mark. (Trade Marks Act, Cap. 332, secs. 44-48.) (*See* Infringement below.)

Definitions A trademark is a mark or symbol with a device, brand, heading, label, ticket, name, signature, word, letter, numeral, or any combination of these that is used or proposed to be used in relation to goods or services to indicate a connection in the course of trade between the goods or services and the person who, as owner or registrant, has a right to use the mark with or without identification of that person. (Trade Marks Act, Cap. 332, sec. 10.) A mark may be registered only if it is adapted to distinguish or is capable of distinguishing the goods or services of the owner or registrant of the mark from those of another person. (Trade Marks Act, Cap. 332, sec. 10.)

Registrable Marks The trademark register contains 42 classes of goods and services. Within those classes, a person may apply for Part A or Part B registration. A person may be permitted to register in both Part A and Part B, provided that separate applications are filed. Registration under Part A gives greater infringement protection than registration under Part B. (*See* Infringement below.)

Registration under Part A is available only if the mark contains or consists of at least one of the following (Trade Marks Act, Cap. 332, sec. 10): a name of a company, individual, or firm represented in a special or particular manner; a signature of the applicant for registration or the applicant's predecessor in business; an invented word or words; a word or words that are without any direct reference to the character or quality of the goods or services for which the mark is used and that do not ordinarily signify a geographical name or a surname; or any other distinctive mark, provided that the registrant provides evidence of distinctiveness and that the Registrar determines that registration is available.

In considering whether registration is available for a particular mark, the Registrar may take into account either or both of the following: (1) the extent to which the trademark inherently distinguishes the goods or services; or (2) the extent to which, because of the manner in which the mark is used or because of any other circumstances, the trademark can in fact distinguish the goods or services. (Trade Marks Act, Cap. 332, sec. 10.)

Registration under Part B is available for a mark in connection with particular goods or services, provided the mark can distinguish those goods or services with which the owner of the mark is or may be connected in the course of trade from other goods or services for which no such connection exists, whether generally or because the trademark is registered, or sought to be registered, for use subject to certain conditions or limits. A trademark that does not on its face appear to be inherently distinctive may be registered under Part B. The Registrar determines the availability of Part B registration for a particular mark using considerations similar to those used for Part A registrations. (Trade Marks Act, Cap. 332, sec. 11.)

Neither Part A nor Part B registration is available for any application to register a mark that contains any of the following: (1) material that could cause confusion or that is deceptive, illegal, immoral, or scandalous; (2) the words, abbreviations, or equivalent expressions to *patent, patented, registered, registered design, registered trade mark, copyright,* or *to counterfeit this is a forgery*; (3) representations or close imitations of the President of Singapore, the Red Cross, or the Geneva Cross; (4) representations of the crest of Singapore, the coat of arms of the President of Singapore, the royal or imperial arms of Singapore, or any other crests, armorial bearings, insignia, or devices of similar items; (5) representations of the royal or imperial crowns or of the royal, imperial, or national flag of Singapore; (6) the words *royal, imperial, presidential, Singapore government,* or any other words that could suggest that the mark is authorized by such persons or entities; or (7) the word *ANZAC.* (Trade Marks Act, Cap. 332, secs. 12-16.)

Registration Process Procedures for registration are burdensome, and therefore the Trade Mark Rules provide an inquiry process by which an applicant may seek preliminary advice from the Registrar as to whether a mark can be registered under Part A or Part B. (See Trade Mark Rules, Form TMR-17.) Once the applicant decides to apply for registration, the applicant must submit four copies of the trademark artwork with the application. (Trade Marks Act, Cap. 332, sec. 73; Trade Mark Rules 19-21.) If any word in the trademark is not in Roman characters, the application must include a transliteration and a translation that satisfies the Registrar, unless the Registrar otherwise directs. (Trade Marks Rule 26.) The Registrar will search the Register and will examine the trademark for compliance with the Act. The Registrar will then issue a report on the availability of registration. If the Registrar finds that the mark cannot be registered, the business is allowed two months to overcome the Registrar's objections. If the business does not resubmit a corrected mark, the Registrar will assume that the application is withdrawn unless the business requests an extension of time for compliance. (Trade Mark Rules 23-32.) An application that is not completed within 12 months from the date it is filed because of the fault of the applicant is considered to be abandoned unless it is completed within the time permitted by the Registrar in a written notice. (Trade Marks Act, Cap. 332, sec. 22.)

Once the Registrar approves the application, the applicant may be required to submit additional, corrected photographs or artwork of the mark sufficient for publication in the official government newspaper. (Trade Mark Rules 37-40.) The Registrar will have the registration application advertised in the official government publication. (Trade Marks Act, Cap. 332, sec. 18.) In the absence of any objections, the Registrar will enter the mark in the Register and will issue a certification of registration to the registrant two months after the date of publication. (Trade Marks Act, Cap. 332, sec. 31.)

Any person may file with the Registrar a notice of opposition on the approved form, provided they do so within two months from the date of the advertisement. The Registrar will send a copy of the notice to the applicant, who may send a counter-statement, also on an approved form, specifying grounds that support the application for registration of the mark. After the Registrar sends the counter-statement to the opponent, the opponent may submit evidence to support the opposition. The Registrar will then hold a hearing, decide the issue, and forward a decision within three months. If the opponent fails to submit any evidence, the opponent is deemed to have abandoned the opposition. (Trade Marks Act, Cap. 332, sec. 19; Trade Marks Rules 41-54.) A party who loses its claim before the Registrar may appeal to a court. (Trade Marks Act, Cap. 332, sec. 19.)

Rectification of Register A person who is injured by the entry, omission, wrongful retention, or incorrect or defective entry of a trademark in the Register may apply to the Registrar to have the Register rectified. (Trade Mark Rules 78-85.) If the Registrar does not rectify the Register, the person may bring a legal action to have the court order the Registrar to rectify the Register. (Trade Marks Act, Cap. 332, secs. 39-41.)

Period of Protection A trademark registration is initially valid for 10 years and can thereafter be renewed indefinitely for successive periods of 10 years each, beginning from expiration of the last registration of the trademark. (Trade Marks Act, Cap. 332, secs. 32-36.) A registration may be renewed not more than three months before it expires. (Trade Marks Act, Cap. 332, secs. 32-36; Trade Mark Rule 60.) If not renewed, the registration is removed. (Trade Marks Act, Cap. 332, sec. 35; Trade Mark Rule 63.) However, the registrant may apply to have the registration restored, provided the application is made within 12 months from the expiration date. (Trade Marks Act, Cap. 332, sec. 35A(1); Trade Marks Rule 64.)

Assignment A trademark may be assigned with or without the goodwill of the business in which it is used. The owner may also assign a limited right to use the mark in a specific market and may retain the right to use it in Singapore and other markets. (Trade Marks Act, Cap. 332, secs. 29, 44; Trade Marks Rules 90-97.) An assignee of a mark may apply to the Registrar for registration of the assigned title to the mark. The application procedure for registration is similar to application for an original trademark, in that on receiving proof of the assigned title, the Registrar will have the application advertised, take objections, and then register the assigned title to the mark once any objections are resolved. (Trade Marks Rules 67-73.)

Infringement No infringement action is available for an unregistered trademark. (Trade Marks Act, Cap. 332, sec. 49.)

An owner of a registered trademark may enforce his or her rights in the same manner as for ownership of personal property. Thus, relief for trademark infringement may be sought through a legal action for damages. (Trade Marks Act, Cap. 332, secs. 44, 45.)

For marks registered under Part A, infringement includes, but is not limited to, a person's unauthorized use of a mark in the course of trade, provided: (1) the mark is identical to, or nearly resembling a registered trademark used in relation to goods or services for which it is registered; and (2) the unauthorized mark is used in relation to goods or services in such a way as either (a) to confuse those goods or services with the ones offered by the registered trademark owner, or (b) to suggest that the registered owner is connected with, or has authorized the use of the trademark in relation to, the goods and services offered by the infringer. The authorized use of a trademark must be within the limits entered on the Register, and no infringement occurs if the unauthorized concurrent use of a trademark is in fact beyond the limits or conditions of the registration. (Trade Marks Act, Cap. 332, secs. 45, 48.)

After seven years from the date of an initial Part A registration, a presumption arises that the registration is valid in all respects. The effect of this presumption is that the trademark will be invalidated only on proof that the registration was obtained by fraud, that the mark contains unlawful matter, or that the contestant continuously used the trademark from a date before the earlier of the registrant's use or the registration of the mark. A plaintiff who sues for infringement of a trademark that is presumed valid after seven years will not be permitted to interfere with or restrain another person's use of an identical or nearly identical mark in relation to goods or services to the extent that such use has been continuous from a date before the earlier of the plaintiff's use or the plaintiff's registration of the mark. (Trade Marks Act, Cap. 332, sec. 48.)

For marks registered under Part B, infringement is the same as for marks registered under Part A. However, a defendant may avoid liability if he or she can establish that use of the trademark is not likely to deceive, cause confusion, or be taken as indicating a connection in the course of trade between the goods or services and the registered trademark owner. (Trade Marks Act, Cap. 332, sec. 46.)

In a criminal action, a person who is found guilty of importing or selling goods using counterfeit trademarks may be fined S$10,000 (about US$6,200) for each item to which the mark is falsely applied to a maximum aggregate fine of S$100,000 (about US$62,000), imprisoned for up to 5 years, or both. (Trade Marks Act, Cap. 332, sec. 73.) The false application of a trademark to services is punishable by a fine of up to S$100,000, imprisonment of up to five years, or both. (Trade Marks Act, Cap. 332, sec. 73A.) The act of counterfeiting another's trademark is punishable by a fine of up to S$100,000, imprisonment of up to five years, or both. (Trade Marks Act, Cap. 332, secs. 70, 71.) Falsely representing that a trademark is registered is punishable by a fine not exceeding S$10,000 (about US$6,200). (Trade Marks Act, Cap. 332, sec. 69.)

Financial Institutions

Singapore's development as a financial center has, if anything, more than kept pace with its growing status as a regional trade center. Asian financial competitors are finding that Singapore is rapidly becoming a force to be reckoned with. Banking and financial markets are well developed, operating with state-of-the-art systems and offering a wide variety of sophisticated products. Singaporean financial institutions have been an adjunct to trade, oriented primarily to servicing trade and industrial growth. However, Singapore's growth in financial services has been planned to a great degree, allowing for a level of proficiency on an across-the-board basis that is missing in most other Asian financial markets. Many of these other centers show areas of sophistication side by side with pockets of stunning financial naiveté, and most developing business centers in Asia hope to take over the position of regional financial leader from Hong Kong. However, at present Singapore appears to be the only competitor positioned to actually do so.

Some important factors contributing to Singapore's development as a financial center are its advanced communication and infrastructural system, government tax incentives for foreign investment, free flow of foreign exchange, effective government regulation and monitoring of markets, a heavy presence among foreign banks, the availability of low-interest loans for both foreigners and residents, and strong regional ties between Singaporean companies and businesses in Hong Kong, China, and other Southeast Asian nations.

Companies doing business in Singapore have access to a wide variety of financial services, including full-service commercial banking, stock market capitalization and trading, business loans, trade financing, and venture capital for high-tech development. Foreign individuals and enterprises can freely invest in Singaporean equities, bonds, futures, and venture capital opportunities.

THE BANKING SYSTEM

Singapore's financial system consists of the Monetary Authority of Singapore (MAS), the country's central bank and chief regulatory agency; commercial banks, including foreign bank branches and representative offices; merchant banks, the investment divisions of commercial banks; Asian Currency Units (ACUs), which operate in international foreign exchange and capital markets; various specialized financial institutions; and nonbank financial institutions, such as finance and insurance companies.

The Monetary Authority of Singapore

The Monetary Authority of Singapore, established in 1971, performs all the functions of a central bank except the issuing of currency (currency is the responsibility of the Board of Commissioners of Currency of Singapore, known as the BCCS). The MAS was organized to consolidate under one agency the various functions that had previously been taken care of by different offices. It is administered by a seven-person board of directors with the Minister of Finance as chair. Other board members are appointed by the president of Singapore.

The MAS, which is technically under the authority of the Ministry of Finance (MOF) but retains a large degree of autonomy, is responsible for all matters relating to banks and other financial institutions under the terms of the Banking Act. It licenses and supervises banks, finance companies, insurance companies, money changers, securities dealers, investment advisors, and other financial institutions. The MAS also formulates and implements Singapore's monetary, interest, and exchange rate policies. As with everything else in Singapore, government involvement in determining these policies is heavy, but MAS is a skilled and pragmatic technocratic agency that operates with a great deal of autonomy.

The MAS is generally credited with being a competent watchdog and manager of financial activity. It was one of the few regulators worldwide to catch onto and ban the corrupt Bank of Credit and Commerce International (BCCI) from operating in its domain before the problems of that institution became manifest. In 1992 the MAS refused to allow representatives from the United States Office of the Comptroller of the Currency to examine the books of US banks operating in Singapore, arguing that it was more competent to do so and that it would report to the comptroller's office anything that it needed to know. Singapore's banks are the healthiest in Asia and some of the strongest in the world due to the careful supervision of the MAS.

Commercial Banks

Commercial banks are classified into three groups—full, restricted, and offshore—depending on their type of operating license. There were 131 banks operating in Singapore at the beginning of 1993: 35 full license banks (of which 13 were local), 14 commercial banks with restricted licenses (all foreign owned), and 82 banks with offshore licenses. Virtually all banks are members of the Association of Banks in Singapore, a trade group that helps set policies and standards. Bank applications are approved based on reputation, asset size, extent of international operations, and potential contribution to the growth of Singapore's economy in general and the development of its financial sector in particular.

Full license banks are the stars of the banking world. They have few restrictions on their banking operations and can offer all standard and many specialized banking services to business and individual customers. They accept all kinds of deposits, provide transaction and remittance services, and extend credit (including letters of credit, guarantees, and trade financing). Full license banks can conduct business with Singaporeans and foreigners, although they are discouraged from Singapore dollar-denominated lending to nonresidents or to residents intending to use Singapore dollars outside of the country. Banks are required to consult the MAS before extending Singapore dollar credits in excess of S$5 million (about US$3.1 million) to any nonresident. They can conduct business in foreign currencies if they have a separate Asian Currency Unit license, which virtually all do. Full license banks are subject to strict limitations on branching and the installation of automatic teller machines (ATMs), although most full license banks do maintain an extensive branch network. They can also trade in gold, securities, and various other financial instruments.

Restricted license banks can engage in most banking operations; however, they cannot offer savings accounts, accept time deposits of less than S$250,000 (about US$156,000), or operate more than one banking facility in Singapore. They generally have ACU licenses, meaning that they can accept deposits and make loans in foreign currencies. Most are active in the wholesale corporate market and in interbank lending.

Offshore license banks cannot accept Singapore dollar deposits except from other banks and authorized financial institutions. They are allowed to offer credit denominated in foreign currencies to both residents and nonresidents. However, they are severely restricted in extending credit denominated in Singapore dollars and can only accept fixed-deposit accounts of greater than S$250,000 (about US$156,000) from nonresidents. Offshore license banks have recently been allowed to increase Singapore dollar loans to residents to a maximum of S$100 million (about US$62.3 million). They typically hold ACU licenses and their foreign exchange operations are concentrated on the Asian dollar market.

Banking in Singapore is heavily computerized. An on-line interbank payments system has been operating since 1985 to handle local clearing functions. In 1987 this system was enhanced to handle book entry trading of government securities. Banks, in conjunction with the post office savings system, operate a network of 1,129 ATMs, and most individual banks offer banking by phone and by computer as well.

Foreign Banks

Foreign banks in Singapore have contributed greatly to the nation's financial development and internationalization. Some 36 full and restricted foreign banks operated in Singapore in 1993. Despite suggestions that foreign banks are prevented from effectively competing in the domestic market, as of 1993 foreign banks held 42 percent of all resident deposits and made 55 percent of all loans to residents. They also handled 70 percent of all trade financing in the country.

Singapore has welcomed foreign banks for the last 20 years, although it has kept them on a somewhat short leash. Since 1990 foreigners have been allowed to own up to 40 percent of Singaporean banks, up from a 20 percent maximum in effect before the change (the 20 percent cap remains in place for finance companies). However, a 5 percent limit on ownership by a single foreign individual, firm, or affiliated group of owners keeps foreign ownership spread so thinly that it does not bring control along with ownership. Singapore has encouraged exposure to foreign banking practices in order to learn, but with the reservation that allowing such institutions to operate freely in the relatively limited local market would subject domestic institutions to unfair competition and drive down margins.

Singapore's largest foreign bank presences are from the United States, Japan, and Europe. In the year ending June 1993 foreign banks arranged more than one-third of all wholesale syndicated loans in Singapore, nearly US$900 million. The leading foreign banks in Singapore include Citicorp (United States), Sumitomo (Japan), Hongkong Bank-Midland Group (Hong Kong-United Kingdom), and ABN-AMRO Bank NV (Netherlands).

Most foreign banks operate primarily in the Asian dollar market under ACU licenses, and they are allowed to set up merchant bank subsidiaries. Most operate as branch offices. In 1992 52 representative offices of foreign banks were operating in Singapore. Although such offices, which were authorized in 1968, are not allowed to conduct business in the country, they can gather information, monitor correspondent relationships, make contacts, and assess the opportunities for future penetration of the market.

Domestic Banks

Four domestic banks—the Development Bank of Singapore (DBS), the Overseas-Chinese Banking Corporation (OCBC), the United Overseas Bank (UOB), and the Overseas Union Bank (OUB)—dominate the market in Singapore. Two other domestic banks, the Tat Lee Bank and the Keppel Bank, also rank among Asia's largest 100 banks. Most of these banks raise capital through equity offerings on the stock exchange.

Domestic banks are steadily expanding their operations both within Singapore and throughout Asia. Their total lending increased by 11 percent in 1991 and by 14 percent in 1990, despite the world economic downturn during that period. In 1991 major borrowers included general commercial firms (25 percent of lending), construction companies (24.3 percent), and manufacturers (12.7 percent).

On the international scene, Singapore's domestic banks are becoming players in China and Southeast Asia at a time when banks from many other nations are retreating to their home turf. Offshore operations account for 20 to 30 percent of the income of most Singaporean banks—with Malaysian business accounting for between 12 and 14 percent of business in the large banks. The proportion of overseas business is expected to increase. Because most Singapore bankers are ethnic Chinese who speak one or more Chinese dialects, they are in an excellent position to provide the financial services that tie together Chinese business communities in Southeast Asia, Taiwan, Hong Kong, and mainland China, as well as within Singapore itself.

The OCBC and UOB have the greatest foreign exposure, mostly in neighboring Malaysia. OCBC also has good relations in China and has recently been a leader in financing major development projects in that country. Government-owned DBS is becoming the government's official investment arm in the China-Hong Kong area. In March 1992 it acquired 10 percent of Hong Kong's Wing Lun Bank. UOB has strong ties in Thailand and operations in Hong Kong and in Fujian province in China.

Commercial Bank Services

Large commercial banks offer a variety of banking services to manufacturing firms and other clients. Most banks extend credit for five- to 10- year periods at competitive interest rates, covering up to 50 percent of plant and equipment costs and up to 65 percent of the value of factory buildings. Higher percentages are available for particularly desirable projects and for expansion loans. Loans usually require a mortgage or a guarantee, although some banks will make unsecured loans under certain conditions.

Banks in Singapore are allowed to set their own interest rates on loans and deposits according to market and business conditions. Loans are usually made at prime plus a margin. The prime rate tends to be determined based on interbank lending rates, which reflect market rather than official rates. Because of low inflation and their strategy of buying increased market share of loans made in the region, Singapore banks offered the lowest prime interest rate in Asia—4.8 percent—as of September 1993.

In addition to providing loans, many large Singaporean banks also have subsidiaries that operate in the areas of merchant banking, insurance, property development, securities trading as members of the stock exchange, and the underwriting of government bond issues.

Retail Services

In the retail banking sector, banks compete by issuing credit cards to Singaporean citizens and by providing extensive access through ATMs. Electronic funds transfer became available in 1984 when a payments system for direct transfers to individuals and commercial accounts was set up, allowing customers to make bill payments and other transfers. Beginning in 1985 customers were able to use another point-of-sale system to debit their accounts for purchases in retail outlets. The system is operated by the four largest domestic banks and the post office savings system and has terminals in nearly 4,000 retail outlets, allowing Singapore to function on a largely cashless basis.

By early 1992 more than one million credit cards had been issued in Singapore, which has a total population of only three million. The MAS acted to regulate what it viewed as a dangerous proliferation of credit by requiring that cardholders have minimum annual incomes of S$30,000 (about US$18,700). Maximum credit lines on cards were reduced from the equivalent of three months' salary to two months' salary to rein in the explosive growth of credit.

Asian Currency Units

Asian Currency Units are the departments of commercial banks licensed to carry out foreign currency operations. They were developed to make Singapore the regional center of foreign exchange and at the same time to insulate the domestic financial system from the effects of international fluctuations. Despite its commitment to free convertibility, the government does not wish to see the Singapore dollar internationalized, ostensibly because of the country's small size. To advance Singapore's position as an international financial center while buffering its domestic economy from international market fluctuations, the government requires a strict separation between ACU and domestic banking operations.

Because the use of Singapore dollars is restricted, foreign businesspeople operating in Singapore usually must open accounts with ACUs as a prerequisite to procuring foreign currency-denominated loans. Deposits and loans made through ACUs must be made in foreign currencies such as the US dollar, the yen, or the deutsche mark, although the US dollar is the primary currency for ACU transactions.

Both residents and nonresidents can use Asian Currency Unit services, although ACUs are basically designed to make Singapore a convenient center for the regional operations of foreign businesses. Banks with ACU licenses accept time and demand deposits, provide open and advise letters of credit, and invest in foreign currency-denominated securities and discount bills. Authorized services include offshore loans to foreigners; offshore trade financing; interbank operations involving loans, deposits, bankers' acceptances, bills, and negotiable certificates of deposit; and securities underwriting, dealing, brokering, and investing. Other services include offshore funds management; gold and futures transactions; financial advisory services; provision of guarantees, performance bonds, and standby letters of credit; transfer agent services; offshore remittances; interest rate and currency swaps; foreign exchange trading; and offshore syndications and underwritings. Offshore syndications are generally tax-free provided that they are carried out by onshore licensed firms. To promote Singapore as an offshore treasury center for international operations, such activities carried out through ACUs are also granted tax concessions or outright exemptions.

The government offers tax breaks—a concessionary rate of 10 percent, down from the standard 30 percent—on income derived from ACU operations. This incentive, which applies to a wide range of financial activities, has allowed Singapore to overtake Hong Kong as the preeminent foreign currency trading center in Southeast Asia. Average daily turnover of foreign exchange was US$74 billion in 1991. International banks have taken advantage of ACUs to make Singapore a major source of international funding. About three-quarters of ACU activity occurs between banks, while the remainder involve large multinational corporations, foreign governments, local businesses engaged in activities requiring foreign exchange, and entrepreneurs in other parts of Asia.

Bank Requirements and Restrictions

The MAS sets stringent capital, reserve, and liquidity requirements for banks operating in Singapore. While this strictness ties up assets, banks meeting MAS standards tend to be financially sound by definition.

If Singapore is a bank's primary center of incorporation, the bank must have issued and paid-up capital of at least S$3 million (about US$1.9 million). Overseas banks must have minimum capital of S$6 million (about US$3.75 million), of which at least S$3 million must be held in Singapore in a form approved by the MAS. The required capital adequacy ratio of banks is 12 percent, 4 percent higher the international standard set by the Bank for International Settlements (BIS). All banks must hold a minimum cash reserve balance with the MAS; since 1975 this balance has been 6 percent of liabilities. Minimum liquid assets must be at least 18 percent of total liabilities; 10 percent of this amount must be held in government securities and 8 percent must be in cash, overnight repurchases of government securities to a maximum of 5 percent, or trade bills to a maximum of 4 percent. Domestic banks must maintain an internal capital reserve account funded out of annual earnings at rates of up to 50 percent of net income until the account equals 100 percent of capital. Even then, banks must continue to add 5 percent of annual profits to this account. Foreign banks are exempt from this specific capital reserve requirement as long as the MAS is satisfied with their aggregate international reserve positions.

A bank cannot engage in wholesale or retail nonbank commercial activities except to satisfy debts due or for purposes ancillary to the provision of regular bank services. It cannot own more than 40 percent of a nonbank institution, nor is it allowed to own real property in excess of 40 percent of its capital funds, excluding property held for the conduct of normal bank operations.

The MAS is also considering dropping the single-borrower limit from 30 percent to 25 percent of capital in order to reduce a bank's exposure to large borrowers. Borrowers are resisting this move because it would deprive them of leverage in negotiating terms and interest rates.

Merchant Banks

As of mid-1992 there were 77 merchant banks operating in Singapore. Authorized in 1970, merchant banks are usually either joint ventures between foreign banks and local investors or the wholly owned investment subsidiary arms of banks. They are a diverse lot, varying according to the purposes for which they were set up and the expertise of the founders. Some 76 merchant banks maintain MAS-approved ACU operations. They finance domestic activities with Singapore dollar loans, underwrite stock and bond issues, and deal in gold and foreign exchange. Merchant banks have also developed expertise in restructurings, takeovers, risk management, funds management, and other investment banking areas. They cannot offer retail banking services and can accept deposits only from banks and other financial institutions, not from individual or business clients. Most merchant banks are financed through the Asian dollar market. They tend to invest heavily in securities.

Banking Trends and Prospects

Because Singapore is concerned about being overbanked, no new full banking licenses are expected to be issued in the near future. But outside pressure to open up its financial sector could cause Singapore to relent. There is also speculation that several domestic full license banks could merge with the blessing of government officials. Such mergers would give the resulting bigger and better local banks added heft and free up existing licenses for reissue to placate outsiders. Provisions of the General Agreement on Tariffs and Trade (GATT) require Singapore to open aspects of its financial system to greater foreign participation. Although observers do not expect major moves in this direction in the next five years, most consider that within 10 years financial system reform will be an accomplished fact.

Nevertheless, Singapore's main concern is not with outsiders coming in but with services going out. Despite continued growth, the Singaporean economy has just about reached the saturation point for domestic banking services. Demand for loans is slowing, and deposits are growing, outstripping growth in lending by a 3 to 5 percent margin. Although bank profitability has remained good, it has largely come through margin expansion rather than from real growth in volume, and margins are more likely to contract than to expand or stay the same in the future. Of necessity, Singaporean banks will have to look outside to put their growing deposits to work. Outside operations already account for between 30 and 40 percent of large domestic bank income, although most overseas operations have been limited to trade financing rather than lending or other services.

Because of the stringent requirements of the MAS, Singaporean banks have a strong capital position. In this regard they are ahead of the ailing Japanese banks, the one-time giants in regional affairs, but they lack the expertise to manage as opposed to participate in overseas ventures. Singaporean banks have been rushing to do a variety of deals to gain experience, name recognition, and market share, even if many of the deals have resulted in little profit.

NONBANK FINANCIAL INSTITUTIONS

Specialized Financial Institutions

The government-controlled Development Bank of Singapore (DBS) is the country's largest bank in terms of total capitalization and assets. At the end of 1992 DBS had total capital of nearly US$2.5 billion and assets in excess of US$23 billion, making it the ninth largest bank in Asia. Its shares are listed on the Stock Exchange of Singapore (SES), and the government is its largest shareholder.

DBS is Singapore's largest commercial bank, but it is also the major investment and financial services policy arm of the government. It competes directly with local and international commercial banks. Through subsidiaries and associated companies, DBS offers services including venture capital investment, investment management, merchant banking, funds management, stock brokering, computer consulting and software services, credit card processing, factoring, local business financing, property management, and insurance services. It is also Singapore's main source of medium- and long-term loans.

DBS is heavily involved in retail banking, operating 42 local branches in Singapore and more than 130 ATMs. It offers savings accounts, Singapore dollar and ACU current accounts, safe-deposit boxes, remittance services, traveler's checks, and the buying and selling of gold. Customer finance services include housing loans, personal loans, overdrafts, banker's guarantees, tuition loans, and special accounts for senior citizens.

DBS has offices in Hong Kong, Jakarta, London, Los Angeles, New York, Osaka, Seoul, Taipei, and Tokyo. It maintains ties with more than 700 correspondent banks in 75 countries.

Postal Savings Bank

Authorized in 1972, the Postal Savings Bank, or POSBank, is the top domestic consumer banking and savings institution in Singapore. At the beginning of 1992 it had more than four million savings accounts—Singapore has a total population of about three million—with a balance of about S$15.5 billion (about US$9.6 billion). Accounts can only be held by individuals, not businesses. Deposits are guaranteed by the government, and interest earned on deposits is tax-free.

POSBank operates 151 branches and 540 ATMs—

almost half of the country's ATMs—throughout Singapore. It offers passbook savings, payroll deduction plans, and current accounts, as well as a variety of consumer retail banking services, such as 24-hour automated service centers. POSBank's widely used electronic transfer system allows depositors to pay household bills, insurance premiums, taxes, and school fees through regular authorized deductions from their accounts. Depositors can also make payments at department stores and retail outlets using their POSBank ATM card, dramatically reducing residents' need for cash.

Funds held in POSBank accounts provide the Singaporean government with investment capital for national development projects. The government can also use these funds to buy stocks and bonds to regulate the local securities market. Credit POSB Pte. Ltd., a subsidiary of POSBank, offers loans for home improvement, private housing purchases, public housing development (through the Housing and Urban Development Company), and commercial property development.

Finance Companies

At the beginning of 1993 there were 27 finance companies with 103 branches in Singapore. All are locally incorporated, and about half are owned by commercial banks. Finance companies offer small loans that are below the minimum limits of those usually considered by commercial banks. Products include installment loans for car purchases, small mortgages for housing purchases and home improvements, and financing for the purchase of domestic durable goods, such as major appliances. Some finance companies also invest in shipping and general commercial endeavors, as well as leasing, receivables financing, and factoring.

Finance companies are allowed to accept deposits and lend money, but they cannot operate demand deposit and transaction accounts, deal in foreign exchange or gold, or extend unsecured loans in amounts greater than S$5,000 (about US$3,100). They can accept fixed accounts and savings accounts denominated in Singapore dollars. The rates charged on loans are generally higher than bank lending rates. Finance companies must maintain a minimum cash reserve of 6 percent and a liquidity ratio of 10 percent.

Insurance Companies

There were 60 direct insurers, 31 professional reinsurers, and 45 captive insurers in Singapore at the beginning of 1992. Of the free insurers, 11 dealt only in life insurance, 69 were property and casualty insurers, and 11 operated in both lines. Captive insurance subsidiaries were introduced in 1982 to allow large companies or other organizations to self-insure. Domestic and foreign life insurance companies are permitted to operate in Singapore, and, to encourage the development of the sector, the government taxes their insurance earnings at a concessional 10 percent rate. Insurers offer a source of medium- to long-term financing for capital investments.

With the 1986 amendment to the insurance act, regulation of the industry was tightened and policyholders were given greater rights. However, insurers were also granted greater freedom to invest their funds and to make longer-term investments. Agents must register with the trade association operating in their area of specialization, meet certain training standards, be certified by examination, and be licensed to do business.

The life insurance business in Singapore has grown by more than 20 percent each year during the past decade, and prospects for continued growth remain good. Total life insurance premiums collected in 1992 stood at US$1.04 billion, against US$90 million in 1981, and total insurance assets snowballed from US$.30 million to US$3.9 billion over the same period. Offshore business accounts for about 45 percent of the total.

Compared to developed countries, Singapore is underinsured, although popular interest in insurance is growing steadily. Only 57 percent of Singaporeans had life insurance as of mid-1993. This figure is lower than in most Western countries, although it is higher than the insured rate in other Southeast Asian nations. Singapore had an estimated per capita life insurance coverage of about US$16,000 in 1991, compared with US$89,000 in Japan and US$39,000 in the United States.

ECICS Ltd., a subsidiary of the state-run Temasek Holdings, writes commercial and country risk insurance for exporters incorporated in and operating out of Singapore. It can also underwrite bonds and guarantees. Policies are either comprehensive, covering short-term repetitive sales, or specific, covering longer-term special sales, usually of capital equipment. ECICS writes guarantees including letters of assignment, unconditional guarantees to banks, preshipment credit guarantees, and buyer credit guarantees that indemnify the financing bank if the overseas buyer defaults. It also offers policies to cover other risks for which commercial insurance is not available to domestic firms with overseas operations.

Pension Funds

Singapore's Central Provident Fund (CPF), introduced in 1955 and reorganized in its current form in 1968, is a pension scheme that covers virtually all entities operating in the country. The CPF essentially maintains a monopoly over pensions in Singapore. Contributions are mandatory. Employee and employer contributions are calculated on a complex sliding scale, and total savings through the fund can amount to as much as 66 percent of wages, although the more usual combined rate is 38 percent, and there is a cap on contributions. This pension scheme has resulted in the highest savings rate in the world and an immense holding of funds for investment. The Central Provident Fund Board manages these funds, which are generally invested in government paper. The government uses the funds to implement its investment policies, primarily in construction of infrastructure and other capital projects.

Individuals are allowed to tap their pension fund accounts for cash to use in various ways. Up until recently they could use such funds primarily to buy government-built housing (more than 80 percent of Singaporeans now own or are buying their apartments). In an attempt to invigorate the stock market, the government has authorized the CPF Board to invest up to S$22 billion (about US$13.7 billion) in listed equities. CPF account holders are now allowed to invest 80 percent of their funds in excess of US$30,000 in approved financial instruments. About 1.1 million accounts are eligible for investment in the stock market, but individuals must specifically direct the CPF to issue out funds and direct the stock investment. As of late 1993 only about 50,000 have taken the plunge to date, although observers expect this number to jump as individuals become more accustomed to the idea of stock investing.

Other Specialized Financing and Guarantee Arrangements

Singapore has concluded investment guarantee agreements providing for reciprocal indemnification of loss due to war or other noncommercial risk with Canada, France, the Netherlands, Germany, Switzerland, the United Kingdom, the United States, Belgium-Luxembourg, the PRC, Taiwan, Sri Lanka, and the member nations of ASEAN. Payments, to be based on market value at time of loss, are to be made either directly or to the claimant's home government. The agreement with the United Kingdom is funded with assets equal to 105 percent of total eligible assets, providing an additional source of funds in the system. Other country agreements are unfunded.

The MAS operates a rediscounting scheme for preexport and export financing for firms incorporated in Singapore. Funds are advanced through the exporter's bank at a rate comparable to the prevailing money market rate for periods of up to three months. Banks can charge up to 1.5 percent on these funds.

ECICS Ltd., which also provides export insurance, extends medium- and long-term loans at preferential rates to qualified local small- and medium-sized firms for plant and equipment, working capital, and factoring. The related International Factors Ltd. can arrange to advance up to 90 percent of the value of a contract and indemnify qualified sellers against recourse.

The Singapore Economic Development Board (EDB) administers a variety of specific assistance schemes that offer incentives and financing to approved targeted investments. These schemes include product development, research and development, local enterprise financing and technical assistance, business development, and market development projects. The EDB also administers a S$2 billion (about US$1.25 billion) venture capital fund for locally incorporated firms in targeted high-technology areas to invest in and expand their operations. Eligibility requirements vary considerably, and many of these schemes—particularly the venture capital fund—are designed primarily for use by local firms but allow locally incorporated foreign firms with substantial Singaporean equity participation to apply. In any case, the availability of these funds frees up other funds for financing of foreign projects.

Underground Financial Operations

Because of the ready availability of financing and services, a lack of foreign exchange restrictions, and the government's rigid control over virtually every aspect of social and business life, essentially no underground economy or financial sector exists. The Drug Trafficking Act of 1992 made it a criminal offense to participate in money laundering of illegal proceeds. Singapore also participates in G-7 and EC efforts to track and control international flows of funds to hamper illicit transfers and money laundering.

FINANCIAL MARKETS

Authorities have placed a high priority on the swift and orderly development of various capital markets in Singapore and offer incentives to foreign firms that bring in state-of-the-art systems and up-to-date practices. Securities, gold, financial futures, government and corporate bonds, and foreign exchange markets are all available to foreign and domestic investors.

Equities Markets

The Stock Exchange of Singapore The Stock Exchange of Singapore (SES) is the principle securities market in the country. Founded in 1973, the SES grew out of the joint Singaporean-Malaysian market and until 1990 cross-listed shares with the Malaysian exchange. The uncoupling of the two exchanges caused the market capitalization of the SES to fall by about 30 percent, but it has been recovering strongly. At the beginning of 1993 188 companies, five more than a year earlier, with a total capitalization of nearly S$100 billion (about US$61 billion), up nearly 28 percent from a year earlier, were listed on the main board.

About half of that capitalization represents unrestricted shares that are available for purchase by foreigners. Singapore does not ban foreigners from owning shares of local firms except in specific areas of investment, although it does limit foreign institutional holdings to a maximum of 49 percent. However, the bylaws of specific Singaporean companies can and often do restrict foreign holdings in many listed firms. A number of listed companies also have a minimal float of publicly available shares and have dominant shareholders who exercise control, further restricting the attractiveness of shares listed on the market.

Listed companies are classified into industrial and commercial, financial, property, hotel, plantation, and mining sectors. Warrants and options are also traded, and bonds and securitized loans can also be listed, although these are rare. Since 1990 a 30 percent set-aside has been in effect for new listings by small firm applicants, a response to complaints that the SES does not fulfill its function as a capital source for new companies due to its rigid listing requirements. Trading, in lots of 100 shares, takes place in two sessions from 10:00 am to 12:30 pm and from 2:30 to 4:00 pm local time Monday through Friday.

The SES boasts one of the most technologically advanced trading systems in the world, the CLOB (Central Limit Order Book) system. Since introduction of the CLOB in 1990 all trading has been switched over to electronic book entry and paper transfers have been phased out, making the SES the world's first floorless stock exchange.

In 1991 total turnover amounted to 15.6 billion shares with a value of S$30.5 billion (about US$18.4billion). Daily average volume was 62 million shares worth S$121.7 million (about US$73.5 million). In 1992 total turnover was S$27.9 billion (about US$17 billion), down by 8.5 percent. The *Straits Times* Index is the principle gauge of stock performance on the SES. There is a also an SES All-Singapore Index, but it has a lesser following and is less influential. The Straits Times index ended 1993 at 2,425.68, about one point below its all-time high of 2,426.85 and up 60 percent during the year. It had stood at 1,491 at the end of 1992, up 29 percent from a year earlier. This dynamic increase is all the more remarkable because it occurred during a period when the traditional markets for Singaporean goods and services—the United States and Europe—were facing recession and stagnating purchasing power. As funds continue to pour into the country, many SES shares are expected to see continued growth and local analysts are predicting at least a 15 percent rise in 1994, although danger signs are appearing and many observers consider the market to be overbought.

In 1993 the government launched a concerted effort to increase the SES's capitalization. By relaxing rules on how the state pension fund can invest money, the government has freed S$22 billion (about US$13.7 billion) for investment in public equities. In an effort to privatize government holdings, the authorities have promised to list Singapore Telecom on the SES. Singapore Telecom would become the largest company listed on the SES, increasing the market's total capitalization by more than 15 percent. The government has also initiated an advertising campaign designed to lure small investors into the market.

While noting the meteoric rise of the SES, analysts point out that the market is not particularly cheap based on the earnings of the underlying stocks and compared with alternatives and interest rates. Foreign institutional investors also note that problems arise due to the relatively small size of the overall market and of individual companies and the lack of diversity among listed companies. Moreover, many of the largest and most popular issues, such as Singapore Airlines, have limits on foreign ownership. These limits effectively create two classes of stock: one that can be held by foreigners and that trades at a premium and one that can only be held by locals. In fall 1993 the premium on Singapore Airlines shares available to foreigners reached 93 percent, while the average premium on the restricted shares of other companies had risen to 43 percent from 26 percent at the beginning of the year.

Supervision, Regulation, and Operations The MAS monitors the SES and stockbrokers to assure that they comply with the Securities Industry Act and Regulations. The SES is self-governed by a nine-member committee, of which four are elected stockbroker members and five are appointed nonbroker members. The SES has 33 member firms, of which seven are foreign and 10 represent joint ventures. These member firms have a combined paid-up capital of S$451.8 million (about US$281.5 million). Member companies are required to have a minimum paid-up capital of S$10 million (about US$6.25 million). Members must also maintain adjusted net capital of at least S$8 million (about US$5 million) and operate within certain regulatory limits regarding leverage, and exposure to a single client or a single security. Members must also contribute to an exchange reserve fund and are expected to cover any defaults by member firms proportionally according to their capital.

Wholly owned subsidiaries of local banks, foreign brokerage houses, and foreign financial institutions have been admitted to the SES as member companies, although restrictions remained on foreign brokerage operations until 1992 when seven foreign companies were admitted to the exchange as international members. Such international members are permitted to deal freely in SES securities on behalf of nonresident firms and individuals but are only allowed to deal with residents in transactions valued at more than S$5 million (about US$3.1 million). Foreign securities companies operating in Singapore represent France, the United Kingdom, the United States, Japan, Thailand, Malaysia, Australia, and some Middle Eastern countries.

As in other areas of finance, Singaporean SES regulators get generally high marks for their expertise. However, sentiment is growing among the regulated that the regulators have become overzealous and overly conservative, interfering with the free conduct of business. In particular, market watchers complain that regulators have prevented new firms and new products from being traded on the SES, making it a less dynamic institution. There is an active lobby for broadening the SES by allowing short sales and the attendant stock lending, as well as lending using stocks as collateral and a broader options market in shares. A limited options market began in 1993 but remains in the experimental stage.

The cost of doing business on the SES is relatively high—higher than on developed exchanges in Japan and in the West—because of mandated commissions and fees charged for trades on the exchange. Nevertheless, costs are less than those levied on the competing Hong Kong Stock Exchange and on many other emerging market exchanges, and regulators are nudging brokers toward negotiated commissions. Commissions on stock transactions are set on a sliding scale ranging from a high of 1 percent on transactions of less than S$250,000 (about US$156,000) to a minimum of 0.5 percent on transactions exceeding S$1 million (about US$620,000). In addition to broker commissions, a 0.05 percent clearing fee with a maximum charge of S$100 (about US$60) and a contract stamp duty of S$1 per S$1,000 are levied. A transfer stamp duty of 0.2 percent is levied unless the trade is settled on a book entry basis.

Listing Requirements Locally incorporated companies seeking to list their shares on the SES must have a minimum issued and paid-in capital of S$15 million (about US$9.3 million); at least 25 percent of outstanding shares must be owned by more than 500 shareholders, and these large shareholders must own more than 20 percent of the shares outstanding if the company's capital is less than S$50 million (about US$31 million) and 10 percent if its capital is greater than S$100 million (about US$62 million). The company must have been in operation for at least five years and have earned profits in each of the last three years, and the company must be determined to have adequate working capital and proven strength and continuity of management. In practice the listing committee has usually required capital far above the stated minimum.

A foreign company seeking a listing on the SES must already be listed on its home stock exchange; have net tangible assets of at least S$50 million (about US$31 million); and have had net income of at least S$50 million (about US$31 million) in each of the last three years or net income of at least S$20 million (about US$12.5 million) for one of the last three years.

Firms seeking to list debt securities must have at least S$750,000 (about US$467,000) of debt outstanding in the class of securities for which SES quotation is desired. There must be a minimum of 100 holders of these securities, and the securities must conform to stipulations of issue and be structured as trust deeds.

Investment funds denominated in foreign currencies are allowed to list on the SES if they are determined to be managed by a reputable and established company that has been in operation for at least five years. They must also have minimum capital of US$30 million, and at least 20 percent of their shares must be held by the public.

Listed companies must file semiannual reports and a preliminary annual report within three months and a final annual report within six months after the close of the fiscal year. These statements must include certain items of mandated information for public release.

Secondary Stock Exchanges In addition to the SES, two other stock markets operate in Singapore. The Stock Exchange of Singapore Dealing and Automated Quotation System (SESDAQ) was established in 1987 to allow small- and medium-sized companies to raise funds through equities issues. In 1991 SESDAQ traded 167.9 million shares valued at S$157.5 million (about US$98 million). At the end of 1992 SESDAQ listed 25 companies—up from 14 in 1990—with a total capitalization of S$1 billion (about US$620 million). The more volatile SESDAQ offers investors the prospect of higher returns to offset the higher risks associated with the type of unseasoned small companies that list through it.

SESDAQ listing requirements are less stringent than those of the SES. No minimum capital is required, but at least 500,000 shares, or 15 percent, but not more than 50 percent, of the company's shares must be publicly owned. A moratorium exists on disposal of shares by major shareholders for one year following listing. Companies are not required to be profitable in order to be listed, but they normally must have been operating for at least three years. Domestic and foreign-owned companies incorporated in Singapore may list on SESDAQ. The costs of listing on SESDAQ are about half of what they are to list on the SES, and regulation is somewhat looser.

Introduced in 1990, CLOB International is basically a computerized clearing house that matches orders for securities traded on foreign markets, mostly in Malaysia. This market is an extension of the use of the automated system used to settle trades on the SES. At the end of 1992 CLOB International listed 126 issues—113 Malaysian stocks, 10 Hong Kong stocks, and 3 other foreign stocks—down from a total of 144 when it was introduced. Total capitalization for these companies stood at S$129.5 billion (about US$79 billion), down from S$158.3 billion (about US$96 billion) in 1991, due mainly to the increasing separateness of the Singaporean and Malaysian exchanges.

The SES has also attempted a linkup between the United States National Association of Securities Dealers (NASD) to give local investors more direct access US and European markets. So far cooperation has been limited to experimental transmissions of data on closing quotes and trading volume in 24 US stocks rather than as a mechanism for actual trading of issues in foreign markets.

International Funds Management Investment funds managed in Singapore in 1992 totaled US$22.9 billion, four times more than in 1989. The MAS reports that more than 80 percent of those funds came from foreign investors, and that half of those funds were invested in equities. At the end of 1992 113 such fund managers were operating in Singapore. Now that such funds are approved for CPF investment, domestic demand for such accounts is expected to pick up.

Debt and Money Markets

Government Securities and Bonds Since 1987 the Singapore government has auctioned government securities ranging in maturity from three months to seven years. Government paper includes 91-day, 182-day, and 364-day bills, as well as two-year, five-year, and seven-year bonds. The seven-year bonds were first issued in 1992. The government uses this market primarily to deal with excess liquidity from the CPF. Purchases help banks meet their reserve and liquidity requirements.

Prior to the establishment of this market, the government placed long-term debt with a few banks, insurance companies, wealthy individuals, and nonprofit organizations. Sales in the revamped government debt market are carried out by seven primary dealers who act as government securities clearing houses. Applications to purchase government securities at auction must be made through these seven dealers. There are 36 secondary dealers owned by banks, merchant banks, and stock brokerage firms and an additional 70 securities purchasing accounts held by banks for their own investment accounts.

Dealer prices for government securities are quoted in lots of S$3 million (about US$1.9 million) for new issues and S$1 million (about US$620,000) for existing issues. Finance companies, insurance firms, public corporations, and individuals may purchase notes and bonds in minimum amounts of S$1,000 (about US$620) and treasury bills in minimum amounts of S$10,000 (about US$6,200).

All trading is on a book entry basis. Average daily turnover has stabilized around S$200 million (about US$125 million) since 1990, with repurchase agreements adding another S$200 million to that daily figure. At the beginning of 1992 the total capitalization of the government securities market was S$11.2 billion (about US$6.75 billion), double the amount outstanding when the market was reorganized in 1987.

Corporate Debt The primary issuance and subsequent trading of corporate debt is limited, although a fairly active market for rights and warrants and securitized loans exists. A few straight and convertible corporate bonds have been issued, but debt financing has yet to take off in Singapore. In 1992 a total of S$428 million (about US$261 million) was raised in the private debt market. A total of 161 debt issues were listed on the SES in 1992, but liquidity is very limited, and most bonds are traded over the counter or through private placements. The corporate bond market has yet to reach critical mass. However, more international firms are establishing bond-trading capabilities in Singapore.

Money Markets The money market is a nexus of informal trading activity consisting of two closely linked spheres: the interbank and discount markets. Interbank lending is conducted exclusively in Singapore dollars, with foreign currency trading being kept strictly separate. Merchant banks are the largest and most frequent borrowers because they are not required to reserve for such borrowings. Most interbank lending is accomplished through money brokers rather than through direct bank-to-bank deals. The discount market began in the early 1970s with the rise of discount houses to mediate between the MAS and the banks to adjust the money in circulation. The actual discount houses were replaced by the government securities dealers when the government market was formalized in 1987. Instruments traded consist primarily of commercial bills, short-term government paper, and negotiable bank certificates of deposit.

Futures Markets

The futures market in Singapore began with the Gold Exchange of Singapore, which was established in 1978. In 1983 this exchange was reorganized to include trading in financial futures and renamed the Singapore International Monetary Exchange (SIMEX). The SIMEX was the first formal financial futures exchange in Asia, trading its first contract, a gold future, in 1984. Options trading began in 1987.

The SIMEX has an agreement with the Chicago Mercantile Exchange (CME), on which the SIMEX operations and procedures are modeled, allowing contracts opened on the CME to be closed on the SIMEX and vice versa, without additional transaction costs. The relationship between these two exchanges allows global futures trading to go on 24 hours a day. Eurodollar, deutsche mark, yen, and pound sterling futures contracts can be traded through this system.

At the beginning of 1992 SIMEX listed 11 futures and five options contracts. The futures contracts include gold, three interest rate contracts (the three-month Eurodollar, three-month Euroyen, and three-month Euromark); three foreign currency contracts (the US$-DM, US$-¥, and US$-£); three petroleum contracts (the gas oil, high sulfur fuel oil, and Dubai crude oil); and the Japanese Nikkei Average Stock Index (in 1986 the SIMEX became the first market in the world to offer a contract based on the Nikkei). The options contracts are based on the three-month Eurodollar interest rate, the three-month Euroyen interest rate, the US$-¥, the US$-DM contracts, and the Nikkei stock average. A new futures contract based on Hong Kong's Hang Seng Stock Index was scheduled to begin trading in late 1993.

Trading volume on the SIMEX has increased steadily since its inception. Average daily volume increased from 1,579 contracts in 1984 to 24,175 in 1991, doubling to 49,065 in 1992. A total of 5.7 million contracts were traded in 1990, with 6.1 million being traded in 1991, while 12.2 million contracts were traded in 1992. In 1992 the three-month Eurodollar contract accounted for 46 percent of all trading, while trading in Nikkei futures accounted for 27 percent. Trading set new records again in 1993.

At the end of 1992 the SIMEX had 40 corporate clearing members, who guarantee the market's obligations; 32 corporate nonclearing members; 12 commercial associates, who trade solely in oil contracts; and 479 individual members. The SIMEX is authorized to issue special permits to nonmembers who wish to trade in a specific contract or type of contract. It was granted a 10 year tax holiday in 1984, and members' trades with nonresidents are subject to the concessionary 10 percent rate.

Commodities Markets

Singapore has long been a center for trading in commodities such as cocoa, palm oil, tin, coffee, and pepper. Oddly enough, however, it has had no organized or officially regulated commodities market. Transactions are carried out by private traders through a telecommunications network. Most trading is on a relatively small retail spot-contract basis, although some market makers will write futures contracts as hedges for large buyers.

The exception to this largely informal system is the Rubber Association of Singapore (RAS), the world's leading rubber exchange. Reorganized as the RAS Commodities Exchange, it began formal trading in three rubber contracts in 1992. The exchange, which is run through a computerized market making system to allow for immediate information availability, has plans to begin offering formal contracts in other commodities in the future.

To encourage the development of this market, the government offers tax concessions to firms dealing in approved commodities that meet certain criteria—mainly annual turnover of S$200 million (about US$125 million). A similar arrangement is available to approved oil traders doing at least S$100 million per year (about US$62 million).

Asian Dollar Market and Asian Dollar Bond Market

Most banks operating in Singapore are permitted to deal in foreign currencies through Asian Currency Unit licenses. With the rapid development of infrastructure and businesses in Asian countries, ACUs serve to channel international funds, primarily from the United States and Europe, to Asian countries, and to a far lesser extent to channel investment funds from Asia to the developed world. The Asian dollar market, which was formally set up in 1968, is analogous to the more developed Eurodollar market. Largely due to the incentives offered by the government in its bid to gain this business, the Singaporean dollar market has surpassed similar operations in Tokyo and Hong Kong to become the primary market in Asia, and the fourth largest foreign exchange market in the world. Growing from a strictly interbank market, it has become a major source of investment funding in Asia. Banks usually extend credit for foreign currencies on a three- or six- month rollover basis with floating interest rates based on the Singapore interbank offering rate (SIBOR) or the internationally recognized London interbank offering rate (LIBOR).

Asian dollar market borrowers can also raise medium- and long-term capital by entering the Asian dollar bond market. This market consists of foreign currency-denominated bonds managed by an Asian institution and listed on the SES. The first of these bonds was issued in 1971 by DBS Bank. Most issues are in US dollars, but debt has also been issued in deutsche marks, Australian dollars, yen, European currency units (ECUs), and special drawing rights (SDRs). In addition to straight bond issues, the market is also used for floating-rate notes and convertible bond issues. The market was about US$390 million in 1970, growing to US$280 billion in 1988. In 1991 the gross size of the Asian dollar market was US$357.7 billion. Size and volume were down from the previous year due to the global recession.

Gold Markets

Singapore supports gold-trading activity ranging from the standardized to the highly informal. Futures trading was done on the Gold Exchange of Singapore until that institution was replaced in 1984 by the Singapore International Monetary Exchange (SIMEX), which continues to trade a gold contract. There are also two spot markets: the Loco-London market is operated for contracts in which delivery is taken in London, and the Loco-Singapore market, also known as the physical gold market, for contracts in which actual delivery is taken locally. Singapore is an active center for trade in gold for delivery rather than as a commodity speculation or hedge.

Precious metals have traditionally been in great demand as a store of wealth by both Singapore's Chinese and Indian populations. Singapore maintains no controls on the import or export of precious metals. In 1991 imports of gold amounted to about 263 metric tons, down from 300 metric tons in 1990. Prices for gold follow the rate set on the London Metals Exchange (LME), the foremost such market worldwide.

FURTHER READING

This discussion is provided as a basic guide to money, finances, financial institutions, and financial markets in Singapore. Those interested in current developments may wish to consult the *Far Eastern Economic Review* and *Asia Money* both of which frequently cover economic and financial developments in Singapore.

Currency & Foreign Exchange

INTERNATIONAL PAYMENT INSTRUMENTS

Singapore is an international trade and transshipment center, and as such its financial and business institutions are well versed in the use of internationally recognized payment instruments. Letters of credit (L/Cs) and sight drafts are the preferred means of handling transactions. L/C payments are normally for periods of 90, 120, and 180 days. Singapore does not usually authorize countertrade arrangements; however, the Singapore Trade Development Board (STDB) can make exceptions when a trading partner lacks foreign exchange to carry out a proposed deal.

CURRENCY

The currency in use is the Singapore dollar (S$), which replaced the Straits dollar in 1967. Singapore issues coins in denominations of S$0.01, S$0.05, S$0.10, S$0.20, S$0.50, and S$1, and bills of S$1, S$5, S$10, S$50, S$100, S$1,000, and S$10,000. As of the end of 1993 S$1.605 was equal to US$1. The S$50 bill was worth about US$31. The S$10,000 note (about US$6,230) is not commonly used and few are in circulation. The government also issues commemorative and special numismatic coins and notes that are legal tender but not found in general circulation. The government reserves the right to restrict the use of coins for transactions of more than a few dollars: S$0.50 coins cannot be used for transactions of more than S$10, and smaller coins cannot be used for transactions of more than S$2.

Money from Brunei is also in use and accepted as interchangeable at face value with Singapore dollars according to treaty terms between the two countries. Malaysian notes are no longer accepted. The US dollar is commonly used, and businesspeople should verify whether the unit in question is the S$ or the US$, because both use the "$" sign and are often referred to interchangeably.

As with everything else, Singapore takes its money seriously: it is an offense to destroy, mutilate, or deface coins or currency, and representations used in advertising, art, or other media are strictly controlled.

Currency issue is the responsibility of the Board of Commissioners of Currency of Singapore (BCCS), which oversees financial service delivery, issues currency, monitors currency demand, and manages the currency reserve fund. All currency is fully backed by gold and foreign hard currency reserves. The International Monetary Fund (IMF) rates the Singapore dollar as one of the world's strongest currencies.

REMITTANCE AND EXCHANGE CONTROLS

Singapore removed all exchange controls in 1978, allowing residents and nonresidents, whether individuals or businesses, to move funds into or out of the country at will. There are no restrictions on the movement of investment capital into or out of Singapore, nor are there limits on repatriation of profits. No formalities, approvals, or registrations of flows of funds are required.

Government policy does limit the use of the Singapore dollar as the unit for international transactions, placing limits on financial institution lending in Singapore dollars to nonresidents inside the country and to residents for use outside the country, but it does allow free conversion. Bank accounts can be denominated in local currency or in virtually any other currency. Restrictions do exist on the use of the South African rand, but these largely political rules are likely to be dropped in the future.

Because there are no exchange controls, there is no black market, although exchange rates do vary depending on the outlet where the transaction occurs. Businesses will almost always operate through a bank and may be able to negotiate wholesale rates, although the reduction in the spread will be marginal. For smaller transactions, the airport offers good rates and banks remain the primary outlet. Rates vary from bank to bank, and some negate favorable rates with minimum or sliding scale service fees. Traveler's checks usually get a slightly better rate than cash.

Money changers, mostly Indians located around Change Alley in the financial district, usually offer better rates than do the banks. They deal mainly in US dollars, but will do exchanges in a variety of currencies, including some restricted currencies (Indian rupees are common). They take traveler's checks as well as cash. In addition to favorable rates, money changers can also offer entertainment: visitors report that it is worth the price of admission (nil) to see them do rapid, complex cross-rate calculations, often to several decimal places and in their heads.

Hotels change money at relatively unfavorable rates. Most establishments, especially in the main tourist areas, accept foreign currency and traveler's checks at a rate that amounts to a slight discount. They accept international credit cards with varying degrees of gracefulness and usually offer a 3 to 5 percent discount for cash.

Foreigners can often get money from home by using automatic teller machines (ATMs). Some of these will even accept foreign ATM or debit cards, and the rest will usually issue a cash advance on a recognized credit card, as will banks. You will need to know the systems that your bank belongs to and your personal identification number (PIN). The ease of using these procedures is offset by the costs—interest and transfer fees—although the institutions backing the cards usually offer decent exchange rates on such transactions. Money transfers are available from most large banks; branch-to-branch transfers within the same bank are quickest and easiest, but transfers can also be arranged via correspondent banks.

FOREIGN EXCHANGE OPERATIONS

The business of trading money is huge in Singapore. Singapore has surpassed Hong Kong and Tokyo as the premier foreign exchange (forex) market in the region, especially for Asian dollar transactions. It is also the fourth largest foreign exchange market worldwide. In 1991 average daily turnover was US$74 billion, down from US$83 billion in 1990 due to the global recession but still higher than the US$62 billion posted in 1989. Singapore's geographic location between North America and Europe allows traders to use it as a link in order to be able to trade 24 hours a day.

Activity is centered in the major currencies—the US dollar, the Japanese yen, the German deutsche mark, and the English pound sterling—but Singapore traders also make markets in lesser-volume currencies such as the Swiss franc and the French franc, and trading in other Asian currencies is growing rapidly. Singapore is also busy developing and using various forex hedging and derivative instruments and facilitates active speculative and arbitrage trading.

Trading in the Asian dollar market, which began in 1968, supports a regional and international money and capital market. Financial institutions must set up separate Asian Currency Units to trade in this market, which uses foreign exchange as its sole medium. ACUs are allowed to provide foreign currency-denominated banking and investment services as well as forex trading for their own and client accounts. In 1991 gross assets in the ACU market were US$357.7 billion, down by 8 percent from 1990 due to recessionary pressures.

Singapore's Foreign Exchange Rates - Year End Actual
Singapore Dollar (S$) to United States Dollar (US$)

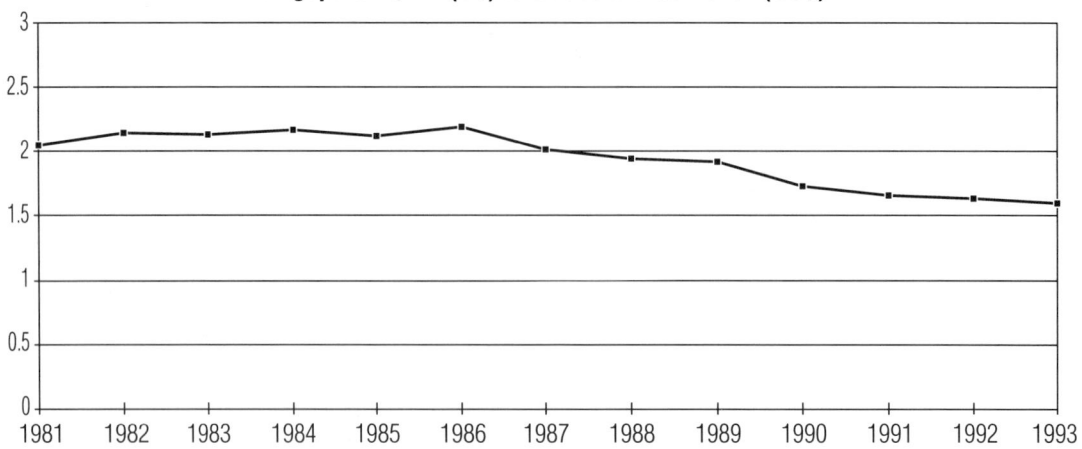

January 1, 1994 US$1 = $S1.60

RATES OF EXCHANGE

Since 1973 the Singapore dollar has floated against a trade-weighted basket of currencies of the country's major trading partners. Between 1981 and late 1993 the Singapore dollar has traded as high as 1.594 and as low as 2.258 to the US dollar, with the high point coming in November 1993. The Singapore dollar has been appreciating steadily against the US dollar since reaching its low point due to recession in 1985. During 1993 it rose by 1.75 percent against the US dollar—from 1.623 in January to 1.594 in November. This general trend is expected to continue.

FOREIGN RESERVES

At the end of 1989 foreign reserves stood at US$20.3 billion. In March 1993 reserves were US$41.4 billion, having more than doubled in less than four years and continuing the 20 percent average annual growth that they have maintained since the late 1980s. Most observers suggest that despite a merchandise trade deficit, Singapore's foreign reserve position will continue to grow because of the country's burgeoning surplus in intangibles trade, derived from its service exports and revenues from its booming financial services sector.

Singapore has essentially no medium- or long-term public debt because it funds its activities from current budget surpluses and internal savings. Its short-term debt consists of government money market paper designed to assure liquidity for the financial system, manage the money supply, and control interest rates.

FURTHER READING

This discussion is provided as a basic guide to money, finances, financial institutions, and financial markets in Singapore. Those interested in current developments may wish to consult the *Far Eastern Economic Review* and *Asia Money* both of which frequently cover economic and financial developments in Singapore.

Exchange Rates—S$/US$

	Jan	Feb	Mar	Apr	May	Jun	Jul	Aug	Sep	Oct	Nov	Dec
1981	2.081	2.087	2.099	2.110	2.149	2.144	2.158	2.170	2.144	2.098	2.061	2.053
1982	2.061	2.109	2.121	2.133	2.088	2.138	2.146	2.159	2.167	2.198	2.212	2.152
1983	2.077	2.076	2.085	2.101	2.092	2.120	2.129	2.142	2.142	2.135	2.133	2.132
1984	2.131	2.128	2.089	2.085	2.101	2.112	2.147	2.147	2.163	2.167	2.155	2.173
1985	2.201	2.256	2.258	2.220	2.223	2.229	2.211	2.219	2.227	2.139	2.108	2.121
1986	2.129	2.140	2.165	2.188	2.216	2.223	2.186	2.160	2.168	2.178	2.192	2.190
1987	2.151	2.141	2.142	2.134	2.120	2.118	2.118	2.182	2.092	2.089	2.044	2.013
1988	2.026	2.018	2.013	2.004	2.011	2.028	2.046	2.042	2.041	2.020	1.962	1.944
1989	1.940	1.928	1.941	1.950	1.957	1.957	1.959	1.960	1.977	1.962	1.959	1.918
1990	1.887	1.864	1.878	1.878	1.859	1.847	1.819	1.790	1.767	1.726	1.710	1.727
1991	1.745	1.718	1.759	1.769	1.769	1.778	1.755	1.727	1.700	1.694	1.671	1.656
1992	1.634	1.636	1.660	1.657	1.641	1.624	1.614	1.608	1.599	1.608	1.634	1.639
1993	1.652	1.646	1.644	1.623	1.614	1.617	1.623	1.610	1.597	1.570	1.602	1.605

Source: US Federal Reserve System

International Payments

International transactions add an additional layer of risk for buyers and sellers that are familiar only with doing business domestically. Currency regulations, foreign exchange risk, political, economic, or social upheaval in the buyer's or seller's country, and different business customs may all contribute to uncertainty. Ultimately, however, the seller wants to make sure he gets paid and the buyer wants to get what he pays for. Choosing the right payment method can be the key to the transaction's feasibility and profitability.

There are four common methods of international payment, each providing the buyer and the seller with varying degrees of protection for getting paid and for guaranteeing shipment. Ranked in order of most security for the supplier to most security for the buyer, they are: Cash in Advance, Documentary Letters of Credit (L/C), Documentary Collections (D/P and D/A Terms), and Open Account (O/A).

Cash in Advance

In cash in advance terms the buyer simply prepays the supplier prior to shipment of goods. Cash in advance terms are generally used in new relationships where transactions are small and the buyer has no choice but to pre-pay. These terms give maximum security to the seller but leave the buyer at great risk. Since the buyer has no guarantee that the goods will be shipped, he must have a high degree of trust in the seller's ability and willingness to follow through. The buyer must also consider the economic, political and social stability of the seller's country, as these conditions may make it impossible for the seller to ship as promised.

Documentary Letters of Credit

A letter of credit is a bank's promise to pay a supplier on behalf of the buyer so long as the supplier meets the terms and conditions stated in the credit. Documents are the key issue in letter of credit transactions. Banks act as intermediaries, and have nothing to do with the goods themselves.

Letters of credit are the most common form of international payment because they provide a high degree of protection for both the seller and the buyer. The buyer specifies the documentation that he requires from the seller before the bank is to make payment, and the seller is given assurance that he will receive payment after shipping his goods so long as the documentation is in order.

Documentary Collections

A documentary collection is like an international cash on delivery (COD), but with a few twists. The exporter ships goods to the importer, but forwards shipping documents (including title document) to his bank for transmission to the buyer's bank. The buyer's bank is instructed not to transfer the documents to the buyer until payment is made (Documents against Payment, D/P) or upon guarantee that payment will be made within a specified period of time (Documents against Acceptance, D/A). Once the buyer has the documentation for the shipment he is able to take possession of the goods.

D/P and D/A terms are commonly used in ongoing business relationships and provide a measure of protection for both parties. The buyer and seller, however, both assume risk in the transaction, ranging from refusal on the part of the buyer to pay for the documents, to the seller's shipping of unacceptable goods.

Open Account

This is an agreement by the buyer to pay for goods within a designated time after their shipment, usually in 30, 60, or 90 days. Open account terms give maximum security to the buyer and greatest risk to the seller. This form of payment is used only when the seller has significant trust and faith in the buyer's ability and willingness to pay once the goods have been shipped. The seller must also consider the economic, political and social stability of the buyer's country as these conditions may make it impossible for the buyer to pay as promised.

DOCUMENTARY COLLECTIONS (D/P, D/A)

Documentary collections focus on the transfer of documents such as bills of lading for the transfer of ownership of goods rather than on the goods themselves. They are easier to use than letters of credit and bank service charges are generally lower.

This form of payment is excellent for buyers who wish to purchase goods without risking prepayment and without having to go through the more cumbersome letter of credit process.

Documentary collection procedures, however, entail risk for the supplier, because payment is not made until after goods are shipped. In addition, the supplier assumes the risk while the goods are in transit and storage until payment/acceptance take place. Banks involved in the transaction do not guarantee payments. A supplier should therefore only agree to a documentary collection procedure if the transaction includes the following characteristics:

- The supplier does not doubt the buyer's ability and willingness to pay for the goods;
- The buyer's country is politically, economically, and legally stable;
- There are no foreign exchange restrictions in the buyer's home country, or unless all necessary licenses for foreign exchange have already been obtained; and
- The goods to be shipped are easily marketable.

Types of Collections

The three types of documentary collections are:
1. Documents against Payment (D/P)
2. Documents against Acceptance (D/A)
3. Collection with Acceptance (Acceptance D/P)

All of these collection procedures follow the same general step-by-step process of exchanging documents proving title to goods for either cash or a contracted promise to pay at a later time. The documents are transferred from the supplier (called the remitter) to the buyer (called the drawee) via intermediary banks. When the supplier ships goods, he presents documents such as the bill of lading, invoices, and certificate of origin to his representative bank (the remitting bank), which then forwards them to the buyer's bank (the collecting bank). According to the type of documentary collection, the buyer may then do one of the following:

- With Documents against Payment (D/P), the buyer may only receive the title and other documents after paying for the goods;
- With Documents against Acceptance (D/A), the buyer may receive the title and other documents after signing a time draft promising to pay at a later date; or
- With Acceptance Documents against Payment, the buyer signs a time draft for payment at a latter date. However, he may only obtain the documents after the time draft reaches maturity. In essence, the goods remain in escrow until payment has been made.

In all cases the buyer may take possession of the goods only by presenting the bill of lading to customs or shipping authorities.

In the event that the prospective buyer cannot or will not pay for the goods shipped, they remain in legal possession of the supplier, but he may be stuck with them in an unfavorable situation. Also, the supplier has no legal basis to file claim against the prospective buyer. At this point the supplier may:

- Have the goods returned and sell them on his domestic market; or
- Sell the goods to another buyer near where the goods are currently held.

If the supplier takes no action the goods will be auctioned or otherwise disposed of by customs.

Documentary Collection Procedure

The documentary collection process has been standardized by a set of rules published by the International Chamber of Commerce (ICC). These rules are called the Uniform Rules for Collections (URC) and are contained in ICC Publication No. 322. (See the last page of this section for ICC addresses and list of available publications.)

The following is the basic set of steps used in a documentary collection. Refer to the illustration on the following page for a graphic representation of the procedure.

(1) The seller (remitter, exporter) ships the goods.
(2) and (3) The seller forwards the agreed upon documents to his bank, the remitting bank, which in turn forwards them to the collecting bank (buyer's bank).
(4) The collecting bank notifies the buyer (drawee, importer) and informs him of the conditions under which he can take possession of the documents.
(5) To take possession of the documents, the buyer makes payment or signs a time deposit.
(6) and (7) If the buyer draws the documents against payment, the collecting bank transfers payment to the remitting bank for credit to the supplier's account. If the buyer draws the documents against acceptance, the collecting bank sends the acceptance to the remitting bank or retains it up to maturity. On maturity, the collecting bank collects the bill and transfers it to the remitting bank for payment to the supplier.

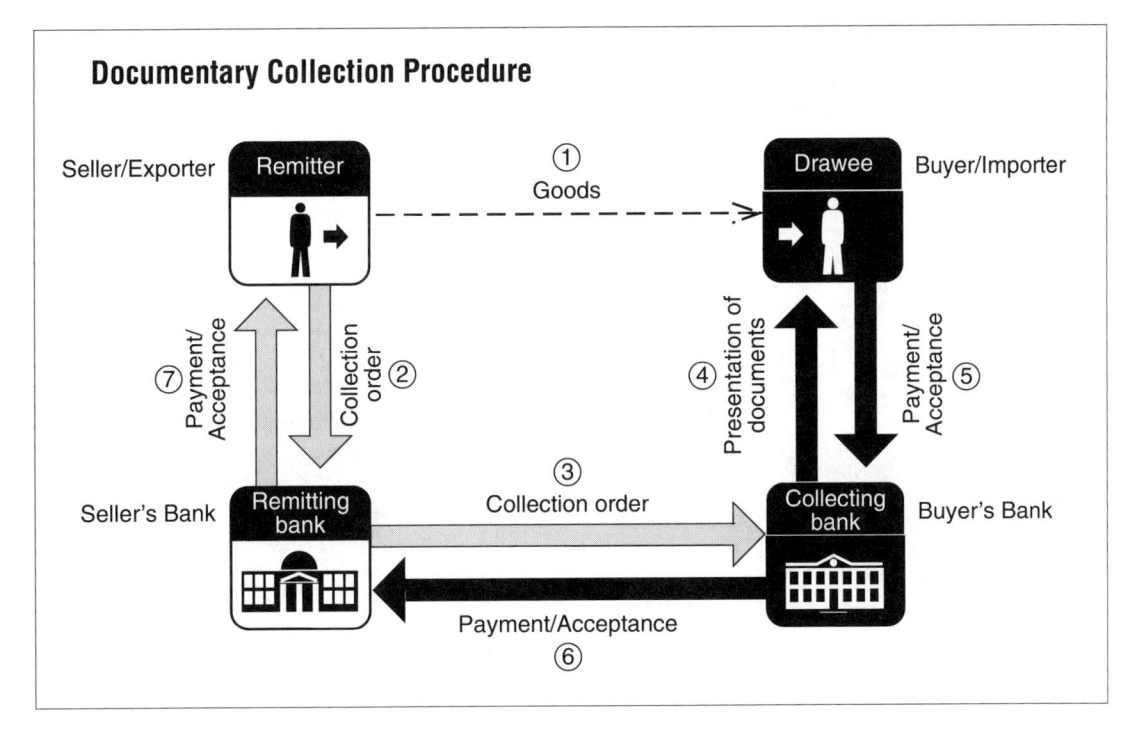

Documentary Collection Procedure

Seller/Exporter — Remitter
① Goods →
Drawee — Buyer/Importer

② Collection order
⑦ Payment/Acceptance
④ Presentation of documents
⑤ Payment/Acceptance

Seller's Bank — Remitting bank
③ Collection order →
Collecting bank — Buyer's Bank
⑥ Payment/Acceptance

TIPS FOR BUYERS

1. The buyer is generally in a secure position because he does not assume ownership or responsibility for goods until he has paid for the documents or signed a time draft.
2. The buyer may not sample or inspect the goods before accepting and paying for the documents without authorization from the seller. However, the buyer may in advance specify a certificate of inspection as part of the required documentation package.
3. As a special favor, the collecting bank can allow the buyer to inspect the documents before payment. The collecting bank assumes responsibility for the documents until their redemption.
4. In the above case, the buyer should immediately return the entire set of documents to the collecting bank if he cannot meet the agreed payment procedure.
5. The buyer assumes no liability for goods if he refuses to take possession of the documents.
6. Partial payment in exchange for the documents is not allowed unless authorized in the collection order.
7. With documents against acceptance, the buyer may receive the goods and resell them for profit before the time draft matures, thereby using the proceeds of the sale to pay for the goods. The buyer remains responsible for payment, however, even if he cannot sell the goods.

TIPS FOR SUPPLIERS

1. The supplier assumes risk because he ships goods before receiving payment. The buyer is under no legal obligation to pay for or to accept the goods.
2. Before agreeing to a documentary collection, the supplier should check on the buyer's creditworthiness and business reputation.
3. The supplier should make sure the buyer's country is politically and financially stable.
4. The supplier should find out what documents are required for customs clearance in the buyer's country. Consulates may be of help.
5. The supplier should assemble the documents carefully and make sure they are in the required form and endorsed as necessary.
6. As a rule, the remitting bank will not review the documents before forwarding them to the collecting bank. This is the responsibility of the seller.
7. The goods travel and are stored at the risk of the supplier until payment or acceptance.
8. If the buyer refuses acceptance or payment for the documents, the supplier retains ownership. The supplier may have the goods shipped back or try to sell them to another buyer in the region.
9. If the buyer takes no action, customs authorities may seize the goods and auction them off or otherwise dispose of them.
10. Because goods may be refused, the supplier should only ship goods which are readily marketable to other sources.

LETTERS OF CREDIT (L/C)

A letter of credit is a document issued by a bank stating its commitment to pay someone (supplier/exporter/seller) a stated amount of money on behalf of a buyer (importer) so long as the seller meets very specific terms and conditions. Letters of credit are often called documentary letters of credit because the banks handling the transaction deal in documents as opposed to goods. Letters of credit are the most common method of making international payments, because the risks of the transaction are shared by both the buyer and the supplier.

STEPS IN USING AN L/C

The letter of credit process has been standardized by a set of rules published by the International Chamber of Commerce (ICC). These rules are called the Uniform Customs and Practice for Documentary Credits (UCP) and are contained in ICC Publication No. 400. (See the last page of this section for ICC addresses and list of available publications.) The following is the basic set of steps used in a letter of credit transaction. Specific letter of credit transactions follow somewhat different procedures.

- After the buyer and supplier agree on the terms of a sale, the buyer arranges for his bank to open a letter of credit in favor of the supplier.
- The buyer's bank (the issuing bank), prepares the letter of credit, including all of the buyer's instructions to the seller concerning shipment and required documentation.
- The buyer's bank sends the letter of credit to a correspondent bank (the advising bank), in the seller's country. The seller may request that a particular bank be the advising bank, or the domestic bank may select one of its correspondent banks in the seller's country.
- The advising bank forwards the letter of credit to the supplier.
- The supplier carefully reviews all conditions the buyer has stipulated in the letter of credit. If the supplier cannot comply with one or more of the provisions he immediately notifies the buyer and asks that an amendment be made to the letter of credit.
- After final terms are agreed upon, the supplier prepares the goods and arranges for their shipment to the appropriate port.
- The supplier ships the goods, and obtains a bill of lading and other documents as required by the buyer in the letter of credit. Some of these documents may need to be obtained prior to shipment.
- The supplier presents the required documents to the advising bank, indicating full compliance with the terms of the letter of credit. Required documents usually include a bill of lading, commercial invoice, certificate of origin, and possibly an inspection certificate if required by the buyer.
- The advising bank reviews the documents. If they are in order, the documents are forwarded to the issuing bank. If it is an irrevocable, confirmed letter of credit the supplier is guaranteed payment and may be paid immediately by the advising bank.
- Once the issuing bank receives the documents it notifies the buyer who then reviews the documents himself. If the documents are in order the buyer signs off, taking possession of the documents, including the bill of lading, which he uses to take possession of the shipment.
- The issuing bank initiates payment to the advising bank, which pays the supplier.

The transfer of funds from the buyer to his bank, from the buyer's bank to the supplier's bank, and from the supplier's bank to the supplier may be handled at the same time as the exchange of documents, or under terms agreed upon in advance.

Parties to a Letter of Credit Transaction

Buyer/Importer	Buyer	Issuing bank	Buyer's bank
Seller/Supplier/Exporter	Seller	Advising bank	Seller's bank

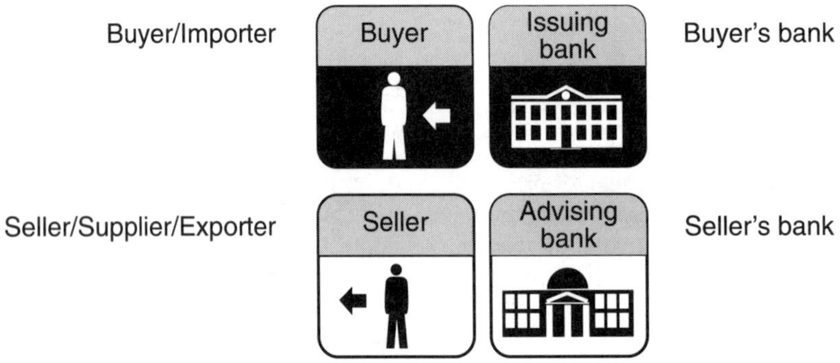

INTERNATIONAL PAYMENTS

Issuance

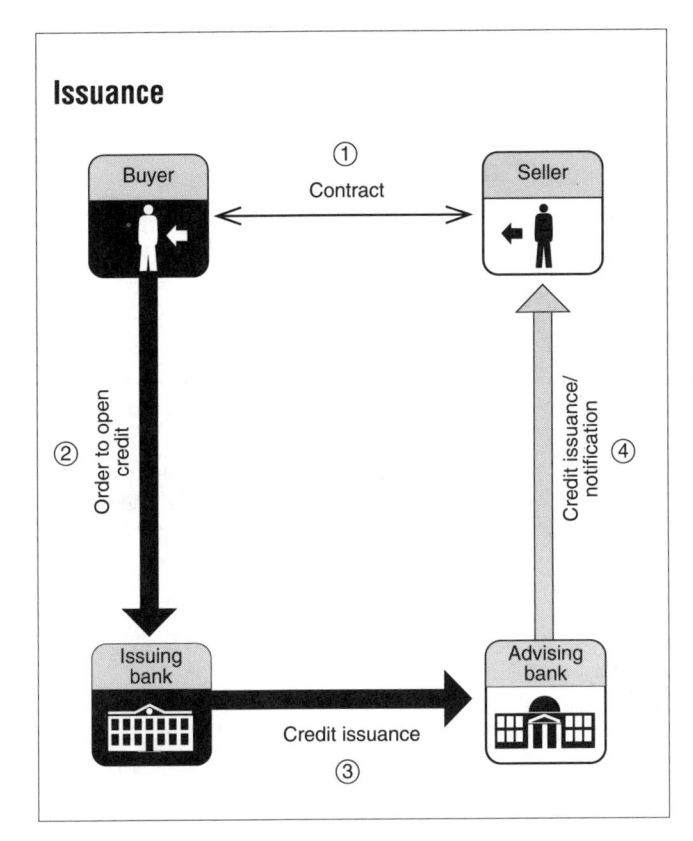

Issuance of a Letter of Credit

① Buyer and seller agree on purchase contract.
② Buyer applies for and opens a letter of credit with issuing ("buyer's") bank.
③ Issuing bank issues the letter of credit, forwarding it to advising ("seller's") bank.
④ Advising bank notifies seller of letter of credit.

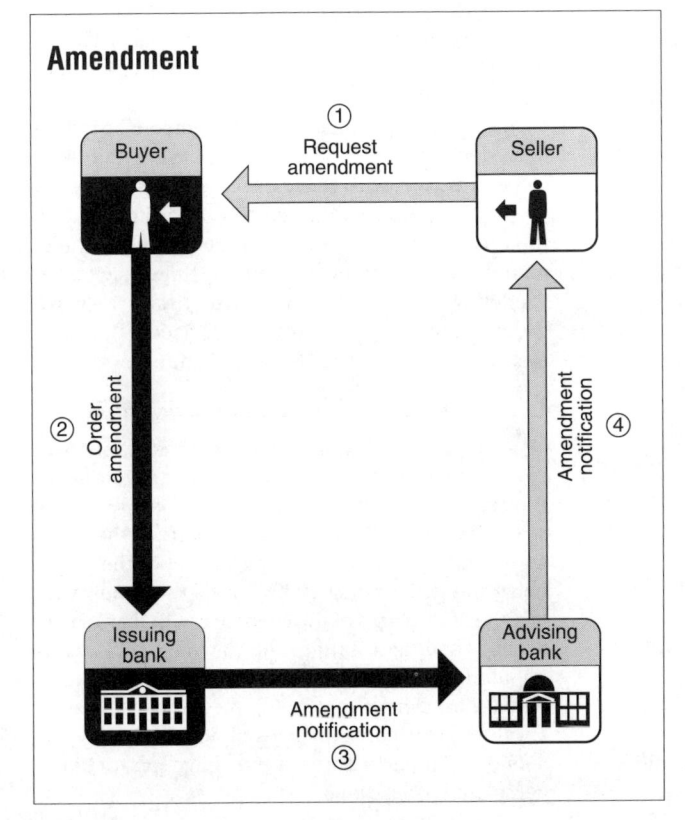

Amendment

Amendment of a Letter of Credit

① Seller requests (of the buyer) a modification (amendment) of the terms of the letter of credit. Once the terms are agreed upon:
② Buyer issues order to issuing ("buyer's") bank to make an amendment to the terms of the letter of credit.
③ Issuing bank notifies advising ("seller's") bank of amendment.
④ Advising bank notifies seller of amendment.

Utilization

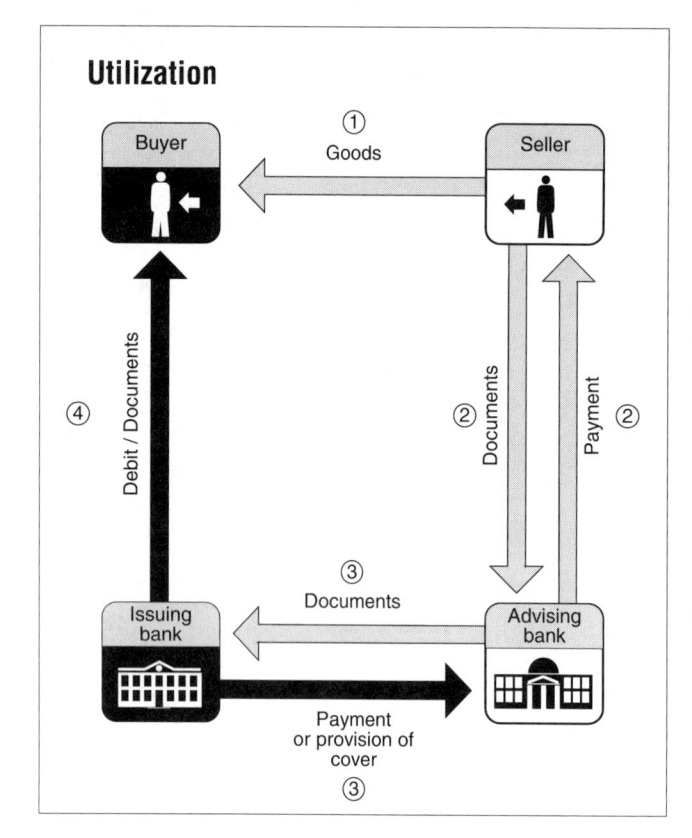

Utilization of a Letter of Credit
(irrevocable, confirmed credit)

① Seller ships goods to buyer.
② Seller forwards all documents (as stipulated in the letter of credit) to advising bank. Once documents are reviewed and accepted, advising bank pays seller for the goods.
③ Advising bank forwards documents to issuing bank. Once documents are reviewed and accepted, issuing bank pays advising bank.
④ Issuing bank forwards documents to buyer. Seller's letter of credit, or account, is debited.

COMMON PROBLEMS IN LETTER OF CREDIT TRANSACTIONS

Most problems with letter of credit transactions have to do with the ability of the supplier to fulfill obligations the buyer establishes in the original letter of credit. The supplier may find the terms of the credit difficult or impossible to fulfill and either tries to do so and fails, or asks the buyer for an amendment to the letter of credit. Observers note that over half of all letters of credit involving parties in East Asia are amended or renegotiated entirely. Since most letters of credit are irrevocable, amendments to the original letter of credit can only be made after further negotiations and agreements between the buyer and the supplier. Suppliers may have one or more of the following problems:

- Shipment schedule stipulated in the letter of credit cannot be met.
- Stipulations concerning freight cost are deemed unacceptable.
- Price is insufficient due to changes in exchange rates.
- Quantity of product ordered is not the expected amount.
- Description of product to be shipped is either insufficient or too detailed.
- Documents stipulated in the letter of credit are difficult or impossible to obtain.

Even when suppliers accept the terms of a letter of credit, problems often arise at the stage where banks review, or negotiate, the documents provided by the supplier against the requirements specified in the letter of credit. If the documents are found not to be in accord with those specified in the letter of credit, the bank's commitment to pay is invalidated. In some cases the supplier can correct the documents and present them within the time specified in the letter of credit. Or, the advising bank may ask the issuing bank for authorization to accept the documents despite the discrepancies found.

Limits on Legal Obligations of Banks

It is important to note once again that banks *deal in documents and not in goods*. Only the wording of the credit is binding on the bank. Banks are not responsible for verifying the authenticity of the documents, nor for the quality or quantity of the goods being shipped. As long as the *documents* comply with the specified terms of the letter of credit, banks may accept them and initiate the payment process as stipulated in the letter of credit. Banks are free from liability for delays in sending messages caused by another party, consequences of Acts of God, or the acts of third parties whom they have instructed to carry out transactions.

TYPES OF LETTERS OF CREDIT

Basic Letters of Credit

There are two basic forms of letters of credit: the Revocable Credit and the Irrevocable Credit. There are also two types of irrevocable credit: the Irrevocable Credit not Confirmed, and the Irrevocable Confirmed Credit. Each type of credit has advantages and disadvantages for the buyer and for the seller. Also note that the more the banks assume risk by guaranteeing payment, the more they will charge for providing the service.

1. Revocable credit This credit can be changed or canceled by the buyer without prior notice to the supplier. Because it offers little security to the seller revocable credits are generally unacceptable to the seller and are rarely used.

2. Irrevocable credit The irrevocable credit is one which the issuing bank commits itself irrevocably to honor, provided the beneficiary complies with all stipulated conditions. This credit cannot be changed or canceled without the consent of both the buyer and the seller. As a result, this type of credit is the most widely used in international trade. Irrevocable credits are more expensive because of the issuing bank's added liability in guaranteeing the credit. There are two types of irrevocable credits:

a. The Irrevocable Credit not Confirmed by the Advising Bank (Unconfirmed Credit) This means that the buyer's bank which issues the credit is the only party responsible for payment to the supplier, and the supplier's bank is obliged to pay the supplier only after receiving payment from the buyer's bank. The supplier's bank merely acts on behalf of the issuing bank and therefore incurs no risk.

b. The Irrevocable, Confirmed Credit In a confirmed credit, the advising bank adds its guarantee to pay the supplier to that of the issuing bank. If the issuing bank fails to make payment the advising bank will pay. If a supplier is unfamiliar with the buyer's bank which issues the letter of credit, he may insist on an irrevocable confirmed credit. These credits may be used when trade is conducted in a high risk area where there are fears of outbreak of war or social, political, or financial instability. Confirmed credits may also be used by the supplier to enlist the aid of a local bank to extend financing to enable him to fill the order. A confirmed credit costs more because the bank has added liability.

Special Letters of Credit

There are numerous special letters of credit designed to meet specific needs of buyers, suppliers, and intermediaries. Special letters of credit usually involve increased participation by banks, so financing and service charges are higher than those for basic letters of credit. The following is a brief description of some special letters of credit.

1. Standby Letter of Credit This credit is primarily a payment or performance guarantee. It is used primarily in the United States because US banks are prevented by law from giving certain guarantees. Standby credits are often called non-performing letters of credit because they are only used as a backup payment method if the collection on a primary payment method is past due.

Standby letters of credit can be used, for example, to guarantee the following types of payment and performance:

- repayment of loans;
- fulfillment by subcontractors;
- securing the payment for goods delivered by third parties.

The beneficiary to a standby letter of credit can draw from it on demand, so the buyer assumes added risk.

2. Revolving Letter of Credit This credit is a commitment on the part of the issuing bank to restore the credit to the original amount after it has been used or drawn down. The number of times it can be utilized and the period of validity is stated in the credit. The credit can be cumulative or noncumulative. Cumulative means that unutilized sums can be added to the next installment whereas noncumulative means that partial amounts not utilized in time expire.

3. Deferred Payment Letter of Credit In this credit the buyer takes delivery of the shipped goods by accepting the documents and agreeing to pay his bank after a fixed period of time. This credit gives the buyer a grace period, and ensures that the seller gets payment on the due date.

4. Red Clause Letter of Credit This is used to provide the supplier with some funds prior to shipment to finance production of the goods. The credit may be advanced in part or in full, and the buyer's bank finances the advance payment. The buyer, in essence, extends financing to the seller and incurs ultimate risk for all advanced credits.

5. Transferable Letter of Credit This allows the supplier to transfer all or part of the proceeds of the letter of credit to a second beneficiary, usually the ultimate producer of the goods. This is a common financing tactic for middlemen and is used extensively in the Far East.

6. Back-to-Back Letter of Credit This is a new credit opened on the basis of an already existing, non-transferable credit. It is used by traders to make payment to the ultimate supplier. A trader receives a letter of credit from the buyer and then opens another letter of credit in favor of the supplier. The first letter of credit is used as collateral for the second credit. The second credit makes price adjustments from which come the trader's profit.

OPENING A LETTER OF CREDIT

The wording in a letter of credit should be simple but specific. The more detailed an L/C is, the more likely the supplier will reject it as too difficult to fulfill. At the same time, the buyer will wish to define in detail what he is paying for.

Although the L/C process is designed to ensure the satisfaction of all parties to the transaction, it cannot be considered a substitute for face-to-face agreements on doing business in good faith. It should therefore contain only those stipulations required from the banks involved in the documentary process.

L/Cs used in trade with East Asia are usually either irrevocable unconfirmed credits or irrevocable confirmed credits. In choosing the type of L/C to open in favor of the supplier, the buyer should take into consideration generally accepted payment processes in the supplier's country, the value and demand for the goods to be shipped, and the reputation of the supplier.

In specifying documents necessary from the supplier, it is very important to demand documents that are required for customs clearance and those that reflect the agreement reached between the buyer and the supplier. Required documents usually include the bill of lading, a commercial and/or consular invoice, the bill of exchange, the certificate of origin, and the insurance document. Other documents required may be copies of a cable sent to the buyer with shipping information, a confirmation from the shipping company of the state of its ship, and a confirmation from the forwarder that the goods are accompanied by a certificate of origin. Prices should be stated in the currency of the L/C, and documents should be supplied in the language of the L/C.

THE APPLICATION

The following information should be included on an application form for opening an L/C.

(1) **Beneficiary** The seller's company name and address should be written completely and correctly. Incomplete or incorrect information results in delays and unnecessary additional cost.

(2) **Amount** Is the figure a maximum amount or an approximate amount? If words like "circa," "ca.," "about," etc., are used in connection with the amount of the credit, it means that a difference as high as 10 percent upwards or downwards is permitted. In such a case, the same word should also be used in connection with the quantity.

(3) **Validity Period** The validity and period for presentation of the documents following shipment of the goods should be sufficiently long to allow the exporter time to prepare his documents and ship them to the bank. Under place of validity, state the domicile of either the advising bank or the issuing bank.

(4) **Beneficiary's Bank** If no bank is named, the issuing bank is free to select the correspondent bank.

(5) **Type of Payment Availability** Sight drafts, time drafts, or deferred payment may be used, as previously agreed to by the supplier and buyer.

(6) **Desired Documents** Here the buyer specifies precisely which documents he requires. To obtain effective protection against the supply of poor quality goods, for instance, he can demand the submission of analysis or quality certificates. These are generally issued by specialized inspection companies or laboratories.

(7) **Notify Address** An address is given for notification of the imminent arrival of goods at the port or airport of destination. Damage of goods in shipment is also cause for notification. An agent representing the buyer may be used.

(8) **Description of Goods** Here a short, precise description of the goods is given, along with quantity. If the credit amount carries the notation "ca.," the same notation should appear with the quantity.

(9) **Confirmation Order** It may happen that the foreign beneficiary insists on having the credit confirmed by the bank in his country.

INTERNATIONAL PAYMENTS   231

Sample Letter of Credit Application

Sender American Import-Export Co., Inc. 123 Main Street San Francisco, California Our reference AB/02	**Instructions** **to open a Documentary Credit** San Francisco, 30th September 19.. Place / Date
Please open the following [X] irrevocable [ ] revocable documentary credit	**Domestic Bank Corporation** Documentary Credits P.O. Box 1040 San Francisco, California
Beneficiary ① Singapore Trading Corporation 435 Orchard Road. Singapore	Beneficiary's bank (if known) ④ Singapore Commercial Bank Central Office Singapore
Amount ② US$70,200.--	Please advise this bank [ ] by letter [X] by letter, cabling main details in advance [ ] by telex / telegram with full text of credit
Date and place of expiry ③ 25th November 19.. in San Francisco	
Partial shipments Transhipment [X] allowed [ ] not allowed [ ] allowed [X] not allowed	Terms of shipment (FOB, C & F, CIF) CIF San Francisco
Despatch from / Taking in charge at For transportation to Singapore San Francisco	Latest date of shipment Documents must be presented not later than 10th Nov. 19.. ③ 15 days after date of despatch
Beneficiary may dispose of the credit amount as follows [X] at sight upon presentation of documents ⑤ [ ] after .... days, calculated from date of	[ ] by a draft due .... drawn on [ ] you [ ] your correspondents which you / your correspondents will please accept

against surrender of the following documents ⑥
[X] invoice   (...3... copies)
Shipping document
[X] sea: bill of lading, to order, endorsed in blank
[ ] rail: duplicate waybill
[ ] air: air consignment note
[ ]
	[X] insurance policy, certifcte (......... copies)
	  covering the following risks:
	  "all risks" including war up to
	[ ] Additional documents final destination in
	  the USA
	[X] Confirmation of the carrier that the
	  ship is not more than 15 years old
	[X] packing list (3 copies)

Notify address in bill of lading / goods addressed to American Import-Export Co., Inc. ⑦ 123 Main Street San Francisco, California	Goods insured by [ ] us [X] seller

Goods ⑧
1'000 "Record players ANC 83 as per proforma invoice
       no. 74/1853 dd 10th September 19.."
     at US$70.20 per item	

Your correspondents to advise beneficiary [ ] adding their confirmation [X] without adding their confirmation ⑨ Payments to be debited to our U.S. Dollars .... account no 10-32679150	

NB. The applicable text is marked by [X]

E 6801 N 1/2 3.81 5000

American Import-Export Co., Inc.

Signature _____

For mailing please see overleaf

TIPS FOR PARTIES TO A LETTER OF CREDIT

Buyer

1. Before opening a letter of credit, the buyer should reach agreement with the supplier on all particulars of payment procedures, schedules of shipment, type of goods to be sent, and documents to be supplied by the supplier.
2. When choosing the type of L/C to be used, the buyer should take into account standard payment methods in the country with which he is doing business.
3. When opening a letter of credit, the buyer should keep the details of the purchase short and concise.
4. The buyer should be prepared to amend or renegotiate terms of the L/C with the supplier. This is a common procedure in international trade. On irrevocable L/Cs, the most common type, amendments may be made only if all parties involved in the L/C agree.
5. The buyer can eliminate exchange risk involved with import credits in foreign currencies by purchasing foreign exchange on the forward markets.
6. The buyer should use a bank experienced in foreign trade as the L/C issuing bank.
7. The validation time stated on the L/C should give the supplier ample time to produce the goods or to pull them out of stock.
8. The buyer should be aware that an L/C is not failsafe. Banks are only responsible for the documents exchanged and not the goods shipped. Documents in conformity with L/C specifications cannot be rejected on grounds that the goods were not delivered as specified in the contract. The goods shipped may not in fact be the goods ordered and paid for.
9. Purchase contracts and other agreements pertaining to the sale between the buyer and supplier are not the concern of the issuing bank. Only the terms of the L/C are binding on the bank.
10. Documents specified in the L/C should include those the buyer requires for customs clearance.

Supplier

1. Before signing a contract, the supplier should make inquiries about the buyer's creditworthiness and business practices. The supplier's bank will generally assist in this investigation.
2. The supplier should confirm the good standing of the buyer's bank if the credit is unconfirmed.
3. For confirmed credit, the supplier should determine that his local bank is willing to confirm credits from the buyer and his bank.
4. The supplier should carefully review the L/C to make sure he can meet the specified schedules of shipment, type of goods to be sent, packaging, and documentation. All aspects of the L/C must be in conformance with the terms agreed upon, including the supplier's address, the amount to be paid, and the prescribed transport route.
5. The supplier must comply with every detail of the L/C specifications, otherwise the security given by the credit is lost.
6. The supplier should ensure that the L/C is irrevocable.
7. If conditions of the credit have to be modified, the supplier should contact the buyer immediately so that he can instruct the issuing bank to make the necessary amendments.
8. The supplier should confirm with his insurance company that it can provide the coverage specified in the credit, and that insurance charges in the L/C are correct. Insurance coverage often is for CIF (cost, insurance, freight) value of the goods plus 10 percent.
9. The supplier must ensure that the details of goods being sent comply with the description in the L/C, and that the description on the invoice matches that on the L/C.
10. The supplier should be familiar with foreign exchange limitations in the buyer's country which may hinder payment procedures.

GLOSSARY OF DOCUMENTS IN INTERNATIONAL TRADE

The following is a list and description of some of the more common documents importers and exporters encounter in the course of international trade. For the importer/buyer this serves as a checklist of documents he may require of the seller/exporter in a letter of credit or documents against payment method.

Bill of Lading A document issued by a transportation company (such as a shipping line) to the shipper which serves as a receipt for goods shipped, a contract for delivery, and may serve as a title document. The major types are:

Straight (non-negotiable) Bill of Lading Indicates that the shipper will deliver the goods to the consignee. The document itself does not give title to the goods. The consignee need only identify himself to claim the goods. A straight bill of lading is often used when the goods have been paid for in advance.

Order (negotiable or "shippers order") Bill of Lading This is a title document which must be in the possession of the consignee (buyer/importer) in order for him to take possession of the shipped goods. Because this bill of lading is negotiable, it is usually made out "to the order of" the consignor (seller/exporter).

Air Waybill A bill of lading issued for air shipment of goods, which is always made out in straight non-negotiable form. It serves as a receipt for the shipper and needs to be made out to someone who can take possession of the goods upon arrival—without waiting for other documents to arrive.

Overland/Inland Bill of Lading Similar to an Air Waybill, except that it covers ground or water transport.

Certificate of Origin A document which certifies the country of origin of the goods. Because a certificate of origin is often required by customs for entry, a buyer will often stipulate in his letter of credit that a certificate of origin is a required document.

Certificate of Manufacture A document in which the producer of goods certifies that production has been completed and that the goods are at the disposal of the buyer.

Consular Invoice An invoice prepared on a special form supplied by the consul of an importing country, in the language of the importing country, and certified by a consular official of the foreign country.

Dock Receipt A document/receipt issued by an ocean carrier when the seller/exporter is not responsible for moving the goods to their final destination, but only to a dock in the exporting country. The document/receipt indicates that the goods were, in fact, delivered and received at the specified dock.

Export License A document, issued by a government agency, giving authorization to export certain commodities to specified countries.

Import License A document, issued by a government agency, giving authorization to import certain commodities.

Inspection Certificate An affidavit signed by the seller/exporter or an independent inspection firm (as required by the buyer/importer), confirming that merchandise meets certain specifications.

Insurance Document A document certifying that goods are insured for shipment.

Invoice/Commercial Invoice A document identifying the seller and buyer of goods or services, identifying numbers such as invoice number, date, shipping date, mode of transport, delivery and payment terms, and a complete listing and description of the goods or services being sold including prices, discounts, and quantities. The commercial invoice is usually used by customs to determine the true cost of goods when assessing duty.

Packing List A document listing the merchandise contained in a particular box, crate, or container, plus type, dimensions, and weight of the container.

Phytosanitary (plant health) Inspection Certificate A document certifying that an export shipment has been inspected and is free from pests and plant diseases considered harmful by the importing country.

Shipper's Export Declaration A form prepared by a shipper/exporter indicating the value, weight, destination, and other information about an export shipment.

GLOSSARY OF TERMS OF SALE

The following is a basic glossary of common terms of sale in international trade. Note that issues regarding responsibility for loss and insurance are complex and beyond the scope of this publication. The international standard of trade terms of sale are "Incoterms," published by the International Chamber of Commerce (ICC), 38, Cours Albert Ier, F-75008 Paris, France. Other offices of the ICC are British National Committee of the ICC, Centre Point, 103 New Oxford Street, London WC1A 1QB, UK and US Council of the ICC, 1212 Avenue of the Americas, New York, NY 10010, USA.

C&F (Cost and Freight) Named Point of Destination The seller's price includes the cost of the goods and transportation up to a named port of destination, but does not cover insurance. Under these terms insurance is the responsibility of the buyer/importer.

CIF (Cost, Insurance, and Freight) Named Point of Destination The seller's price includes the cost of the goods, insurance, and transportation up to a named port of destination.

Ex Point of Origin ("Ex Works" "Ex Warehouse" etc.) The seller's price includes the cost of the goods and packing, but without any transport. The seller agrees to place the goods at the disposal of the buyer at a specified point of origin, on a specified date, and within a fixed period of time. The buyer is under obligation to take delivery of the goods at the agreed place and bear all costs of freight, transport and insurance.

FAS (Free Alongside Ship) The seller's price includes the cost of the goods and transportation up to the port of shipment alongside the vessel or on a designated dock. Insurance under these terms is usually the responsibility of the buyer.

FOB (Free On Board) The seller's price includes the cost of the goods, transportation to the port of shipment, and loading charges on a vessel. This might be on a ship, railway car, or truck at an inland point of departure. Loss or damage to the shipment is borne by the seller until loaded at the point named and by the buyer after loading at that point.

Ex Dock—Named Port of Importation The seller's price includes the cost of the goods, and all additional charges necessary to put them on the dock at the named port of importation with import duty paid. The seller is obligated to pay for insurance and freight charges.

GLOSSARY OF INTERNATIONAL PAYMENT TERMS

Advice The forwarding of a letter of credit or an amendment to a letter of credit to the seller, or beneficiary of the letter of credit, by the advising bank (seller's bank).

Advising Bank The bank (usually the seller's bank) which receives a letter of credit from the issuing bank (the buyer's bank) and handles the transaction from the seller's side. This includes: validating the letter of credit, reviewing it for internal consistency, forwarding it to the seller, forwarding seller's documentation back to the issuing bank, and, in the case of a confirmed letter of credit, guaranteeing payment to the seller if his documents are in order and the terms of the credit are met.

Amendment A change in the terms and conditions of a letter of credit, usually to meet the needs of the seller. The seller requests an amendment of the buyer who, if he agrees, instructs his bank (the issuing bank) to issue the amendment. The issuing bank informs the seller's bank (the advising bank) who then notifies the seller of the amendment. In the case of irrevocable letters of credit, amendments may only be made with the agreement of all parties to the transaction.

Back-to-Back Letter of Credit A new letter of credit opened in favor of another beneficiary on the basis of an already existing, nontransferable letter of credit.

Beneficiary The entity to whom credits and payments are made, usually the seller/supplier of goods.

Bill of Exchange A written order from one person to another to pay a specified sum of money to a designated person. The following two versions are the most common:

Draft A financial/legal document where one individual (the drawer) instructs another individual (the drawee) to pay a certain amount of money to a named person, usually in payment for the transfer of goods or services. Sight Drafts are payable when presented. Time Drafts (also called usance drafts) are payable at a future fixed (specific) date or determinable (30, 60, 90 days etc.) date. Time drafts are used as a financing tool (as with Documents against Acceptance D/P terms) to give the buyer time to pay for his purchase.

Promissory Note A financial/legal document wherein one individual (the issuer) promises to pay another individual a certain amount.

Collecting Bank (also called the presenting bank) In a Documentary Collection, the bank (usually the buyer's bank) that collects payment or a time draft from the buyer to be forwarded to the remitting bank (usually the seller's bank) in exchange for shipping and other documents which enable the buyer to take possession of the goods.

Confirmed Letter of Credit A letter of credit which contains a guarantee on the part of both the issuing and advising bank of payment to the seller so long as the seller's documentation is in order and terms of the credit are met.

Deferred Payment Letter of Credit A letter of credit where the buyer takes possession of the title documents and the goods by agreeing to pay the issuing bank at a fixed time in the future.

Discrepancy The noncompliance with the terms and conditions of a letter of credit. A discrepancy may be as small as a misspelling, an inconsistency in dates or amounts, or a missing document. Some discrepancies can easily be fixed; others may lead to the eventual invalidation of the letter of credit.

D/A Abbreviation for "Documents against Acceptance."

D/P Abbreviation for "Documents against Payment."

Documents against Acceptance (D/A) *See* Documentary Collection

Documents against Payment (D/P) *See* Documentary Collection

Documentary Collection A method of effecting payment for goods whereby the seller/exporter instructs his bank to collect a certain sum from the buyer/importer in exchange for the transfer of shipping and other documentation enabling the buyer/importer to take possession of the goods. The two main types of Documentary Collection are:
  Documents against Payment (D/P) Where the bank releases the documents to the buyer/importer only against a cash payment in a prescribed currency; and
  Documents against Acceptance (D/A) Where the bank releases the documents to the buyer/importer against acceptance of a bill of exchange guaranteeing payment at a later date.

Draft *See* Bill of exchange.

Drawee The buyer in a documentary collection.

Forward Foreign Exchange An agreement to purchase foreign exchange (currency) at a future date at a predetermined rate of exchange. Forward foreign exchange contracts are often purchased by buyers of merchandise who wish to hedge against foreign exchange fluctuations between the time the contract is negotiated and the time payment is made.

Irrevocable Credit A letter of credit which cannot be revoked or amended without prior mutual consent of the supplier, the buyer, and all intermediaries.

Issuance The act of the issuing bank (buyer's bank) establishing a letter of credit based on the buyer's application.

Issuing Bank The buyer's bank which establishes a letter of credit in favor of the supplier, or beneficiary.

Letter of Credit A document stating commitment on the part of a bank to place an agreed upon sum of money at the disposal of a seller on behalf of a buyer under precisely defined conditions.

Negotiation In a letter of credit transaction, the examination of seller's documentation by the (negotiating) bank to determine if they comply with the terms and conditions of the letter of credit.

Open Account The shipping of goods by the supplier to the buyer prior to payment for the goods. The supplier will usually specify expected payment terms of 30, 60, or 90 days from date of shipment.

Red Clause Letter of Credit A letter of credit which makes funds available to the seller prior to shipment in order to provide him with funds for production of the goods.

Remitter In a documentary collection, an alternate name given to the seller who forwards documents to the buyer through banks.

Remitting Bank In a documentary collection, a bank which acts as an intermediary, forwarding the remitter's documents to, and payments from the collecting bank.

Sight Draft *See* Bill of Exchange.

Standby Letter of Credit- A letter of credit used as a secondary payment method in the event that the primary payment method cannot be fulfilled.

Time Draft *See* Bill of Exchange.

Validity The time period for which a letter of credit is valid. After receiving notice of a letter of credit opened on his behalf, the seller/exporter must meet all the requirements of the letter of credit within the period of validity.

Revocable Letter of Credit A letter of credit which may be revoked or amended by the issuer (buyer) without prior notice to other parties in the letter of credit process. It is rarely used.

Revolving Letter of Credit A letter of credit which is automatically restored to its full amount after the completion of each documentary exchange. It is used when there are several shipments to be made over a specified period of time.

FURTHER READING

For more detailed information on international trade payments, refer to the following publications of the International Chamber of Commerce (ICC), Paris, France.

Uniform Rules for Collections This publication describes the conditions governing collections, including those for presentation, payment and acceptance terms. The Articles also specify the responsibility of the bank regarding protest, case of need and actions to protect the merchandise. An indispensable aid to everyday banking operations. (A revised, updated edition will be published in 1995.) ICC Publication No. 322.

Documentary Credits: UCP 500 and 400 Compared This publication was developed to train managers, supervisors, and practitioners of international trade in critical areas of the new UCP 500 Rules. It pays particular attention to those Articles that have been the source of litigation. ICC Publication No. 511.

The New ICC Standard Documentary Credit Forms Standard Documentary Credit Forms are a series of forms designed for bankers, attorneys, importers/exporters, and anyone involved in documentary credit transactions around the world. This comprehensive new edition, prepared by Charles del Busto, Chairman of the ICC Banking Commission, reflects the major changes instituted by the new "UCP 500." ICC Publication No. 516.

The New ICC Guide to Documentary Credit Operations This new Guide is a fully revised and expanded edition of the "Guide to Documentary Credits" (ICC publication No. 415, published in conjunction with the UCP No. 400). The new Guide uses a unique combination of graphs, charts, and sample documents to illustrate the Documentary Credit process. An indispensable tool for import/export traders, bankers, training services, and anyone involved in day-to-day Credit operations. ICC Publication No. 515.

Guide to Incoterms 1990 A companion to "Incoterms," the ICC "Guide to Incoterms 1990" gives detailed comments on the changes to the 1980 edition and indicates why it may be in the interest of a buyer or seller to use one or another trade term. This guide is indispensable for exporters/importers, bankers, insurers, and transporters. ICC Publication No. 461/90.

These and other relevant ICC publications may be obtained from the following sources:

ICC Publishing S.A.
International Chamber of Commerce
38, Cours Albert Ier
75008 Paris, France
Tel: [33] (1) 49-53-28-28 Fax: [33] (1) 49-53-28-62
Telex: 650770

International Chamber of Commerce
Borsenstrasse 26
P.O. Box 4138
8022 Zurich, Switzerland

British National Committee of the ICC
Centre Point, New Oxford Street
London WC1A QB, UK

ICC Publishing, Inc.
US Council of the ICC
156 Fifth Avenue, Suite 820
New York, NY 10010, USA
Tel: [1] (212) 206-1150 Fax: [1] (212) 633-6025

Corporate Taxation

AT A GLANCE

Corporate Income Tax Rate (%)	27 (a)
Capital Gains Tax Rate (%)	0
Branch Tax Rate (%)	27 (a)
Withholding Tax (%) (c)	
Dividends	0 (b)
Interest	27
Royalties from Patents, Know-how, etc.	27
Branch Remittance Tax	0
Net Operating Losses (Years)	
Carryback	0
Carryforward	Unlimited (d)

(a) Numerous tax exemptions and reductions are available. See Taxes on Corporate Income and Gains.
(b) See Taxes on Corporate Income and Gains.
(c) See Treaty Withholding Tax Rates.
(d) See Determination of Trading Income.

TAXES ON CORPORATE INCOME AND GAINS

Corporate Income Tax

Income tax is charged on all income derived from sources in Singapore, together with income from sources outside Singapore if received in Singapore. A nonresident company carrying on business in Singapore is similarly taxed on Singapore-source income and on foreign-source income received in Singapore if that income is effectively connected with a Singapore permanent establishment. A company is resident in Singapore if its management and control are exercised in Singapore; the place of incorporation is of no relevance.

Rates of Income Tax

For year of assessment 1994 (*see* Administration), the rate of tax is 27 percent, applicable to both resident and nonresident companies.

The following tax exemptions and tax reductions are available.

Approved Financial Institutions Approved banks, insurance companies, and leasing companies pay a 10 percent tax on their off-shore profits.

Pioneer Industries A pioneer enterprise is exempt from company tax for five to 10 years from the date it commences commercial production.

Pioneer Service Companies For pioneer service companies, the tax exemption period is five to 10 years.

Post-Pioneer Companies The maximum tax relief period for a post-pioneer company is 10 years. During this period, the rate of tax levied on profits from the qualifying activities is reduced, but not below 10 percent.

Expansion of Established Enterprises Holders of an expansion certificate are exempt from tax on the increase in income resulting from the expansion for a period of up to five years. To qualify, the expansion must involve over S$10 million of capital expenditures for manufacture of an approved product.

Expanding Service Companies Certain companies expanding their activities may apply for approval as an expanding service company. On approval, tax exemption is given on the incremental profits made during a period not exceeding five years.

Export Incentives For a period of five years, 90 percent of a company's qualifying export profits exceeding a predetermined base is exempt from tax. This time limit may be extended to 15 years for projects with fixed capital expenditures exceeding S$150 million, provided 50 percent of the company's paid-up capital is held by permanent residents of Singapore. There is no local shareholding requirement if the expenditure is to exceed S$1 billion.

Investment Allowance This is available as an alternative to pioneer status and export incentives.

Note: *This section is courtesy of and © Ernst & Young from their Worldwide Corporate Tax Guide, 1994 Edition. This material should not be regarded as offering a complete explanation of the taxation matters referred to. Ernst & Young is a leading international professional services firm with offices in 120 countries, including Singapore. Refer to the "Important Addresses" chapter for addresses and telephone numbers of the Ernst & Young offices in Singapore.*

Tax-exempt profits are limited to specified percentages (up to 50 percent) of actual fixed investment on certain specialized factory buildings and new productive equipment for an approved project.

Warehousing and Servicing Incentives Half of a company's qualifying export profits is exempt from tax for an initial period of five years. To calculate export profits, total profits are divided in the proportion of export and domestic sales. Profits from domestic sales and services are taxed at the full corporate tax rate of 27 percent.

International Consultancy Services Incentive This incentive provides for one-half of the qualifying income to be exempt from tax for five years, provided the qualifying consultancy services have an expected annual income of at least S$1 million.

International Trade Incentive This incentive provides for one-half of the qualifying income to be exempt from tax for five years. It applies to trading companies that export locally manufactured goods or nontraditional commodities.

Approved Royalties, Fees, Development Contributions An application may be made to relieve the recipient from tax on such payments.

Operational Headquarters Approved operational headquarters (OHQs) for multinational groups of companies are taxed at a concessionary rate of 10 percent on their service income for up to 10 years (with provision for renewal). Other income, such as interest and royalties received by an OHQ from its overseas subsidiaries and associated companies, may also be eligible for tax concessions. OHQs pay only Singapore or foreign tax, whichever is higher. If the OHQ entity owns equity in the overseas affiliates, dividend flows may be exempt from Singapore tax for a period of 10 years (with provision for renewal).

Treasury and Finance Center Incentive Income from foreign exchange transactions, offshore investments and financial services provided to related companies is subject to a 10 percent tax.

Approved International Trader Incentive Income from trading activities involving foreign suppliers and customers with a minimum trading volume of S$200 million is subject to a concessionary 10 percent rate of tax for five years.

Approved International Shipping Enterprise Income from the operation of ships in international waters is exempt from tax.

Approved Oil Trader Income from trading in oil and oil-based products with nonresidents and other approved oil traders is subject to tax at a rate of 10 percent.

Venture Capital and Regional Funds Incentive Gains on the disposal of approved local and overseas investments and interest on convertible loan stocks derived from approved overseas investments are exempt from tax for a period of up to 10 years.

Art and Antiques Incentive Various tax reliefs and tax reductions are granted to certain businesses and transactions related to the art and antiques market. A tax rate of 10 percent applies for the first five years, with possible extensions.

Capital Gains

There is no taxation of capital gains. However, in certain circumstances the Singapore Revenue views the acquisition and disposal of real estate or stocks and shares as derived from the carrying on of a trade.

Special rules apply to the taxation of gains arising from investment sales by unit trusts and approved investment companies. Under the rules, such transactions are regarded as trading activities subject to relief based on the holding period and reduced tax rates.

Administration

The tax year, known as a year of assessment, runs from January 1 to December 31. The period for which profits are identified for assessment is called the basis year. Thus, income earned during the basis year 1993 is assessed for taxation in the 1994 tax year. If a company doing business in Singapore adopts an accounting period other than the calendar year, the assessable profits are those for the 12-month accounting period ending in the year preceding the year of assessment.

A company is required to file an income tax return within 21 days from date of issue. However, it is possible to obtain extensions—currently, up to July 31 of the year of assessment. The Singapore Revenue requires an estimate of the taxable income of the

---

Singapore Tax Code Changes—1994

Singapore made some basic changes in its tax code effective April 1, 1994. These changes involved the institution of a goods and services tax—similar to a value-added tax—of 3 percent on all goods and services at each stage of production and including final consumption, except on products designated specifically for export. The 3 percent levy also applies to imported goods and services.

The corporate income tax rate was simultaneously lowered to 27 percent from 30 percent to compensate for this new approach to taxation. Businesses with annual sales of less than S$1 million are exempt from this levy, and exemptions are also available on transactions involving residential real estate. Personal income tax rates were also lowered to 30 percent from 33 percent to offset the effect of the new tax on goods and services. (Refer to the "Personal Taxation" chapter.)

company to support an application for extension. If an estimate is not given, the Singapore Revenue may prepare an assessment based on its own estimate.

Income tax is due for payment within 30 days of the date of issue of the notice of assessment. Companies are permitted to pay tax in monthly installments, up to a maximum of 10, the first installment, being payable one month after the end of the accounting period.

Companies with an accounting year that ends on or before September 30 are required to furnish an estimate of taxable income within three months of the end of the accounting period. An advance assessment notice will be prepared based on the estimate submitted; tax assessed must be paid as set out above.

A late payment penalty of 5 percent of the tax due is imposed if the tax is not paid by the due date. If the tax is not paid within 60 days of the imposition of the penalty, an additional penalty of 1 percent of the tax is levied for each complete month that the tax remains outstanding, up to a maximum of 12 percent.

Late filings of a tax return are subject to penalties of up to S$1,000.

Dividends

Dividends paid by a resident company are subject to a withholding tax of 27 percent. Because Singapore operates a full imputation system, this withholding tax may be treated as paid (franked) out of the corporate income tax that has already been paid on taxable profits. Only if the tax withheld on dividends exceeds the cumulative tax on the company's profits does the company have to pay the excess amount to the tax authority. This payment is treated as an advance payment of corporate income tax and can be used to offset future corporate tax liabilities. A nonresident company may distribute after-tax profits without incurring any additional liability. The 27 percent withholding tax satisfies the Singapore tax liability of a nonresident shareholder; for a resident shareholder, the credit is applied towards that shareholder's tax liability, and any excess of the credit over the tax liability is refundable.

Foreign Tax Relief

Singapore has concluded double taxation treaties with 28 countries, but notably not with the United States. Under the treaties, foreign tax relief is limited to the lower of the foreign tax paid and the Singapore tax payable on that income; credit is computed on a source-by-source basis. A unilateral tax credit system, similar to treaty relief, is also available for certain professional, consulting, and service income remitted from specified countries, including the United States.

For dividends received by residents on or after January 1, 1993, a unilateral tax credit is granted for dividend withholding tax suffered. In addition, unilateral relief is granted for underlying foreign taxes paid by a 25 percent-owned foreign subsidiary on profits out of which dividends are paid.

In other circumstances, foreign tax relief is limited to the lower of the foreign tax paid in a British Commonwealth country and one-half of the Singapore tax paid. Even in these circumstances, the relief is restricted to those countries that grant corresponding relief.

DETERMINATION OF TRADING INCOME

General

An assessment is based on the audited financial statements prepared under generally accepted accounting principles, subject to certain adjustments and provisions. A non-resident company trading in Singapore, in practice, prepares the financial statements of its Singapore branch in accordance with the Singapore Companies Act; these would be accepted as representing that concept.

For expenses to be deductible, they must be incurred wholly and exclusively in the production of income. Expenses attributable to foreign-source income are not deductible unless the foreign-source income is received in Singapore and is therefore subject to taxation. Losses incurred in the production of foreign-source income are not deductible against Singapore-source income. No deduction is allowed for the book depreciation of fixed assets, although tax depreciation (capital allowances) is granted according to statutory rates. Special rules govern the deductibility of various categories of expenses for investment companies.

Inventories

Trading inventory is valued at the lower of cost or net realizable value. Cost must be determined on a first-in, first-out (FIFO) basis; the last-in, first-out (LIFO) basis is not accepted.

Provisions

Provisions for doubtful debts are allowable deductions but only to the extent that the debts arose from the trade and became bad or are estimated to have become bad during the basis period. Other provisions are deductible if they are accurately estimated. Provisions of a general nature are not allowable.

Capital Allowances
(Tax Depreciation)

Plant and Machinery Depreciation allowances are given for capital expenditures incurred on the acquisition of plant and machinery used for the purposes of a trade or business. The expenditure may be written off in equal amounts over three years. The cost of computers, office automation equipment, or robotic machinery may be written off in the year of acquisition. Expenditures on automobiles and commercial vehicles (other than privately owned vehicles for which no capital allowances are available) are subject to an initial allowance of 20 percent in the first year. The balance of the expenditure after the initial allowance may be written off as annual allowances in equal amounts over six years, beginning in the year of acquisition. Allowances on passenger cars are limited to a total of S$35,000.

Industrial Buildings An initial allowance is given on industrial buildings amounting to 25 percent of the qualifying expenditure; a depreciation allowance of 3 percent a year on a straight-line basis can also be claimed. There is no depreciation allowance for commercial buildings or hotels other than for approved hotels on the island of Sentosa.

Other Matters Capital allowances are also granted on capital expenditure incurred to acquire approved patent rights and know-how for use in an entity's manufacturing trade or business. Allowances are generally subject to recapture on the sale of a qualifying asset if the sales proceeds exceed the tax-depreciated value. If sales proceeds are less than the tax-depreciated value, an additional corresponding allowance is given.

Relief for Trading Losses Trading losses may be offset against all other chargeable income of the same year. Unused losses may be carried forward indefinitely and offset against any other chargeable income. Excess depreciation allowances can also be offset against other chargeable income of the same year and carried forward indefinitely, provided the trade giving rise to the capital allowances is still carried on.

The carryforward of losses and excess capital allowances is subject to the shareholders remaining substantially the same (50 percent or more) at the end of the year in which the losses or capital allowances arose and on the first day of the year of assessment in which relief is claimed. If the shareholder of the loss company is itself another company, look-through provisions apply through the corporate chain to the final beneficial shareholder. It is frequently difficult to determine continuity of shareholding when the losses or allowances are incurred by a subsidiary of a publicly listed company.

The Singapore Revenue has the authority to allow companies to deduct their tax losses notwithstanding a substantial change in ownership at the relevant dates if the change is not motivated by tax considerations (such as when the change is caused by nationalization or privatization of industries or if the shareholding of the company or its parent changes substantially as a result of the shares being widely traded on recognized exchanges). These losses, however, may be carried forward for offset only against profits from the same business. In practice, the discretion is rarely exercised except in cases of nationalization or privatization.

Groups of Companies There are no group relief provisions in Singapore. Thus, the losses of one company within a group cannot be used to reduce profits of another company in the same group.

The Singapore Revenue will examine each set of accounts and computations separately, and intercompany transactions are scrutinized to ensure that they are conducted at arm's length.

OTHER SIGNIFICANT TAXES

The table below summarizes other significant taxes.

Nature of Tax	Rate (%)
Goods and services tax, on any supply of goods and services, except for an exempt supply, made in Singapore by a taxable person (a business is taxable if its annual supplies exceed S$1 million) in the course of business (effective April 1, 1994)	0/3
Social security contributions (Central Provident Fund) For employees under age 55, on monthly ordinary wages (lower rates apply if employee is age 55 or older)	
Employer (limited to S$1,110 a month)	18.5
Employee (limited to S$1,290 a month)	21.5
Contributions on bonuses and nonregular payments (if total wages exceed S$100,000, generally limited to contributions payable on 40 percent of total annual ordinary wages)	
Employer	18.5
Employee	21.5
Skills development levy, payable by employer for employees earning S$750 or less a month	1

MISCELLANEOUS MATTERS

Foreign Exchange Controls

Singapore does not impose any restrictions on the remittance and repatriation of funds in and out of Singapore. All transactions take place at fluctuating exchange rates.

CORPORATE TAXATION

Antiavoidance Legislation

Legislation permits the Singapore Revenue to substitute an arm's length price if transactions with a foreign affiliate are entered into at an artificial price. Despite the considerable number of related-party transactions taking place between residents and foreign companies, these provisions are seldom applied in practice provided a reasonable profit accrues to residents.

TREATY WITHHOLDING TAX RATES

Withholding taxes (normally at 27 percent) are imposed on interest and royalties paid to a nonresident. Withholding tax is also levied on payments for the use of, or the right to use, scientific, technical, industrial, or commercial knowledge. The tax also applies to rents or other payments under any agreement or arrangement for the use of any movable property and also to payments made for the management or assistance in management of any trade, business, or profession. If the technical service or management assistance is rendered wholly outside Singapore, and the payment is demonstrably at arm's length, it may be exempt from the tax.

Singapore, at present, does not levy a separate withholding tax on dividends, although dividends are "franked" by the underlying corporate income tax rate of 27 percent.

The rates of tax may be reduced under the terms of a double tax treaty and details of the rates applicable to treaty countries are set out below.

	Interest %	Royalties (l) %
Australia	10	10
Bangladesh	10	10
Belgium	15	0
Canada	15 (b)	15
China	10 (c)(d)	10
Denmark	15 (d)	15
Finland	10 (d)	10
France	10 (d)(e)	0
Germany	10 (d)	0
India	30	30
Indonesia	10	15
Israel	15	0
Italy	12.5 (d)	15 (a)
Japan	15 (d)(e)	10
Korea	10 (d)	15
Malaysia	30	30
Netherlands	10 (d)	0
New Zealand	15	15
Norway	15 (d)	15
Pakistan	12.5	10
Papua New Guinea	10	10
Philippines	15	25 (f)
Sri Lanka	10 (k)	15 (g)
Sweden	15 (h)	0
Switzerland	10 (i)	5 (i)
Taiwan	30	15
Thailand	25 (d)(j)	15
United Kingdom	15 (i)	15 (i)
Nontreaty countries	27	27

(a) The withholding rate is 20 percent for royalties on copyrights of literary or artistic works or for motion picture films, videotapes, etc.
(b) Interest paid to the Export Development Corporation of Canada is exempt.
(c) 7 percent if received by a bank or other financial institution.
(d) Interest received by a contracting state or subdivision, or by certain state institutions, is exempt.
(e) Interest paid by an industrial undertaking is exempt.
(f) Exempt if royalties are approved under the Economic Expansion Incentives Act.
(g) If new technology transferred; otherwise 27 percent.
(h) The rate is 10 percent for interest paid by an "industrial undertaking" to a financial institution in Sweden.
(i) No withholding applies to interest or royalties on certain "approved" transactions.
(j) 10 percent if received by a financial institution; 25 percent otherwise.
(k) Exempt if payable to the government or a banking or financial institution.
(l) The reduced treaty rates often do not apply to royalties on copyrights of literary or artistic works including motion picture or cinematographic films, and tapes and films for radio or television broadcasting. Reference should be made to the applicable tax treaty.

Personal Taxation

AT A GLANCE—MAXIMUM RATES

Income Tax Rate (%)	30
Capital Gains Tax Rate (%)	0
Net Worth Tax Rate (%)	0
Estate Tax Rate (%)	10

INCOME TAXES—EMPLOYMENT

Who Is Liable

A person is subject to tax on employment income for services performed in Singapore, regardless of whether the remuneration is paid in or outside Singapore. Residents are also liable for tax on income from sources outside Singapore that is received in Singapore.

Individuals are resident for tax purposes if employed or physically present in Singapore for a total period of 183 days or more during the calendar year preceding a year of assessment. In practice, an individual whose stay extends into three or more consecutive years of assessment is, by concession, considered resident for all years, even if fewer than 183 days were spent in Singapore in the first and last years of stay.

Taxable Income

Taxable employment income includes cash remuneration such as wages, salary, leave pay, fees, commissions, bonuses, gratuities, perquisites, and allowances received as compensation for income. Benefits in kind derived from employment, such as home-leave passage, employer-provided housing, and employer-provided automobiles, are also taxable. Certain of these benefits receive special tax treatment. Compulsory statutory contributions made by employers to the Central Provident Fund (see Social Security Taxes) for an individual performing services in Singapore do not constitute taxable income. Effective January 1, 1993, contributions made by an employer to any provident pension fund located outside Singapore are taxable as income when the contributions are paid.

Income Tax Rates

A person who is a tax resident in Singapore is taxed on assessable income after personal deductions at the following rates for the 1994 assessment year (income from the 1993 calendar year).

Married persons are taxed either jointly or separately, at the taxpayers' election, on all types of income. For rates applicable to nonresidents, see page 245.

Assessable Income Exceeding S$	Not Exceeding S$	Tax on Lower Amount* S$	Rate on Excess* %
0	5,000	0	2.5
5,000	7,500	125	5
7,500	10,000	250	6
10,000	15,000	400	7
15,000	20,000	750	8
20,000	25,000	1,150	11
25,000	35,000	1,700	13
35,000	50,000	3,000	15
50,000	75,000	5,250	19
75,000	100,000	10,000	22
100,000	150,000	15,500	24
150,000	200,000	27,500	25
200,000	400,000	40,000	28
400,000	—	96,000	30

*A S$700 rebate is given for the 1994 year of assessment (income year 1993). This rebate will be reduced by S$50 each year until it reaches S$500 in the 1998 year of assessment, after which it will remain at S$500. The figures quoted are before the rebate.

Note: This section is courtesy of and © Ernst & Young from their Worldwide Personal Tax Guide, 1994 Edition. This material should not be regarded as offering a complete explanation of the taxation matters referred to. Ernst & Young is a leading international professional services firm with offices in 120 countries, including Singapore. Refer to the "Important Addresses" chapter for addresses and telephone numbers of the Ernst & Young offices in Singapore.

Deductible Expenses

Expenses incurred wholly and exclusively in the production of income qualify for deduction, but the deductions available against employment income are limited in practice. The general view taken by the Inland Revenue Authority of Singapore is that an employer normally pays all the necessary expenses incurred by an employee in the course of discharging the duties of office. Employees must be able to prove to the Inland Revenue Authority of Singapore that expenses claimed were necessarily incurred in performing their duties.

Personal Deductions and Allowances

Personal deductions are granted to individuals resident in Singapore. The deductions for the 1994 assessment year are summarized in the following table.

Self	S$3,000
Wife relief	S$1,500
Earned income	
Under age 55	S$1,000
Age 55 to 59	S$3,000
Age 60 and above	S$4,000
Under 55 years and blind, or	
physically or mentally disabled	S$2,000
Child relief	
1st, 2nd, and 3rd child	S$1,500 each
4th and 5th child	
(born before Aug. 1, 1973)	S$ 300 each
4th child	
(born on or after Jan. 1, 1988)	S$1,500
(For children receiving education	
abroad, these amounts may be	
doubled.)	
Dependent relative deduction	S$3,500

Singapore Tax Code Changes—1994

On April 1, 1994 the tax rate on personal income was lowered to 30 percent from 33 percent. At the same time, the tax rate on corporate income dropped to 27 percent from 30 percent and Singapore instituted a 3 percent goods and services tax, similar to a value-added tax. (Refer to the "Corporate Taxation" chapter.) This tax is levied on businesses, not individuals. However, it effectively raises the cost of virtually all items they purchase. Foreign visitors buying a minimum of S$320 of goods in Singapore may apply for a refund. Applicants are forewarned that they must note that they intend to seek a refund when buying the goods in order to receive the necessary paperwork, which must be presented at departure.

Enhanced child reliefs are available for married women working in Singapore. Special tax rebates are available for parents who are Singapore citizens. Special deductions are available for reservists.

The following deductions for life insurance premiums and contributions to approved pension funds are granted:

- For an employee, the total of life insurance premiums and amounts contributed to designated approved pension funds other than the Central Provident Fund (CPF), to the extent CPF contributions are less than S$5,000;
- For an individual carrying on a trade, business, profession, or vocation, 18 percent of assessable trade income or S$12,960, whichever is lower; and
- A maximum of S$6,000 for contributions made to the taxpayer's or the taxpayer's parents' CPF retirement account.

In addition, fees for approved courses may be deducted, up to a S$2,000 maximum.

INCOME TAX— SELF-EMPLOYMENT/ BUSINESS INCOME

Who Is Liable

Individuals who carry on a trade, business, profession, or vocation in Singapore are taxed on their profits.

Whether an individual is carrying on a trade is a question of fact, to be determined by the circumstances of each case.

Taxable Income

Income subject to tax is based on financial accounts prepared under generally accepted accounting principles. Adjustments are then made to the profits or losses to conform with tax law. Business income is aggregated with other types of income to determine taxable income, which is taxed at rates described in the Income Tax Rates table.

Deductible Expenses

Expenses incurred are deductible in ascertaining taxable profits if they were expended wholly and exclusively in producing income and are not specifically disallowed. Expenses specifically not deductible include personal expenses, income taxes paid in and outside Singapore, contributions to unapproved provident funds, and certain automobile expenses. Book depreciation of fixed assets is not an allowable deduction. Tax depreciation (capital allowances) is granted instead.

DIRECTORS' FEES

Directors' fees are considered to be gains or profits from employment and are taxed in the same manner as salaries. The source of directors' fees is the site of the company's residence, which is deemed to be the site from which it is managed and controlled.

INVESTMENT INCOME

Dividend Income Dividend income is taxed at the source at a rate of 27 percent, unless the dividend is declared from profits that are specifically exempt from tax or enjoy concessionary tax rates under the tax laws. Nonexempt dividends are aggregated with other sources of income and taxed at the rates described in the Income Tax Rates table. The tax deducted at source from the dividend is available for offset against the tax liability for that assessment year, and any excess is refunded to the taxpayer.

Dividends declared from exempt profits are tax-free in the hands of the recipient.

Interest Income Except for interest income derived from the national savings bank (Post Office Savings Bank) and certain tax-free government bonds, all other forms of interest are taxed with other income described in Income Tax Rates. However, nonresident individuals receiving interest from deposits with approved banks and from Asian dollar bonds are exempt from paying tax on such income.

RELIEF FOR LOSSES

Losses incurred and capital allowances granted in the operation of a trade, business, profession, or vocation are available for setoff against any other income of the individual liable for tax for the same year of assessment. Any unabsorbed losses and capital allowances may be carried forward without any time limitation for setoff against future income from all sources. This relief is allowed only for a loss incurred in connection with a trade, business, profession, or vocation. Relief for a qualifying loss is mandatory and may not be deferred.

CAPITAL GAINS AND LOSSES

Singapore has no capital gains tax. However, in certain circumstances the tax authorities may treat gains from purchases and sales of real estate and stocks or shares as revenue subject to income tax if the taxpayer is in the business of dealing in real estate and shares.

ESTATE TAX

Estate duty is charged on the market value of all movable and immovable property in Singapore, settled or not settled, which passes on the death of the owner. If the deceased was domiciled in Singapore at the time of death, all movable property outside Singapore is also subject to estate duty.

The rates of duty, which are normally not changed from year to year, are 5 percent for the first S$10 million of the estate and 10 percent on the excess.

The following items are exempt from estate duty:

- Any number of residential properties up to an aggregate value of S$3 million; and

---

NONRESIDENTS

The rates of tax applied to income of nonresident individuals are set forth in the following table.

Income Category	Rate (%)
Income from employment (other than from directorships)	15%
(Or the tax payable as a resident, whichever is greater. However, if a nonresident individual is employed in Singapore for no more than 60 days in a calendar year, the employment income is exempt from Singapore tax.)	
Income from employment as a company director	27
Income from a trade, business, profession, or vocation	27
Interest (other than from approved banks or loans on which the interest is tax-exempt)	27
Royalties, rents, or dividends	27
Income of public entertainers *(net of expenses)*	15
Other sources	27

Foreign-source income received in Singapore by a nonresident is specifically tax-exempt. Nonresidents must file tax returns with the Inland Revenue Authority of Singapore by April 15 of each assessment year.

- Taxable property up to S$500,000 in value or the balance in the Central Provident Fund account, if greater.

Singapore does not impose a gift tax.

SOCIAL SECURITY TAXES

No social security taxes are currently levied in Singapore. However, a statutory savings scheme known as the Central Provident Fund provides for an employee's old age retirement. Both employees and employers are required to contribute to the scheme. Expatriate employees may be exempt from the scheme if they satisfy certain conditions.

The rates of contribution are 21.5 percent of ordinary monthly wages up to a maximum contribution of S$1,290 for employees less than 55 years of age and 18.5 percent for employers, up to a maximum contribution of S$1,110. These rates are current until changed by subsequent legislation.

Contributions on bonuses and nonregular wages are subject to limits if the employee's total wages for the year exceed S$100,000. If total annual wages exceed S$100,000 and annual ordinary wages do not exceed S$72,000, contributions on additional wages are payable on an amount up to a limit of S$100,000 less annual ordinary wages. If annual ordinary wages exceed S$72,000, the contributions on the additional wages may not exceed the amount that would be payable on 40 percent of annual ordinary wages.

Lower rates of contribution apply to employees 55 years of age or older. Individuals who carry on a trade, business, profession, or vocation may also participate in the scheme to a limited extent.

On reaching the age of 55, an employee is entitled to withdraw free of all taxes the accumulated contribution up to a certain limit, plus accrued interest. If the employee permanently leaves Singapore (and Malaysia) at an earlier time, the funds may be withdrawn. The employee's balance may also be withdrawn for certain specified purposes, such as the acquisition of residential property, investment in shares, and the payment of certain hospital expenses for the entire family.

ADMINISTRATION

The tax year is the assessment year, and tax is levied on a preceding-year basis. For example, in the 1994 assessment year, tax is levied on income from the 1993 calendar year. Individuals must file returns by April 15 of the assessment year. Sole proprietors and partners must attach their certified financial statements.

An individual may pay the tax due for each year of assessment in one lump sum within one month of the issuance of an assessment. Alternatively, the tax may be paid in installments, up to a maximum of 12.

DOUBLE TAX RELIEF/ DOUBLE TAX TREATIES

Relief for double taxation is granted on income from accounting, legal, medical, engineering, architectural, computer-related, technical, and other consulting services performed in certain countries that do not have a tax treaty with Singapore.

Double taxation relief is also available for foreign taxes levied on income tax in Singapore if Singapore has a tax treaty with the country concerned and the individual is resident in Singapore for tax purposes.

Singapore has concluded tax treaties with the following countries.

Australia	Netherlands
Bangladesh	New Zealand
Belgium	Norway
Canada	Pakistan
China	Papua New Guinea
Denmark	Philippines
Finland	Saudi Arabia (1)
France	Sri Lanka
Germany	Sweden
India	Switzerland
Indonesia	Taiwan
Israel	Thailand
Italy	United Arab Emirates (1)
Japan	United Kingdom
Korea	United States (2)
Malaysia	

(1) Applies to income only from international air transport.
(2) Applies to income only from international sea and air transport.

Double taxation relief provided by the tax treaties is generally computed by the following method.

- If the foreign tax is less than the Singapore effective tax payable on the foreign income, the relief granted is the full amount of the foreign tax paid.
- If the foreign tax is greater than the Singapore effective tax payable on the foreign income, the relief is restricted to the Singapore tax payable.

Individuals who receive employment income in Singapore and are tax residents of countries that have concluded tax treaties with Singapore may be exempt from Singapore income tax if their period of employment in Singapore does not exceed a certain number of days (usually 183) in a calendar year and they satisfy certain additional criteria specified in the treaties.

Ports & Airports

The late 1980s and early 1990s have seen a major increase in the amount of air and shipping traffic passing through East Asia. Planners saw this coming some years back, and have been scrambling to expand and improve facilities at ports and airports throughout the region; many are currently at or over capacity. Air cargo traffic has been growing faster in Asia than anywhere else in the world, and passenger traffic has increased by leaps and bounds. However, with all the new facilities opening in the near future, there are estimates that by 1997 airport capacity will actually exceed demand. This may mean that airlines and cargo carriers will schedule more frequent service with smaller aircraft, something which is not currently possible, largely because of the small number of slots available at airport terminals and the many major airports operating with only one runway. Long a major center for shipping, Asia is fast becoming the leader in container port traffic. The largest increases in container traffic worldwide have been at the Asian hub ports, and four of the world's five leading countries in container traffic are located on the western side of the Pacific Rim: Japan, Singapore, Hong Kong, and Taiwan are ranked two through five, respectively. The US, which covers the eastern edge of the Pacific, is ranked number one.

CHANGI INTERNATIONAL AIRPORT

Changi International Airport in Singapore is one of the most efficient and most pleasant airports in the world, and has become even more so since the opening of Terminal 2 in 1990. While new terminals and airports are currently under construction throughout Asia, Singapore already has one of the best facilities anywhere. Both Hong Kong and Bangkok, two other Asian hub airports, are straining to operate above capacity at this time, while Singapore has at least a little bit of room. It is expected that by 1995 Changi will reach its capacity of 24 million passengers annually, but officials believe the airport will operate fairly comfortably until Terminal 3 opens in 1997. And unlike many Asian airports, Changi already has two runways, one of which was recently lengthened; a third runway will not be built unless a fourth terminal is constructed, which is unlikely for the forseeable future. The major missing link at Changi International is the lack of a rail link to the city center. Singapore has an excellent Mass Rapid Transit rail system, but political squabbles have stymied efforts to extend it to the airport, and a solution does not appear to be on the horizon.

Changi is the 13th busiest air cargo facility in the world, handling 721,000 metric tons in 1992, just a shade behind Taipei's Chiang Kai Shek International Airport, and the annual tonnage is increasing at a rate of approximately 10 percent a year. Another telling indicator for the airport's future in cargo is the number and range of cargo carriers. There are now 65 air cargo carriers operating regular flights from Changi, representing every part of the globe, from the US to the Indian sub-continent, from Eastern Europe to East Asia and Oceania, and from the Middle East to South America. The airport has more than 64,000 square meters of warehouse space available for cargo and three cargo agent buildings. Singapore Airlines is the major carrier, while Thai Air, Garuda Indonesia, and Quantas all run a large range of aircraft out of Changi. All-cargo carriers include Asia Pacific Air Cargo, Cargolux, Nippon Cargo, Federal Express, and United Parcel Service, while the major US carriers are United Airlines and Northwest. For addresses and telephone numbers, refer to the Transport section of "Important Addresses" chapter.

Changi International Airport is administered by:

Civil Aviation Authority of Singapore
PO Box 1
Singapore 9181
Tel: [65] 5421122, 5412179 (Air Cargo Dept.)
Fax: [65] 5421231, 5425390 (Air Cargo Dept.)

Domestic Transportation

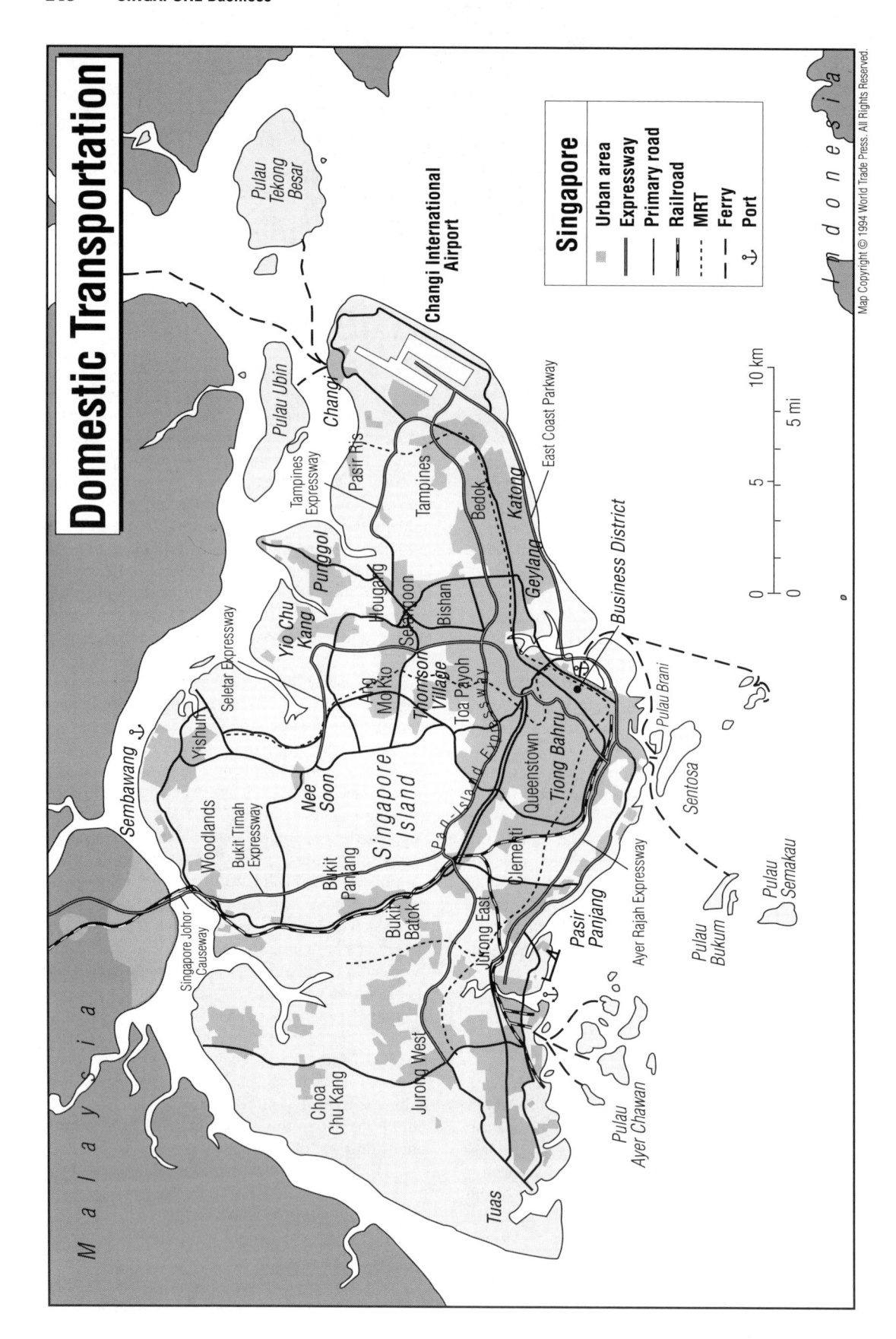

PORT OF SINGAPORE

Singapore is one of the world's busiest ports, and until recently the busiest container port in the world; in this category, it has slipped to a very close second place behind Hong Kong. The main wharves and ports in use are: Tanjong Pagar Container Terminal, Keppel Wharves, Sembawang Wharves, Pasir Panjang Wharves, and Jurong Port. The latter is owned by the Jurong Town Corporation. The Tanjong Pagar Container Terminal handles the majority of container traffic, although the other wharves and ports are equipped for containers to various degrees. A new container terminal is under construction on the island of Pulau Brani, which will be connected to Tanjong Pagar Container Terminal at East Lagoon by a causeway. The first berth opened in early 1992, and when finished, it will have five container berths and three feeder berths, with storage for 51,000 TEU's. New distribution complexes at Pasir Panjang and on Keppel Road have recently begun operation as well. Singapore is the home of the largest ship repair operation in the world, with 21 drydocks and a vast range of services available.

The Singapore port is operated by:

Port of Singapore Authority
460 Alexandra Rd., PSA Bldg.
Singapore 0511
Tel: [65] 2747111 Fax: [65] 2744677
Tlx: 21507 PORT

The Jurong Port is owned by:

Jurong Town Corporation
Port Office, Jurong Port Road
Singapore 2261
Tel: [65] 2650666 Fax: [65] 2656614

Facilities available:

Transportation Service—Truck only.
Cargo Storage—Covered, 2 million square meters. Open, 16,517 square meters. Refrigerated, 27,980 square meters.
Special Cranes—Heavy lift capacity is 120 metric tons. Container, 18 with 36 metric ton capacity.
Air Cargo—Changi Airport is located 20 km from port.
Cargo Handling—Port equipment is adequate to handle most normal cargo traffic. For specialized handling needs, 1 ore and bulk cargo terminal, 6 tanker terminals, 1 liquefied gas terminal and 3 Ro-Ro berths are on hand.
Weather—Weather is extremely wet with over 230 cms of rain yearly.

Business Dictionary

The transliteration system used in this mini-dictionary for Chinese, the primary non-English business language in Singapore, is known as **pin-yin,** the official Chinese phonetic system used in the People's Republic of China and in Singapore. An older transliteration system know as Wade-Giles is still used in Taiwan. No transliteration system is ideal since there are some sounds that cannot be adequately represented using the Latin alphabet. Only through listening to and imitating the pronunciation of native speakers can a truly accurate reproduction of the actual sounds of the language be achieved.

In the "Pronunciation" column, multi-syllabic Chinese words have been joined by a hyphen to assist reading and pronunciation.

TONES

Each Chinese syllable contains one of four tones or is unstressed.

1st tone (high level): is spoken high and the voice does not rise or fall.
2nd tone (rising): starts with the voice lower at the entry point, then finishes at the same level as the 1st tone.
3rd tone (falling-rising): starts with the voice lower than the 2nd tone, then dips and rises to a point just lower than the 1st tone.
4th tone (falling): the voice falls from high to low.

Each syllable is pronounced with one of these tones unless it is unstressed. In such cases the tone distinctions are absent and the unstressed syllable is pronounced light and short.

VOWELS

a	like **a** in j**a**r (but without the r-sound)
e	like **e** in h**e**r (but without the r-sound)
i	(1) like **ee** in f**ee**
	(2) after **c, s, z, ch, sh, r,** like **i** in h**i**t but sounded at the back of the mouth
o	like **aw** as in p**aw**
u	like **oo** as in m**oo**n
ü	similar to German **ü** or **u** in French l**u**ne
ou	like **ou** in s**ou**l
ian	like y**en**
ui	like **way**

In syllables with compound vowels, the pronunciation starts from one vowel and "flows" into the other(s) e.g. **i, ia, iao.**

CONSONANTS

These are pronounced approximately the same as in English, with the following exceptions:

c	like **ts** in hi**ts**
ch	like **ch** in **ch**urch, but the tip of the tongue is curled back to touch the roof of the mouth
g	always as in **g**ap
h	like **ch** in Scottish lo**ch**
j	like **j** in **j**eep (pronounced as near to the front of the mouth as possible)
q	like **ch** in **ch**eer
r	like **r** in English, but the tongue is curled back to touch the roof of the mouth so that it sounds something like the **s** in plea**s**ure.
s	always like **s** in **s**ap
sh	like **sh** in **sh**ip but with the tongue curled back to touch the roof of the mouth
x	like **sh** in **sh**ip but with the lips spread widely
y	like **y** in **y**ellow
z	like **ds** in bi**ds**
zh	like **j** in **j**eep but with the tip of the tongue curled up to touch the roof of the mouth

The consonants **p, t,** and **k** should be pronounced with a strong puff of breath.

252 SINGAPORE Business

English	Mandarin	Transliteration	Pronunciation

GREETINGS AND POLITE EXPRESSIONS

English	Mandarin	Transliteration	Pronunciation
Hello (morning)	早安	zao3 an1	dzow ahn
(daytime)	您好	nin2 hao3	nin how (rhymes with how)
(evening)	晚安	wan3 an1	wahn ahn
Good-bye	再见	zai4 jian4	dzy jen
Please	请	qing3	ching
Pleased to meet you.	很高兴能认识您	hen3 gao1 xing4 neng2 ren4 shi nin2	hun gow-shing nung ren-shi (shi as in ship) nin
Please excuse me.	对不起	dui4 bu qi3	dway boo chee
Excuse me for a moment. (when leaving a meeting)	对不起，请稍等	dui4 bu qi3, qing3 shao1 deng3	dway boo chee, ching show (rhymes with how) dung
Congratulations	祝贺您	zhu4 he4 nin2	jew-her nin
Thank you.	谢谢	xie4 xie	sheh-sheh
Thank you very much.	非常感谢	fei1 chang2 gan3 xie4	fay-chahng gahn-sheh
Thank you for the gift.	谢谢您的礼物	xie4 xie nin2 de li3 wu4	sheh-sheh nin-duh lee-woo
I am sorry. I don't understand Chinese.	对不起，我不懂中文	dui4 bu qi3, wo3 bu4 dong3 zhong1 wen2	dway boo chee, war boo dong jong-wen
Do you speak English?	你会说英语吗？	ni3 hui4 shuo1 ying1 yu3 ma?	nee hway shaw ying-yu (French u as in lune) mah?
My name is...	我叫...	wo3 jiao4...	war jow...
Is Mr./Ms. ... there? on the telephone	...先生/女士在吗？	...xian1-sheng/nu3 shi4 zai4 ma?	...shen-shung/nu (French u) -shi dzigh mah?
Can we meet (tomorrow)?	明天见好吗？	ming2 tian1 jian4 hao3 ma?	ming-tyen jen how mah?
Would you like to have dinner together?	我们一起吃晚饭好吗？	wo3 men yi1 qi3 chi1 wan3 fan4 hao3 ma?	warm'n ee-chee chi (as in chip) wahn-fahn how mah?
Yes	好的	hao3 de	how-duh
No	不行	bu4 xing2	boo-shing

DAY/TIME OF DAY

English	Mandarin	Transliteration	Pronunciation
morning	上午	shang4 wu3	shahng-woo
noon	中午	zhong1 wu3	jong-woo
afternoon	下午	xia4 wu3	shah-woo
evening	傍晚	bang4 wan3	bahng-wahn
night	夜晚	ye4 wan3	yeh-wahn
today	今天	jin1 tian1	jin-tyen
yesterday	昨天	zuo2 tian1	dzaw (as in law)-tyen
tomorrow	明天	ming2 tian1	ming-tyen
Monday	星期一	xing1 qi1 yi1	shing-chee ee
Tuesday	星期二	xing1 qi1 er4	shing-chee err

English	Mandarin	Transliteration	Pronunciation
Wednesday	星期三	xing1 qi1 san1	shing-chee sahn
Thursday	星期四	xing1 qi1 si4	shing-chee si (as in sip)
Friday	星期五	xing1 qi1 wu3	shing-chee woo
Saturday	星期六	xing1 qi1 liu4	shing-chee lyoo
Sunday	星期日	xing1 qi1 ri4	shing-chee ri (as in rip)
holiday	假日	jia4 ri4	jyah ri
New Year's Day	新年，元旦	xin1 nian2 , yuan2 dan4	shin nyen, ywahn dahn
time	时间	shi2 jian1	shi-jen

NUMBERS

English	Mandarin	Transliteration	Pronunciation
one	一	yi1	yee
two	二	er4	err
three	三	san1	sahn
four	四	si4	si
five	五	wu3	woo
six	六	liu4	lyoo
seven	七	qi1	chee
eight	八	ba1	bah
nine	九	jiu3	jew
ten	十	shi2	shi (as in ship)
eleven	十一	shi2 yi1	shi-ee
fifteen	十五	shi2 wu3	shi-woo
twenty	二十	er4 shi2	err-shi
twenty-one	二十一	er4 shi2 yi1	err-shi-ee
thirty	三十	san1 shi2	sahn-shi
thirty-one	三十一	san1 shi2 yi1	sahn-shi-ee
fifty	五十	wu3 shi2	woo-shi
one hundred	一百	yi4 bai3	ee-buy
one hundred one	一百零一	yi4 bai3 ling2 yi1	ee buy ling ee
one thousand	一千	yi4 qian1	ee chen
one million	一百万	yi4 bai3 wan4	ee buy wahn
first	第一	di4 yi1	dee-ee
second	第二	di4 er4	dee-err
third	第三	di4 san1	dee-sahn

English	Mandarin	Transliteration	Pronunciation
GETTING AROUND TOWN			
Where is...?	...在哪里？	...zai4 na3 li3?	...dzigh nah-lee?
Does this train go to ...?	这辆火车去...吗？	zhe4 liang4 huo3 che1 qu4...ma?	jay lyahng whar-cher chu (French u)...mah?
Please take me to (location)	请送我到...（地点）	qing3 song4 wo3 dao4 ...	ching song war dow (ow as in how)...
Where am I?	这是哪里？	zhe4 shi4 na2 li3?	jay shi nah-lee?
airplane	飞机	fei1 ji1	fay-jee
airport	飞机场	fei1 ji1 chang3	fay-jee chahng
bus (public)	汽车（公共汽车）	qi4 che1(gong1 gong4 qi4 che1)	chee cher (gong-gong chee-cher)
taxi	出租车	chu1 zu1 che1	choo-dzoo cher
train	火车	huo3 che1	whar-cher
train station	火车站	huo3 che1 zhan4	whar-cher jahn
ticket	车票	che1 piao4	cher-pyow (ow as in cow)
one-way (single) ticket	单程车票	dan1 cheng2 che1 piao4	dahn-chung cher-pyow
round trip (return) ticket	双程（往返）车票	shuang1 cheng2 (wang2 fan3) che1 piao4	shwahng-chung(wahng-fahn) cher-pyow
PLACES			
airport	飞机场	fei1 ji1 chang3	fay-jee chahng
bank	银行	yin2 hang2	yin-hahng
barber shop	理发店	li3 fa4 dian4	lee-fah-dyen
beauty parlor	美容厅	mei3 rong2 ting1	may-rong ting
business district	商业区	shang1 ye4 qu1	shahng-yeh chu (French u)
chamber of commerce	商会	shang1 hui4	shahng hway
clothes store	服装店	fu2 zhuang1 dian4	foo-jwahng dyen
exhibition	展览	zhan2 lan3	jahn-lahn
factory	工厂	gong1 chang3	gong-chahng
hotel	旅馆	lü2 guan3	lu (French u)-gwahn
hospital	医院	yi1 yuan4	ee-ywahn
market	市场	shi4 chang3	shi-chahng
post office	邮局	you2 ju2	yoe (oe as in Joe)-ju (French u)
restaurant	餐馆	can1 guan3	tsahn-gwahn
rest room/toilet (W.C.)	盥洗室/卫生间	guan4 xi3 shi4/wei4 sheng1 jian1	gwahn-shee-shi, way-shung-jen
sea port	海港	hai2 gang3	high-gahng
train station	火车站	huo3 che1 zhan4	whar-cher jahn

English	Mandarin	Transliteration	Pronunciation
At the bank			
What is the exchange rate?	兑换率是多少？	dui4 huan4 lü4 shi4 duo1 shao3?	dway-hwahn lu (French u) shi daw show (rhymes with how)?
I want to exchange...	我想兑换...	wo2 xiang3 dui4 huan4 ...	war shyahng hwahn...
Australian dollar:	澳元	ao4 yuan2	ow (Rhymes with now) ywahn
British pound	英磅	ying1 bang4	ying bahng
Chinese yuan (PRC)	人民币元（中华人民共和国货币）	ren2 min2 bi4 yuan2 (zhong1 hua2 ren2 min2 gong4 he2 guo2 huo4 bi4)	ren-min-bee ywahn (jong hwah ren-min gong- her-gwar hwar-bee)
French franc	法郎	fa4 lang2	fah-lahng
German mark	德国马克	de2 guo2 ma3 ke4	der-gwor mah-ker
Hong Kong dollar	港币	gang3 bi4	gahng-bee
Indonesia rupiah	印度尼西亚卢比	yin4 du4 ni2 xi1 ya4 lu2 bi3	yin-doo-nee-she-ya loo-bee
Japanese yen	日元	ri4 yuan2	ri ywahn
Korean won	韩国圆	han2 guo2 yuan2	hahn-gwar ywahn
Malaysia ringgit	马来西亚林吉特	ma3 lai2 xi1 ya4 lin2 ji2 te4	mah-ly-shee-ya lin-jee-ter
Philippines peso	菲律宾比索	fei1 lü4 bin1 bi3 suo3	fay-lu (French u)-bin bee-saw
Singapore dollar	新加坡元	xin1 jia1 po1 yuan2	shin-jah-paw ywahn
New Taiwan dollar (ROC)	新台币	xin1 tai2 bi4	shin tigh bee
Thailand baht	泰国铢	tai4 guo2 zhu1	tigh-gwaw joo
U.S. dollar	美元	mei3 yuan2	may ywahn
Where should I sign?	我在哪里签字？	wo3 zai4 na2 li3 qian1 zi4?	war dzigh nar-lee chen dzi (i as in zip)?
Traveler check	旅行支票	lü3 xing2 zhi1 piao4	lu (French u)-shing ji-pyow
Bank draft	银行汇票	yin2 hang2 hui1 piao4	yin-hahng hway-pyow
At the hotel			
I have a reservation.	我已经预订了房间	wo2 yi3 jing1 yu4 ding4 le fang2 jian1	war ee-jing yu (French u)-ding-le fahng-jen
Could you give me a single/ double room?	能给我订一个单人/双人房间吗？	neng2 gei3 wo ding4 yi2 ge4 dan1 ren2/shuang1 ren2 fang2 jian1 ma?	nung gay waw ding ee-guh dahn-ren/shwahng-ren fahng-jen mah?
Is there...?	有没有... ?	you3 mei2 you3...?	yoe may-yoe...?
air-conditioning	空调	kong1 tiao2	kong-tyow
heating	暖气	nuan3 qi4	nwahn chee
private toilet	专用盥洗室	zhuan1 yong4 guan4 xi3 shi4	jwahn-yong gwahn-shee shi (i as in ship)
hot water	热水	re4 shui3	rer shway
May I see the room?	我能看一下房间吗？	wo3 neng2 kan4 yi2 xia4 fang2 jian1 ma?	war nung kahn ee-sha fahng-jen mah?
May I have my bill?	请给我帐单	qing3 gei2 wo3 zhang4 dan1	ching gay war jahng-dahn

256 SINGAPORE Business

English	Mandarin	Transliteration	Pronunciation
At the store			
Do you sell...?	这里有没有...？	zhe4 li you3 mei2 you3...?	jer-lee yoe may-yoe...?
Do you have anything less expensive?	有便宜些的吗？	you3 pian2 yi xie1 de ma?	yoe pyen-ee sheh-duh mah?
I would like (quantity).	我想要（数量）...	wo2 xiang3 yao4...	war shyahng-yow (rhymes with how)...
I'll take it.	我要这件	wo3 yao4 zhei4 jian4	war yow jay jen
I want this one.	我想要这个	wo2 xiang3 yao4 zhei4 ge4	war shyahng yao jay-guh
When does it open/close?	什么时候开/关门？	shen2 mo shi2 hou4 kai1/ guan1 men2?	shemma shi-hoe ky/gwahn mun?

COUNTRIES

America (USA)	美国	mei3 guo2	may gwaw
Australia	澳大利亚	ao4 da4 li4 ya4	ow-da-lee-ya
China (PRC)	中华人民共和国	zhong1 hua2 ren2 min2 gong4 he2 guo2	jong-hwah ren-min gong her gwaw
France	法国	fa4 guo2	fah-gwaw
Germany	德国	de2 guo2	der-gwaw
Hong Kong	香港	xiang1 gang3	shyahng-gahng
Indonesia	印度尼西亚	yin4 du4 ni2 xi1 ya4	yin-doo-nee-shee-yah
Japan	日本	ri4 ben3	ri-bun
Korea	韩国	han2 guo2	hahn-gwaw
Malaysia	马来西亚	ma3 lai2 xi1 ya4	mah-ly-shee-yah
Philippines	菲律宾	fei1 lü4 bin1	fay-lu (French u)-bin
Singapore	新加坡	xin1 jia1 po1	shin jyah paw
Taiwan (ROC)	台湾	tai2 wan1	tigh wahn
Thailand	泰国	tai4 guo2	tigh gwaw
United Kingdom	英国	ying1 guo2	ying gwaw

EXPRESSIONS IN BUSINESS

1) General business- related terms

accounting	会计	kuai4 ji4	kwyh-jee
additional charge	额外收费	e2 wai4 shou1 fei4	er-wy fay-yong
advertise	登广告	deng1 guang3 gao4	dung gwahng-gow (as in how)
advertisement	广告	guang3 gao4	gwahng-gow (as in how)
bankrupt	破产	po4 chan3	paw-chahn
brand name	商标, 牌子	shang1 biao1, pai2 zi	shahng-byow, pie-dzi
business	生意	sheng1-yi4	shung-ee

BUSINESS DICTIONARY

English	Mandarin	Transliteration	Pronunciation
buyer	买方	mai3 fang1	my fahng
capital (money)	资金	zi1 jin1	dzi (i as in zip)-jin
cash	现金	xian4 jin1	shyen-jin
charge	记帐	ji4 zhang4	jee-jahng
check	支票	zhi1 piao4	ji-pyow
claim	索赔	suo3 pei2	saw-bay
collect	收帐	shou1 zhang4	show (as in low) jahng
commission	佣金	yong1 jin1	yong jin
company	公司	gong1 si1	gong-si
copyright	版权	ban3 quan2	bahn-chwahn
corporation	股份有限公司	gu3 fen4 you3 xian4 gong1 si	goo-fun yoe-shen gong-si
cost (expense)	费用	fei4 yong4	fay-yong
currency	货币	huo4 bi4	hwaw-bee
customer	客户	ke4 hu4	ker-hoo
D/A (documents against acceptance)	承兑交单	cheng2 dui4 jiao1 dan1	chung-dway jyow-dahn
D/P (documents against payment)	付款交单	fu4 kuan3 jiao1 dan1	foo-kwahn jyow dahn
deferred payment	延期付款	yan2 qi1 fu4 kuan3	yen-chee foo-kwahn
deposit	存款,押金	cun2 kuan3, ya1-jin1	tsoun (ou as in could)-kwahn, yah-jin
design	设计	she4 ji4	sher-jee
discount	折扣	zhe2 kou4	jer-kow (as in low)
distribution	分配	fen1 pei4	fun-pay
dividends	红利	hong2 li4	hong-lee
documents	文件	wen2 jian4	wun-jen
due date	到期日	dao4 qi1 ri4	dow chee ri
exhibit	展览	zhan2 kan3	jahn-lahn
ex works	工厂交货	gong1 chang3 jiao1 huo4	gong-chahng jyow-hwaw
facsimile (fax)	传真	chuan2 zhen1	chwahn-jun
finance	财务，金融	cai2 wu4, jin1 rong2	tsigh-woo, jin-rong
foreign businessman	外商	wai4 shang1	wigh-shahng
foreign capital	外资	wai4 zi1	wigh-dzi
foreign currency	外汇	wai4 hui4	wigh-hway
foreign trade	对外贸易	dui4 wai4 mao4 yi4	dway-wigh mow (as in how)-yee
government	政府	zheng4 fu3	jung-foo

English	Mandarin	Transliteration	Pronunciation
industry	工业	gong1 ye4	gong-yeh
inspection	检查	jian3 cha2	jen-chah
insurance	保险	bao2 xian3	bow (as in cow)-shen
interest	利息	li4 xi1	lee-shee
international	国际的	guo2 ji4 de	gwaw-jee-duh
joint venture	合资	he2 zi1	her-dzi
label	标签	biao1 qian1	byow-chen
letter of credit	信用证	xin4 yong4 zheng4	shin-yong-jung
license	许可证	xu2 ke3 zheng4	shu (French u)-ker jung
loan	贷款	dai4 kuan3	digh-kwahn
model (of a product)	产品模型	chan2 pin3 mo2 xing2	chahn-pin more-shing
office	办公室	ban4 gong1 shi4	bahn-gong-shi
patent	专利	zhuan1 li4	jwahn-lee
pay	支付	zhi1 fu4	ji-foo
payment for goods	物品付款	wu4 pin3 fu4 kuan3	woo-pin foo-kwahn
payment by installment	分期付款	fen1 qi1 fu4 kuan3	fun-chee foo-kwahn
permit	许可	xu2 ke3	shu (French u)-ker
principal	本金	ben3 jin1	bun-jin
private (not government)	私营（非政府性）	si1 ying2 (fei1 zheng4 fu3 xing4)	si (as in sip)-ying (fay jung-foo-shing)
product	产品	chan2 pin3	chahn-pin
profit margin	利润幅度	li4 run4 fu2 du4	lee-roun (ou as in could) foo-doo
registration	注册	zhu4 ce4	joo-tser
report	报告	bao4 gao4	bow-gow (both ow as in how)
research and development (R&D)	研究与发展	yan2 jiu1 yu3 fa1 zhan3	yen-jyew yu (French u) fah-jahn
return (on investment)	（投资）收入	(tou2 zi1) shou1 ru4	(toe-dzi)show (as in low)-roo
sample	样品	yang4 pin3	yahng-pin
seller	卖方	mai4 fang1	migh-fahng
settle accounts	结帐	jie2 zhang4	jyeh-jahng
service charge	服务费	fu2 wu4 fei4	foo-woo-fay
sight draft	即期汇票	ji2 qi1 hui4 piao4	jee-chee hway-pyow (as in how)
tax	税	shui4	shway
telephone	电话	dian4 hua4	dyen-hwah
telex	电传	dian4 chuan2	dyen-chwahn
trademark	商标	shang1 biao1	shahng-byow (as in how)
Visa	Visa信用卡	Visa xin4 yong4 ka3	Visa shin-yong kar

English	Mandarin	Transliteration	Pronunciation

2) Labor

compensation	薪水	xin1 shui3	shin-shway
employee	雇员	gu4 yuan2	goo-ywahn
employer	雇主	gu4 zhu3	goo-joo
fire, dismiss	解雇	jie3 gu4	jyeh-goo
foreign worker	外籍工人	wai4 ji2 gong1 ren2	wigh-jee gong-ren
hire	雇用	gu4 yong4	goo-yong
immigration	移民	yi2 min2	ee-min
interview	面试	mian4 shi4	myen-shi
laborer:	工人	gong1 ren2	gong-ren
skilled	熟练工	shu2 lian4 gong1	shoo-lyen-gong
unskilled	非熟练工	fei1 shu2 lian4 gong1	fay shoo-lyen-gong
labor force	劳动力	lao2 dong4 li4	low (as in how)-dong lee
labor shortage	劳力短缺	lao2 li4 duan3 que1	low-lee dwahn-chweh
labor stoppage	停工	ting2 gong1	ting-gong
labor surplus	人工过剩	ren2 gong1 guo4 sheng4	ren-gong gwaw-shung
minimum wage	最低工资	zui4 di1 gong1 zi1	dzway-dee gong-dzi (i as in zip)
profession/ occupation	职业	zhi2 ye4	ji-yeh
salary	薪水	xin1 shui3	shin-shway
strike	罢工	ba4 gong1	bah-gong
training	培训	pei2 xun4	pay-shune (French u)
union	工会	gong1 hui4	gong-hway
wage	工资	gong1 zi1	gong-dzi (i as in zip)

3) Negotiations (Buying / Selling)

agreement	协议	xie2 yi4	shyeh-ee
arbitrate	仲裁	zhong4 cai2	jong-tsigh
brochure, pamphlet	手册，小册子	shou3 ce4, xiao3 ce4 zi	show (as in low)-tser, shyow (rhymes with how) tser-dzi
buy	买	mai3	migh
confirm	确认	que4 ren4	chweh-ren
contract	合同，契约	he2 tong2, qi4 yue1	her-tong, chee-yweh
cooperate	合作	he2 zuo4	her-dzaw
cost	价值	jia4 zhi2	jah-ji
counteroffer	还价	huan2 jia4	hwahn-jah
countersign	会签	hui4 qian1	hway-chen
deadline	截止日期	jie2 zhi3 ri4 qi1	jyeh-ji ri-chee

English	Mandarin	Transliteration	Pronunciation
demand	要求	yao1 qiu2	yow (as in how)-chyow (as in low)
estimate	估计	gu1 ji4	goo-jee
guarantee	保证	bao3 zheng4	bow (as in how)-jung
label	标签	biao1 qian1	byow (as in how)-chen
license	许可证	xu2 ke3 zheng4	shu (French u)-ker-jung
market	市场	shi4 chang3	shi (i as in ship)-chahng
market price	市场价	shi4 chang3 jia4	shi-chahng jah
minimum quantity	最低量	zui4 di1 liang4	dzway dee lyahng
negotiate	谈判	tan2 pan4	tahn-pahn
negotiate payment	付款谈判	fu4 kuan3 tan2 pan4	foo-kwahn tahn-pahn
order	订单	ding4 dan1	ding-dahn
packaging	包装	bao1 zhuang1	bow (as in how)-jwahng
place an order	发出订单	fa1 chu1 ding4 dan1	fah-choo ding-dahn
price	价格	jia4 ge2	jyah-ger
price list	价格表	jia4 ge2 biao3	jyah-ger-byow (rhymes with how)
product features	产品特点	chan2 pin3 te4 dian3	chahn-pin ter-dyen
product line	产品系列	chan2 pin3 xi4 lie4	chahn-pin shee-lyeh
quality	质量，品质	zhi4 liang4 , pin3 zhi4	ji (short i as in zip)-lyahng, pin-ji
quantity	数量	shu4 liang4	shoo-lyahng
quota	配额	pei4 e2	pay-er
quote (offer)	报价	bao4 jia4	bow (as in how)-jyah
sale	销售	xiao1 shou4	shyow (as in how)-show (as in low)
sales confirmation	销售确认书	xiao1 shou4 que4 ren4 shu1	shyow-show chweh-ren-shoo
sell	销售	xiao1 shou4	shyow-show
sign	签署	qian1 shu3	chen-shoo
signature	签字	qian1 zi4	chen-dzi
specifications	规范	gui1 fan4	gway-fahn
standard (quality)	标准（质量）	biao1 zhun3 (zhi4 liang4)	byow-djoun (ou as in would)ji-lyahng
superior (quality)	优质	you1 zhi4	yow (as in low)-ji
trade	贸易	mao4 yi4	mow (as in how)-ee
unit price	单价	dan1 jia4	dahn-jyah
value	价值	jia4 zhi2	jyah-ji
value added	增值	zeng1 zhi2	dzung-ji
warranty (and services)	保证书（及服务）	bao3 zheng4 shu1 (ji2 fu2 wu4)	bow (as in how)-jung shoo (ji foo-woo)

English	Mandarin	Transliteration	Pronunciation
The price is too high.	价钱太贵	jia4 qian2 tai4 gui4	jyah-chen tigh-gway
We need a faster delivery.	我们需要尽快供货	wo3 men xu1 yao4 jin4 kuai4 gong4 huo4	warm'n shu (French u)-yao jin kwigh gong-hwaw
We need it by...	我们需在...之前收到	wo3 men xu1 zai4 ...zhi1 qian2 shou1 dao4	warm'n shu (French u)-yao dzigh..ji-chen show (as in low)-dow (as in how)
We need a better quality.	我们要比这个质量更好的	wo3 men yao4 bi3 zhe4 ge4 zhi4 liang4 geng4 hao3 de	warm'n yow (as in how) bee jay-guh ji-lyahng gung how (as in cow)-da
We need it to these specifications.	我们要符合这个规范的产品	wo3 men yao4 fu2 he2 zhei4 ge4 gui1 fan4 de chan2 pin3	warm'n yow foo-her jay-guh gway-fahn-da-chahn-pin
I want to pay less.	我想要便宜些的	wo2 xiang3 yao4 pian2 yi2 xie1 de	war shyahng-yao pyen-ee sheh-da
I want the price to include..	我希望这个价钱包括...	wo3 xi1-wang4 zhei4 ge4 jia4 qian2 bao1 kuo4...	war shee-wahng jay-guh jyah-chen bow(as in how)-kaw
Can you guarantee delivery?	您能保证交货时间吗？	nin2 neng2 bao3 zheng4 jiao1 huo4 shi2 jian1 ma?	nin nung bow (as in how)-jung jyow-hwaw shi-jyen mah?

4) Products/ Industries

English	Mandarin	Transliteration	Pronunciation
aluminum	铝	lü3	lu (French u)
automobile	汽车	qi4 che1	chee cher
automotive accessories	汽车零件	qi4 che1 ling2 jian4	chee-cher ling-jyen
biotechnology	生物工艺学	sheng1 wu4 gong1 yi4 xue2	shung-woo gong-ee-shweh
camera	照相机	zhao4 xiang4 ji1	jow (as in how)-shyahng-jee
carpets	地毯	di4 tan3	dee-tahn
cement	水泥	shui3 ni2	shway-nee
ceramics	瓷器	ci2 qi4	tsi-chee
chemicals	化学品	hua4 xue2 pin3	hwah-shweh-pin
clothing	服装	fu2 zhuang1	foo-jwahng
for women	女装	nu3 zhuang1	nu (French u)-jwahng
for men	男装	nan2 zhuang1	nahn-jwahng
for children	童装	tong2 zhuang1	tong-jwahng
coal	煤	mei2	may
computer	电脑	dian4 nao3	dyen-now (as in how)
computer hardware	电脑硬件	dian4 nao3 ying4 jian4	dyen-now ying-jen
computer software	电脑软件	dian4 nao3 ruan3 jian4	dyen-now rwahn-jen
construction	施工	shi1 gong1	shi-gong
electrical equipment	电器设备	dian4 qi4 she4 bei4	dyen-chee sher-bay
electronics	电子	dian4 zi3	dyen-dzi
engineering	工程	gong1 cheng2	gong-chung
fireworks	鞭炮	bian1 pao4	byen-pow (as in how)

English	Mandarin	Transliteration	Pronunciation
fishery products	渔业产品	yu2 ye4 chan2 pin3	yu (French u)-yeh chahn-pin
food products	食品	shi2 pin3	shi-pin
footwear	鞋类	xie2 lei4	shyeh-lay
forestry products	林业产品	lin2 ye4 chan2 pin3	lin-yeh chahn-pin
fuel	燃料	ran2 liao4	rahn-lyow (as in now)
furniture	家俱	jia1 ju4	jyah-ju (French u)
games	游戏	you2 xi4	yow (as in low)-shee
gas	气体	qi4 ti3	chee-tee
gemstone	宝石	bao3 shi2	bow (as in how)-shi
glass	玻璃	bo1 li2	baw-lee
gold	黄金	huang2-jin1	hwahng-jin
hardware	五金器件	wu3 jin1 qi4 jian4	wu-jin chee-jyen
iron	铁	tie3	tyeh
jewelry	珠宝	zhu1 bao3	joo-bow (as in how)
lighting fixtures	灯具	deng1 ju4	dung-ju(French u)
leather goods	皮革制品	pi2-ge2 zhi4 pin3	pee-ger ji-pin
machinery	机械	ji1 xie4	jee-shyeh
minerals	矿物质	kuang4 wu4 zhi4	kwahng-woo ji
musical instruments	乐器	yue4 qi4	yweh-chee
paper	纸张	zhi3 zhang1	ji-jahng
petroleum	石油	shi2 you2	shi-yow (as in low)
pharmaceuticals	药物	yao4 wu4	yow-woo
plastics	塑料	su4 liao4	soo-lyow (as in how)
pottery	陶器	tao2 qi4	tow (as in how)-chee
rubber	橡胶	xiang4 jiao1	shyahng-jyow (as in how)
silk	丝绸	si1 chou2	si-chow (as in low)
silver	银器	yin2 qi4	yin-chee
spare parts	零配件	ling2 pei4 jian4	ling-pay-jen
sporting goods	体育用品	ti3 yu4 yong4 pin3	tee-yu (French u) yong-pin
steel	钢	gang1	gahng
telecommunication equipment	电讯设备	dian4 xun4 she4 bei4	dyen-shune (rhymes with French lune) sher-bay
television	电视	dian4 shi4	dyen-shi
textiles	纺织品	fang3 zhi1 pin3	fahng-ji-pin
tobacco	烟草	yan1 cao3	yen-tsow (as in how)

BUSINESS DICTIONARY

English	Mandarin	Transliteration	Pronunciation
tools:	工具	gong1 ju4	gong-ju (French u)
hand (power)	手动	shou3 dong4	show (as in low)-dong
power	电力	dian4 li4	dyen-lee
tourism	旅游	lü3 you2	lu (French u)-yow (rhymes with low)
toys	玩具	wan2 ju4	wahn-ju (French u)
watches/clocks	手表/钟	shou2 biao3/zhong1	show (as in low)-byow (as in how)
wood	木材	mu4 cai2	moo-tsigh

5) Services

English	Mandarin	Transliteration	Pronunciation
accounting service	会计服务	kuai4 ji4 fu2 wu4	kwigh-jee foo-woo
advertising agency	广告代理商	guang3 gao4 dai4 li3 shang1	gwahng-gow (as in how) digh-lee-shahng
agent	代理人	dai4 li3 ren2	digh-lee-ren
customs broker	报关代理人	bao4 guan1 dai4 li3 ren2	bow (as in how)-gwahn digh-lee-ren
distributor	经销商	jing1 xiao1 shang1	jing-shyow (as in how)-shahng
employment agency	职业介绍所	zhi2-ye4 jie4-shao4-suo3)	ji-yeh jyeh-show(as in how)-saw
exporter	出口商	chu1 kou3 shang1	choo-kow (as in low)-shahng
freight forwarder	货运代理人	huo4 yun4 dai4 li3 ren2	whaw-yune (rhymes with French lune) digh-lee-ren
importer	进口商	jin4 kou3 shang1	jin-kow (as in low)shahng
manufacturer	制造商	zhi4 zao4 shang1	ji-dzow (as in how) shahng
packing service	包装服务	bao1 zhuang1 fu2 wu4	bow (as in how)-jwahng foo-woo
printing company	印刷公司	yin4 shua1 gong1 si1	yin-shwah gong-si
retailer	零售商	ling2 shou4 shang1	ling-show (as in low) shahng
service(s)	服务	fu2 wu4	foo-woo
supplier	供货商	gong4 huo4 shang1	gong hwaw shahng
translation services	翻译服务	fan1 yi4 fu2 wu4	fahn-ee foo-woo
wholesaler	批发商	pi1 fa shang1	pee-fah-shahng

6) Shipping/Transportation:

English	Mandarin	Transliteration	Pronunciation
bill of lading	提单	ti2 dan1	tee-dahn
cost, insurance, freight (CIF)	成本、保险加运费价（到岸价）	cheng2 ben3 , bao2 xian3 jia1 yun4 fei4 jia4 (dao4-an1 jia4)	chung-bun, bow (as in how)-shen jah-yune fay-jah (dow [as in how]-ahn jah)
customs	海关	hai3 guan1	high-gwahn
customs duty	关税	guan1 shui4	gwahn-shway
date of delivery	交货日期	jiao1 huo4 ri4 qi1	jyow (as in how)-hwaw ri-chee
deliver (delivery)	交货	jiao1 huo4	jyow (as in how)-hwaw

English	Mandarin	Transliteration	Pronunciation
export	出口	chu1 kou3	choo kow (as in low)
first class mail	第一类邮件	di4 yi1 lei4 you2 jian4	dee-ee-lay yow (as in low)-jen
free on board (F.O.B.)	船上交货价（离岸价）	chuan2 shang4 jiao1 huo4 jia4 (li2 an4 jia4)	chwahng-shang jyow-hwaw jah (lee ahn jah)
freight	运费	yun4 fei4	yune (French u) fay
import	进口	jin4 kou3	jin-kow (as in low)
in bulk	散装	san3 zhuang1	sahn-jwahng
mail (post)	邮寄	you2 ji4	yow (as in low)-jee
country of origin	原产地	yuan2 chan3 di4	ywahn chahn dee
packing	包装	bao1 zhuang1	bow (as in how)-jwahng
packing list	装箱单	zhuang1 xiang1 dan1	jwahng shyahng dahn
port	港口	gang2 kou3	gahng-kow (as in low)
ship (to send):	发货	fa1 huo4	fah-hwaw
by air	空运	kong1 yun4	kong yune (French u)
by sea	海运	hai3 yun4	high yune (French u)
by train	火车运输	huo3 che1 yun4 shu1	hwaw-cher yune-shoo
by truck	卡车运输	ka3 che1 yun4 shu1	kah-cher yune shoo

WEIGHTS, MEASURES, AMOUNTS

English	Mandarin	Transliteration	Pronunciation
barrel	桶	tong3	tong
bushel	蒲士尔	pu2 shi4 er3	poo-shi-err
centimeter	厘米, 公分	li2 mi3, gong1 fen1	lee-mee, gong-fun
dozen	一打（十二个）	yi1 da2 (shi2 er4 ge4)	ee dah(shi-err-guh)
foot	英尺	ying1 chi3	ying-chi
gallon	加仑	jia1 lun2	jyah-loun (ou as in would)
gram	克	ke4	ker
gross (144 pieces)	罗	luo2	law
gross weight	毛重	mao2 zhong4	mow (as in how)-jong
hectare	公顷	gong1 qing3	gong-ching
hundred (100)	一百	yi4 bai3	ee-bigh
inch	英寸	ying1 cun4	ying-tsoun (ou as in would)
kilogram	公斤	gong1 jin1	gong-jin
kilometer	公里, 千米	gong1 li3, qian1 mi3	gong-lee, chen mee
meter	米	mi3	mee
net weight	净重	jing4 zhong4	jing-jong
mile (English)	英里	ying1 li3	ying-lee
liter	升	sheng1	shung

English	Mandarin	Transliteration	Pronunciation
ounce	盎司	ang4 si1	ahng-si
pint	品脱	pin3 tuo1	pin-taw
pound (weight measure avoirdupois)	磅（常衡重量）	bang4 (chang2 heng2 zhong4 liang4)	bahng (chahng hung jong-lyahng)
quart (avoirdupois)	夸脱	kua4 tuo1	kwah-taw
square meter	平方米	ping2 fang1 mi3	ping fahng mee
square yard	平方码	ping2 fang1 ma3	ping fahng mah
size	尺寸	chi3 cun4	chi-tsoun (ou as in would)
ton	吨	dun1	doun (ou as in would)
yard	码	ma3	mah
jin (Chinese pound)	斤	jin1	jin
liang (Chinese ounce)	两	liang3	lyahng
cun (Chinese inch)	寸	cun4	tsoun (ou as in would)
chi (Chinese foot)	尺	chi3	chi (as in chip)

SINGAPORE-SPECIFIC EXPRESSIONS AND TERMS

English	Mandarin	Transliteration	Pronunciation
You are welcome	不客气	bu2 ke4 qi4	boo ker chee
It doesn't matter	没关系	mei2 guan1 xi4	may gwahn shee
Please do not smoke	请勿吸烟	qing3 wu4 xi1 yan1	ching woo shee yen
Have a nice trip	旅途愉快	lü3 tu2 yu4 kuai4	lu (French u)-too yu (French u) kwigh

COMMON SIGNS

English	Mandarin	Transliteration	Pronunciation
Please do not disturb (sign to put on the door of hotel room)	请勿打扰	qing3 wu4 da2 rao3	ching woo dah-row (as in how)
Enter	入口	ru4 kou3	roo kow (as in low)
Exit	出口	chu1 kou3	choo-kow (as in low)
Men	男厕所（男盥洗室）	nan2 ce4 suo3 (nan2 guan4 xi3 shi4)	nahn tser-saw (nahn gwahn-shee shi)
Women	女厕所（女盥洗室）	nu3 ce4 suo3 (nu3 guan4 xi3 shi4)	nu (French u) tser-saw (nu gwahn shee shi)
No smoking	禁止吸烟	jin4 zhi3 xi1 yan1	jin ji shee yen
Handle with care	小心轻放	xiao3 xin1 qing1 fang4	shyow (as in how) shin ching fahng

Important Addresses

IMPORTANT ADDRESSES TABLE OF CONTENTS

Government	267
Government Agencies	267
Overseas Diplomatic Missions of Singapore	269
Foreign Diplomatic Missions in Singapore	271
Government Run Corporations	274
Trade Promotion Organizations	275
World Trade Center	275
General Trade Associations & Local Chambers of Commerce	275
Foreign Chambers of Commerce & Business Organizations	275
Singapore Economic Development Board	276
Singapore Trade Development Board	277
Industry-Specific Trade Organizations in Singapore	279
Financial Institutions	281
Banks	281
Government Banks	281
Domestic Banks	281
Foreign Banks	281
Insurance Companies	282
Stock & Commodity Exchanges	282
Services	283
Accounting Firms	283
Advertising Agencies	283
Law Firms	284
Translators & Interpreters	285
Transportation	286
Airlines	286
Transportation & Customs Brokerage Firms	287
Publications, Media & Information Sources	290
Directories & Yearbooks	290
Newspapers	292
General Business & Trade Periodicals	292
Industry-Specific Periodicals	293
Radio & Television	297
Libraries	297

GOVERNMENT

GOVERNMENT AGENCIES

Civil Aviation Authority of Singapore
PO Box 1
Singapore 9181
Tel: 5421122 Fax: 5421231 Tlx: 21231
Air cargo Tel: 5412179 Fax: 5425390
Air transport Tel: 5412390 Fax: 5456515

Construction Industry Development Board
National Development
Annex A, 3rd Storey
9 Maxwell Road
Singapore 0106
Tel: 2256711 Fax: 2257301 Tlx: 20818 CONDEB

Customs & Excise Department
1 Maritime Square #03-01 & #10-01
World Trade Centre
Singapore 0409
Tel: 2728222 Fax: 2779090 Tlx: 28817
Telephone Information Service: 2725161

Economic Development Board
250 North Bridge Rd. #24-00
Raffles City Tower
Singapore 0617
Tel: 3362288 Fax: 3396077 Tlx: 26223 SINEDB

Government Publication Division
Singapore National Printers Ltd.
8 Shenton Way #B1-07
Singapore 0106
Tel: 2230834

Industrial Arbitration Court
2nd Storey, City Hall
St. Andrew's Road
Singapore 0617
Tel: 3378191 Fax: 3307215

Information Communication Institute of Singapore
1 Hillcrest Rd. #08-00
Singapore
Tel: 4676000 Fax: 4676601

Jurong Town Corporation
301 Jurong Town Hall Rd., Jurong Town Hall
Singapore 2260
Tel: 5600056 Fax: 5655301 Tlx: 35733 JTC

All addresses and telephone numbers are in the Republic of Singapore unless otherwise noted. The country code for Singapore is [65].

Ministry of Communications
460 Alexandra Rd., PSA Bldg. #39-00
Singapore 0511
Tel: 2707988 Fax: 2799734 Tlx: 25500
Corporate Services & Sea Transport Division
Tel: 2799716

Ministry of Community Development
512 Thomson Rd., MCD Bldg.
Singapore 1129
Tel: 2589595 Tlx: 34361

Ministry of Defence
3 Maritime Sq. #04-00
Singapore 1024
Tel: 4741155 Fax: 7620112 Tlx: 21373

Ministry of Education
Kay Siang Rd.
Singapore 1024
Tel: 4739111 Fax: 4756128 Tlx: 34366

Ministry of Finance
8 Shenton Way, Treasury Bldg.
Singapore 0106
Tel: 2259911 Tlx: 34371
Budget Div. Tel: 3209393 Fax: 3209435
Revenue Fax: 2246847
Commercial Affairs Dept. Tel: 3209438

Ministry of Finance
Central Procurement Office
Depot Road
Singapore 0410
Tel: 2721655 Fax: 2790524

Ministry of Finance
Registry of Companies & Businesses
10 Anson Rd. #05-01/15
Singapore 0207
Tel: 2278551 Fax: 2251676
Public Affairs & Info. Tel: 2283702

Ministry of Foreign Affairs
250 North Bridge Rd. #07-00
Raffles City Tower
Singapore 0617
Tel: 3361177 Fax: 3394330 Tlx: 21242

Ministry of Health
16 College Rd., College of Medicine Bldg.
Singapore 0316
Tel: 2237777 Fax: 2241677 Tlx: 34360

Ministry of Health
Pharmaceutical Dept.
2 Jalan Bukit Merah
Singapore 0316
Tel: 2213014 Fax: 2226797

Ministry of Home Affairs
Phoenix Park, Tanglin Rd.
Singapore 1024
Tel: 2359111 Fax: 7344420 Tlx: 34360

Ministry of Home Affairs
Immigration Department
95 South Bridge Rd.
#08-26 Pidemco Centre
Singapore 0105
Tel: 5322877 Fax: 5301840
Employment Pass Section Tel: 5301866

Ministry of Information and the Arts
Department of Information
460 Alexandra Rd., PSA Bldg.
Singapore 0511
Tel: 2707988 Fax: 2799784

Ministry of Information and the Arts
Department of the Arts
512 Thompson Rd., MCD Bldg.
Singapore 1129
Tel: 2589595 Fax: 3506118

Ministry of Labour
18 Havelock Rd. #07-01
Singapore 0105
Tel: 5341511 Fax: 5344840 Tlx: 34364

Ministry of Law
250 North Bridge Rd. #21-00
Raffles City Tower
Singapore 0617
Tel: 3378191, 3361177 Fax: 3305891 Tlx: 34374

Ministry of National Development
Telok Ayer St., 5/F., MND Bdlg., Annex B
Singapore 0106
Tel: 2221211 Fax: 3226254 Tlx: 34369

Ministry of the Environment
40 Scotts Rd., Environment Bldg.
Singapore 0922
Tel: 7327733 Fax: 7319456

Ministry of Trade and Industry
#33-00 Treasury Bldg.
Singapore 0106
Tel: 2259911 Fax: 3209260 Tlx: 24702
Dept. of Statistics Tel: 2259911

Monetary Authority of Singapore
10 Shenton Way, MAS Bldg.
Singapore 0207
Tel: 2255577 Fax: 2299491 Tlx: 28174 ORCHID

National Computer Board
71 Science Park Dr., NCB Bldg.
Singapore 0511
Tel: 7782211 Fax: 7789641 Tlx: 38610 NCB

National Productivity Board
2 Bukit Merah Central, NPB Bldg.
Singapore 0315
Tel: 2786666 Fax: 2786667 Tlx: 36047
Resources Div. Tel: 2793680
Resource & Information Ctr. Tel: 2793737
International Relations Ctr. Tel: 2793720

National Science & Technology Board
The Pasteur, 16 Science Park Dr. #01-03
Singapore
Tel: 7797066 Fax: 7771711

Office of the Prime Minister
Istana Annexe, Istana
Singapore 0923
Tel: 2358577 Fax: 7324627

Port of Singapore Authority
460 Alexandra Rd., PSA Bldg.
Singapore 0511
Tel: 2747111 Fax: 2744677 Tlx: 21507 PORT

Public Utilities Board
PUB Bldg., 111 Somerset Rd.
Singapore
Tel: 2358888

Singapore Broadcasting Corporation
Caldecott Hill, Andrew Rd.
Singapore
Tel: 2560401

Singapore Trade Development Board
1 Maritime Square #10-40 (Lobby D)
World Trade Centre, Telok Blangah Rd.
Singapore 0409
Tel: 2719388 Fax: 2740770, 2782518
Tlx: 28617 TRADEV
Export Institute Tel: 3342188
Imports & Exports Help Desk Tel: 2790350
Imports & Exports, Changi Airport
Fax: 2724720
Library Tel: 2790433
Market Research Tel: 2790434

Singapore Institute of Standards & Industrial Research (SISIR)
1 Science Park Dr.
Singapore 0511
Tel: 7787777 Fax: 7780086 Tlx: 28499 SISIR

Telecommunication Authority of Singapore
31 Exeter Rd. #05-00 Comcentre
Singapore 0923
Tel: 7343344, 7387788 Fax: 7328428, 7330073
Tlx: 33311 TELECOM

OVERSEAS DIPLOMATIC MISSIONS OF SINGAPORE

Australia
High Commission of the Republic of Singapore
17 Forster Crescent
Yarralumla, ACT 2600, Australia
Tel: [61] (62) 2733944, 2733171
Fax: [61] (62) 2733260 Tlx: 62192

Austria
Consulate of the Republic of Singapore
Raiffeisen Zentralbank
Osterreich AG, Am Stadtpark 9
1030 Wien, Austria
Tel: [43] (222) 71707-1229
Fax: [43] (222) 71707-1656 Tlx: 136989

Belgium
Embassy of the Republic of Singapore
198 Ave Franklin Roosevelt
1050 Brussels, Belgium
Tel: [32] (2) 6602979 Fax: [32] (2) 6608685
Tlx: 26731

Canada
Consulate of the Republic of Singapore
Russell & DuMoulin
#1700-1075 West Georgia St.
Vancouver, BC V6E 3G2, Canada
Tel: [1] (604) 631-4868 Fax: [1] (604) 631-3232

Chile
Consulate of the Republic of Singapore
Amunategui 277, 3/F.
Santiago, Chile
Tel: [56] (2) 6965185, 471345 Fax: [56] (2) 726263
Tlx: 340461

China
Embassy of the Republic of Singapore
1 Xiu Shui Bei Jie
Jianguomenwai
Beijing 100600, PRC
Tel: [86] (1) 5323926, 5323143
Fax: [86] (1) 5322215 Tlx: 22578

Consulate (Shanghai)
400 Wulumuqi Zhong Lu
Shanghai 200031, PRC
Tel: [86] (21) 4370776, 4331362
Fax: [86] (21) 4334150 Tlx: 33540 SINSH

Egypt
Embassy of the Republic of Singapore
40 Sharia Babel St.
Dokki, Cairo 11511
Arab Republic of Egypt
Postal address: ATABA PO Box 356
Cairo, Arab Republic of Egypt
Tel: [20] (2) 704744, 703772, 3495045
Fax: [20] (2) 3481682 Tlx: 21353

France
Embassy of the Republic of Singapore
12 sq. de l'av. Foche
75116 Paris, France
Tel: [33] (1) 45-00-33-61 Fax: [33] (1) 45-00-61-79
Tlx: 630994

Germany
Embassy of the Republic of Singapore
Sudstrasse 133
5300 Bonn 2, Germany
Tel: [49] (228) 312007 Fax: [49] (228) 310527
Tlx: 885642

Greece
Consulate of the Republic of Singapore
10-12 Kifissias Ave.
151 25 Maroussi, Athens, Greece
Tel: [30] (1) 6834875, 6845072 Fax: [30] (1) 6834416
Tlx: 224366

Hong Kong
Commission of the Republic of Singapore
901 Admiralty Centre, Tower I
18 Harcourt Rd.
Hong Kong
Tel: [852] 5272212 Fax: [852] 8613595
Tlx: 73194

All addresses and telephone numbers are in the Republic of Singapore unless otherwise noted. The country code for Singapore is [65].

India
High Commission of the Republic of Singapore
E-6 Chandragupta Marg
Chanakyapuri
New Delhi 110 021, India
Tel: [91] (11) 604162, 608149
Fax: [91] (11) 677798 Tlx: 3172169

Consulate of the Republic of Singapore (Bombay)
No. 94, 9/F., Sakhar Bhawan
230 Nariman Point
Bombay 400 021, India
Tel: [91] (22) 2043205/9 Tlx: 1184026

Consulate of the Republic of Singapore (Madras)
Apex Plaza, 2/F.
3 Nungambakkam High Rd.
Madras 600 034, India
Tel: [91] (44) 476637, 473795, 476393   Tlx: 416108

Indonesia
Embassy of the Republic of Singapore
Blk X/4, Kav No 2
Jln H. R. Rasuna Said, Kuningan
Jakarta 12950, Indonesia
Tel: [62] (21) 5201489, 5201490/1
Fax: [62] (21) 5201486 (Embassy), 5201488 (Commercial) Tlx: 62213

Consulate of the Republic of Singapore (Medan)
3 Jln Tengku Daud
Medan, North Sumatra, Indonesia
Tel: [62] (61) 513366 Fax: [62] (61) 513134

Japan
Embassy of the Republic of Singapore
12-3 Roppongi 5-chome, Minato-ku
Tokyo 106 Japan
Postal address: PO Box 32
Azabu Post Office, Minato-ku
Tokyo 106, Japan
Tel: [81] (3) 3586-9111 Fax: [81] (3) 3582-1085
Tlx: 22404

Consulate-General of the Republic of Singapore (Osaka)
14/F., Osaka Kokusai Bldg.
3-13 Azuchi-machi, 2-chome, Chuo-ku
Osaka 541, Japan
Tel: [81] (6) 261-5131/2, 262-2662
Fax: [81] (6) 261-0338 Tlx: 64596

Korea (South)
Embassy of the Republic of Singapore
7/F., Citicorp Centre Bldg. 89-29
Shinmuon-ro 2-ka, Chongno-ku
Seoul, Korea 110-062
Tel: [82] (2) 722-0442 Fax: [82] (2) 722-5930
Tlx: 24648

Lebanon
Consulate of the Republic of Singapore
Horsh Karam-Independence Ave.
Joe Habis Bldg.
Beirut, Lebanon
Postal address: PO Box 166730
Beirut, Lebanon
Tel: [961] (1) 215998 Tlx: 41689

Malaysia
High Commission of the Republic of Singapore
209 Jalan Tun Razak
50400 Kuala Lumpur, Malaysia
Tel: [60] (3) 2616277, 2616404
Fax: [60] (3) 2616343 Tlx: 30320

Netherlands
Consulate of the Republic of Singapore
Grindweg 88
3055 VD Rotterdam, Netherlands
Postal address: PO Box 4402
3006 AK Rotterdam
Tel: [31] (10) 461-5899 Fax: [31] (10) 461-5895
Tlx: 26083

New Zealand
High Commission of the Republic of Singapore
17 Kabul St., Khandallah
Wellington, New Zealand
Tel: [64] (4) 4792076 Fax: [64] (4) 4792315
Tlx: 3593

Norway
Consulate of the Republic of Singapore
Karl Johansgt 16
Oslo, Norway
Postal address: PO Box 1166 Sentrum
Oslo 0107, Norway
Tel: [47] (2) 485000, 485438   Fax: [47] (2) 484206
Tlx: 71356

Pakistan
Consulate of the Republic of Singapore
Lakson Sq. Bldg., 2 Sarwar Shaheed Rd.
Karachi-1, Pakistan
Tel: [92] (21) 526419, 520141 x99
Fax: [92] (21) 513410 Tlx: 23206, 23280

Philippines
Embassy of the Republic of Singapore
6/F., ODC Bldg., International Plaza
219 Salcedo St., Legaspi Village
Makati, Metro Manila, Philippines
Tel: [63] (2) 8161764/5 Fax: [63] (2) 8184687
Tlx: 63631

Russia
Embassy of the Republic of Singapore
Per Voyevodina 5
Moscow, Russia
Tel: [7] (95) 2413702, 2413913/4
Fax: [7] (95) 2302937 Tlx: 413128

Saudi Arabia
Embassy of the Republic of Singapore
Al Baha St., Oleya District
Riyadh 11693, Kingdom of Saudi Arabia
Postal address: PO Box 94378
Riyadh 11693, Kingdom of Saudi Arabia
Tel: [966] (1) 4657007 Fax: [966] (1) 4652224
Tlx: 406211

Consulate of the Republic of Singapore (Jeddah)
Suite 1021, Corniche Commercial Centre
PO Box 18294
Jeddah 21415, Kingdom of Saudi Arabia
Tel: [966] (2) 6435677, 6437267
Fax: [966] (2) 6430750 Tlx: 605794

Spain
Consulate of the Republic of Singapore
J N Estudio Juridico y Rosellon, 257, 4 B
08008 Barcelona, Spain
Tel: [34] (3) 2378401, 2378501
Fax: [34] (3) 2372407 Tlx: 98096

Spain (Madrid)
Consulate
Huertas No. 13
Madrid 28012 Spain
Tel: [34] (1) 5383719 Fax: [34] (1) 5383718
Tlx: 27307, 43229

Sweden
Consulate-General of the Republic of Singapore
Storgatan 42
11455 Stockholm, Sweden
Tel: [46] (8) 660-0135 Fax: [46] (8) 662-2035

Taiwan
Trade Mission of the Republic of Singapore
9/F., Taiwan First Investment & Trust Bldg.
No. 85 Jen Ai Rd., Section 4
Taipei, Taiwan
Tel: [886] (2) 7721940 Fax: [886] (2) 7721943
Tlx: 27220

Thailand
Embassy of the Republic of Singapore
129 South Sathorn Rd.
Bangkok, Thailand
Tel: [66] (2) 2862111 Fax: [66] (2) 2872578
Tlx: 82930 SINGEMB TH

Turkey
Consulate of the Republic of Singapore
Hilmipasa Sok Irem Apt 30/10
81090 Kozyatagi, Istanbul, Turkey
Tel: [90] (1) 3842348/9 Fax: [90] (1) 3735072

United Kingdom
High Commission of the Republic of Singapore
9 Wilton Crescent
London SW1X 8SA, UK
Tel: [44] (71) 2358315
Fax: [44] (71) 2456583, 2359792 Tlx: 262564

United States of America
Embassy of the Republic of Singapore
3501 International Place NW
Washington DC 20008, USA
Tel: [1] (202) 537-3100 Fax: [1] (202) 537-0876

Consulate of the Republic of Singapore
(Los Angeles area)
2424 SE Bristol #320
Santa Ana Heights, CA 92707, USA
Tel: [1] (714) 476-2330 Fax: [1] (714) 760-1433

Consulate of the Republic of Singapore
(Minnesota)
c/o Hillstrom Bale Anderson Young
Polstein & Pearson
607 Marquette Ave., Suite 400
Minneapolis, MN 55402, USA
Tel: [1] (612) 332-8063 Fax: [1] (612) 332-2089

FOREIGN DIPLOMATIC MISSIONS IN SINGAPORE

Argentina
Embassy
302 Orchard Rd. #10-04, Tong Bldg.
Singapore 0923
Tel: 2354231 Fax: 2354382 Tlx: 23714
Trade Commission
268 Orchard Rd. #13-01, Yen San Bldg.
Singapore 0923
Tel: 7347811 Fax: 7347947

Australia
High Commission
25 Napier Rd.
Singapore 1025
Tel: 7379311, 7317290
Fax: 7337134 (High Commission), 7344265
(Commercial Section) Tlx: 21238

Austria
Embassy
1 Scotts Rd. #22-04, Shaw Centre
Singapore 0922
Tel: 2354087/8/9 Fax: 7371202 Tlx: 21133

Belgium
Embassy
10 Anson Rd. #09-24, International Plaza
Singapore 0207
Tel: 2207677 Fax: 2226976 Tlx: 23301 AMBEL

Brazil
Embassy
302 Orchard Rd. #15-03/04, Tong Bldg.
Singapore 0923
Tel: 7343435/6/7 Tlx: 36204 BRAEMB

Canada
High Commission
80 Anson Rd. #14-00 & #15-00/01, IBM Towers
Singapore 0207
Tel: 2256363 Fax: 2261541 Tlx: 21277 DOMCAN
Canada-ASEAN Centre
Tel: 2257346 Fax: 2227439

Chile
Embassy
105 Cecil St. #14-01, The Octagon
Singapore 0106
Tel: 2238577/8 Fax: 2250677 Tlx: 34187

All addresses and telephone numbers are in the Republic of Singapore unless otherwise noted. The country code for Singapore is [65].

China
Embassy
70-76 Dalvey Rd.
Singapore 1025
Tel: 7343200, 7343273 Fax: 7338590, 7344737
Tlx: 36878 CHICRO

Cyprus
Consulate-General
6 Kung Chong Rd.
Singapore 0315
Tel: 4748473 Fax: 4755623, 4755624 Tlx: 21013

Denmark
Embassy
101 Thomson Rd. #13-01., United Sq.
Singapore 1130
Tel: 2503383 Fax: 2533764 Tlx: 24576

Egypt
Embassy
75 Grange Rd.
Singapore 1024
Tel: 7371811, 7371587 Fax: 7323422
Tlx: 23293 BOUSTAN
Commercial Rep. Bureau
Tel: 2352739 Fax: 7340572

Finland
Embassy
101 Thomson Rd. #21-03, United Sq.
Singapore 1130
Tel: 2544042 Fax: 2534101 Tlx: 21489 FINNS

France
Embassy
5 Gallop Rd.
Singapore 1025
Tel: 4664866/1 Fax: 4663296 Tlx: 21351

Trade Commission
10 Anson Rd. #30-06/07, International Plaza
Singapore 0207
Tel: 2213033 Fax: 2259457 Tlx: 22121 COMATTA

French Financial Agency
10 Collyer Quay #11-04, Ocean Bldg.
Singapore 0104
Tel: 5324755 Fax: 5354628 Tlx: 55478 AGEFI

Germany
Embassy
545 Orchard Rd. #14-00, Far East Shopping Centre
Singapore 0923
Postal address: Tanglin PO Box 94
Singapore 9124
Tel: 7371355 Fax: 7372653 Tlx: 21312

Greece
Consulate-General
51 Anson Rd. #11-51, Anson Centre
Singapore 0207
Tel: 2208622 Fax: 2257870 Tlx: 20000

Maritime Section
19 Keppel Rd. #05-04, Jit Poh Bldg.
Singapore 0208
Tel: 2212364

Hungary
Commercial Section
101 Thomson Rd. #22-05, United Sq.
Singapore 1130
Tel: 2509215, 2504424 Fax: 2534161 Tlx: 53514

India
High Commission
India House, 31 Grange Rd.
Singapore 0923
Postal address: Killiney Rd. PO Box 92
Singapore 0923
Tel: 7376777, 7376809 Fax: 7326909
Tlx: 25526 BHARAT

Indonesia
Embassy
7 Chatsworth Rd.
Singapore 1024
Tel: 7377422 Fax: 7375037 Tlx: 21464 INDON

Ireland
Consulate-General
541 Orchard Rd. #08-02, Liat Towers
Singapore 0923
Tel: 7323430, 7332180 Fax: 7337250

Israel
Embassy
58 Dalvey Rd.
Singapore 1025
Tel: 2350966 Fax: 7337008 Tlx: 21975
Trade Dept. Fax: 7372502

Italy
Embassy
101 Thomson Rd. #27-02/03, United Sq.
Singapore 1130
Tel: 2506022, 2506592 (Consular & Passport)
Fax: 2533301 Tlx: 2117 ITALDIP

Italian Trade Commission
1 Maritime Sq. #12-05, World Trade Centre
Singapore 0409
Tel: 2731444 Fax: 2781954 Tlx: 21865

Japan
Embassy
16 Nassim Rd.
Singapore 1025
Tel: 2358855/9 Fax: 7320781 Tlx: 21353

Korea (South)
Embassy
101 Thomson Rd. #10-03, United Sq.
Singapore 1130
Tel: 2561188 Fax: 2543191 Tlx: 36696 ROKEM

Luxembourg
Consulate-General
65 Chulia St. #41-08, OCBC Centre
Singapore 0104
Tel: 5333444 Fax: 5341443 Tlx: 21772 WEARNES

Malaysia
High Commission
301 Jervois Rd.
Singapore 1024
Tel: 2350111 Fax: 7336135 Tlx: 21406

Trade Commission
150 Orchard Rd. #04-02, Orchard Plaza
Singapore 0923
Tel: 2351605

Investment Office
5 Shenton Way #26-05/07, UIC Bldg.
Singapore 0106
Tel: 2210155

Mexico
Embassy
152 Beach Rd. #06-07/08, Gateway East Tower
Singapore 0718
Tel: 2982678 Fax: 2933484 Tlx: 50469

Trade Commission, SE Asia Regional Office
152 Beach Rd. #09-01, Gateway East Tower
Singapore 0718
Tel: 2968281 Fax: 2985825

Netherlands
Embassy
541 Orchard Rd. #13-01, Liat Towers
Singapore 0923
Tel: 7371155 Fax: 7371940
Tlx: 33815 NEDAMB

New Zealand
High Commission
13 Nassim Rd.
Singapore 1025
Tel: 2359966 Fax: 7339924 Tlx: 21244 TAINUI

Norway
Embassy
16 Raffles Quay #44-01, Hong Leong Bldg.
Singapore 0104
Tel: 2207122, 2221316 (Trade) Fax: 2247079
Tlx: 21225 AMBANOR

Pakistan
High Commission
20-A Nassim Rd.
Singapore 1025
Tel: 7376621, 7376988 Fax: 7374096
Tlx: 36777 PAREP

Panama
Embassy
16 Raffles Quay #41-06, Hong Leong Bldg.
Singapore 0104
Tel: 2218677, 2218678 Fax: 2240892 Tlx: 24524

Peru
Consulate
7 Brookvale Dr. #03-11, Edale Blk
Singapore 2159
Tel: 4670497

Philippines
Embassy
20 Nassim Rd.
Singapore 1025
Tel: 7373977, 7373293 (Commercial Attache),
7346102 (Labor Attache) Fax: 7339544
Tlx: 34445 MABINI

Poland
Embassy
100 Beach Rd. #33-11/12, Shaw Towers
Singapore 0718
Tel: 2942513/4 Fax: 2950016 Tlx: 26355

Portugal
Consulate
9 Malacca St., Rm. 11/12
Singapore 0104
Postal address: Robinson Rd. PO Box 1501
Singapore 9030
Tel: 5353278 Fax: 5334943 Tlx: 20450

Romania
Embassy
48 Jalan Harom Setangkai
Singapore 1025
Tel: 4683424 Fax: 4683425 Tlx: 22184

Russia
Embassy
51 Nassim Rd.
Singapore 1025
Tel: 2351832, 7370048 (Consular Section)
Fax: 7334780 Tlx: 23071 SU POSOL

Trade Representative Office
12 Anguila Park
Singapore 0923
Tel: 7376221

Saudi Arabia
Embassy
10 Nassim Rd.
Singapore 1025
Tel: 7345878 Fax: 7385291 Tlx: 25318
Commercial Office Tel: 2358459 Fax: 7374657

Spain
Consulate
4 Shenton Way #06-02, Shing Kwan House
Singapore 0106
Tel: 2278310 Fax: 2250304 Tlx: 25024 BSSIN

Commercial Section
15 Scotts Rd. #05-08/09, Thong Teck Bldg.
Singapore 0922
Tel: 7329788/9 Fax: 7329780 Tlx: 55047 OFCOM

Sweden
Embassy
111 Somerset Rd. #05-08
PUB Bldg., Devonshire Wing
Singapore 0923
Postal address: Orchard Point PO Box 292
Singapore 9123
Tel: 7342771 Fax: 7322958 Tlx: 23450

Switzerland
Embassy
1 Swiss Club Link
Singapore 1128
Tel: 4685788 Fax: 4668245 Tlx: 21501 AMSWISS

Taiwan
Trade Representative
460 Alexandra Rd. #23-00, PSA Bldg.
Singapore 0511
Tel: 2786511 Fax: 2789962

All addresses and telephone numbers are in the Republic of Singapore unless otherwise noted. The country code for Singapore is [65].

Thailand
Embassy
370 Orchard Rd.
Singapore 0923
Tel: 2354175, 7327769 Fax: 7320778
Tlx: 35981
Commercial Counsellor
Tel: 7373060 Fax: 7322458

Turkey
Embassy
20B Nassim Rd.
Singapore 1025
Tel: 7329211 Fax: 7381786 Tlx: 34668

United Kingdom
High Commission
Tanglin Rd.
Singapore 1024
Tel: 4739333 Fax: 4759706, 4752320

United States of America
Embassy
30 Hill St.
Singapore 0617
Tel: 3380251 Fax: 3384550

GOVERNMENT RUN CORPORATIONS

INTRACO (Import Export Company)
NOL Bldg. #1400
Singapore 0511
Tel: 2780011

Singapore Petroleum Company
DBS Bldg. #42-01, Shenton Way
Singapore 0106
Tel: 2213166

TRADE PROMOTION ORGANIZATIONS

WORLD TRADE CENTER

World Trade Centre Singapore
1 Maritime Square #09-72
Singapore 0409
Tel: 3212783, 3212103, 3212791 Fax: 2740721
Tlx: 34975 WTCS

GENERAL TRADE ASSOCIATIONS & LOCAL CHAMBERS OF COMMERCE

Association of Small & Medium Enterprises
Blk. 139, Kim Tian Rd. #02-00
Singapore 0316
Tel: 2712566 Fax: 2711257

Federation of Merchants' Association
25 Genting Rd. #08-01
Singapore
Tel: 7417822

International Business Women's Association
c/o C K Woo & Company
6001 Beach Rd. #18-01, Golden Mile Tower
Singapore 0719
Postal address: Orchard PO Box 23
Singapore 9123
Tel: 3384070

International Procurement Management
Association of Singapore
111 North Bridge Rd. #20-03
Peninsula Plaza
Singapore 0617
Tel: 7384210 Fax: 3389609

Singapore Chinese Chamber of Commerce & Industry
47 Hill St. #09-00
Singapore 0617
Tel: 3378381 Fax: 3390605 Tlx: 33714

Singapore Federation of Chambers of Commerce & Industry
47 Hill St. #03-01, SCCCI Bldg.
Singapore 0617
Tel: 3389761/2 Fax: 3395630
Tlx: 26228 SFCCI

Singapore Indian Chamber of Commerce
101 Cecil St. #23-01, Tong Eng Bldg.
Singapore 0106
Tel: 2222505 Fax: 2231707 Fax: 22336

Singapore International Chamber of Commerce
6 Raffles Quay #05-00, Denmark House
Singapore 0104
Tel: 2241255 Fax: 2242785
Tlx: 25235 INTCHAM

Singapore Malay Chamber of Commerce
10 Anson Rd. #24-07, International Plaza
Singapore 0207
Tel: 2211066, 2230347 Fax: 2235811
Tlx: 25521 SMCC

Singapore Manufacturers' Association
20 Orchard Rd., SMA House
Singapore 0923
Tel: 3388787 Fax: 3385385 Tlx: 24992

FOREIGN CHAMBERS OF COMMERCE & BUSINESS ORGANIZATIONS

Australia
Singapore Australian Business Council
c/o ANZ Bank
10 Collyer Quay #17-01/05, Ocean Bldg.
Singapore 0104
Tel: 5358355 Fax: 5396111

Belgium
Belgium & Luxembourg Association of Singapore
c/o Royal Embassy of Belgium
37 Chiltern Dr.
Singapore 1335
Tel: 2846701 Fax: 2874050

Canada
Canadian-Singapore Business Association
30 Orange Grove Rd. #07-00
RELC Bldg.
Singapore 0125
Tel: 7389232 Fax: 7389227

Denmark
Danish Business Association of Singapore
c/o Maersk Singapore Pte. Ltd.
21 Collyer Quay #18-00, Hongkong Bank Bldg.
Singapore 0104
Tel: 2250511 Fax: 2251205

European Community
European Community Business Association
c/o Singapore International Chamber of Commerce
6 Raffles Quay #10-01
John Hancock Tower
Singapore 0104
Tel: 2241255 Fax: 2242785

Finland
Finnish Business Council
c/o Embassy of Finland
101 Thomson Rd. #21-03, United Sq.
Singapore 1130
Tel: 2544042 Fax: 2534101

France
French Business Association
30 Orange Grove Rd. #08-802, RELC Bldg.
Singapore 1025
Tel: 2358211 Fax: 7331629

Germany
German Business Association
39A Jalan Pemimpin #05-00, Union Centre
Singapore 2057
Tel: 3536841 Fax: 3536962

All addresses and telephone numbers are in the Republic of Singapore unless otherwise noted. The country code for Singapore is [65]

Indonesia
Indonesia Business Association of Singapore
158 Cecil St. #07-03, Dapenso Bldg.
Singapore 0106
Tel: 2215063 Fax: 2215064

Ireland
Irish Business Association
c/o Irish Trade Board
541 Orchard Rd. #08-02, Liat Towers
Singapore 0923
Tel: 7332180 Fax: 7330291

Israel
Israel Business Association
71 Hillview Ave.
Singapore 2366
Tel: 7605988 Fax: 7621210

Italy
Italian Business Association
c/o Hanna Instruments (Asia Pacific) Pte. Ltd.
Blk 161 Kallang Way #07-08/11
Singapore 1334
Tel: 2967118 Fax: 2916904

Japan
International Business Organization of Osaka
1 Maritime Square #09-30, World Trade Centre
Singapore 0409
Tel: 2734856 Fax: 2740124

Japan
Japanese Chamber of Commerce & Industry
10 Shenton Way #12-04/05/06, MAS Bldg.
Singapore 0207
Tel: 2210541 Fax: 2256197

Netherlands
Dutch Business Association
c/o Rabobank Nederland
50 Raffles Place #32-01, Shell Tower
Singapore 0104
Tel: 2259896 Fax: 2246692

Netherlands
Association of Dutch Businessmen in Singapore
c/o Hollandse Club
22 Camden Park
Singapore 1129
Tel: 4695211 Fax: 4686272

New Zealand
New Zealand Business Council
c/o New Zealand High Commission
13 Nassim Rd.
Singapore 1025
Tel: 2359966 Fax: 7339924, 7325595

Norway
Norwegian Industrial Forum
c/o Norwegian Trade Council
16 Raffles Quay #44-01, Hong Leong Bldg.
Singapore 0104
Tel: 2221316 Fax: 2247079

Sweden
Swedish Business Association in Singapore
c/o Swedish Embassy
111 Somerset Rd. #05-08, PUB Bldg.
Singapore 0923
Tel: 7342771 Fax: 7322958

Switzerland
Swiss Business Association
c/o S.A. Desco Singapore Pte. Ltd.
10 Shenton Way #17-01/02, MAS Bldg.
Singapore 0207
Tel: 2203022 Fax: 2216021

Taiwan
Taipei Business Association in Singapore
47 Hill St. #06-07
Singapore
Tel: 3383916, 3383930

United Kingdom
British Business Association
41 Duxton Rd.
Singapore 0208
Tel: 2277861 Fax: 2277021

United States of America
American Business Council
1 Scotts Rd. #16-07, Shaw Centre
Singapore 0922
Tel: 2350077 Fax: 7325917

SINGAPORE ECONOMIC DEVELOPMENT BOARD

Head Office
250 North Bridge Rd. #24-00
Raffles City Tower
Singapore 0617
Tel: 3362288 Fax: 3396077 Tlx: 26233

France
22 av. Victor Hugo
75116 Paris, France
Tel: [33] (1) 45-00-11-83 Fax: [33] (1) 45-00-61-37
Tlx: 649701 FSEDB

Germany
Untermainanlage 7
6000 Frankfurt/Main 1, Germany
Tel: [49] (69) 233838 Fax: [49] (69) 252882
Tlx: 4189031 SEDB D

Hong Kong
3010-3012, One Pacific Place
88 Queensway
Hong Kong
Tel: [852] 5280608 Fax: [852] 8669287

Indonesia
c/o Embassy of the Republic of Singapore
Block X/4 Kav No. 2
Jalan HR Rasuna Said
Kuningan, Jakarta 12950, Indonesia
Tel: [62] (21) 5201477, 5201489
Fax: [62] (21) 5201486 Tlx: 62213

Italy
via S Pietro all'Orto, 17
21121 Milano, Italy
Tel: [39] (2) 799277 Fax: [39] (2) 780023
Tlx: 334658

Japan
c/o Consulate-General of the Republic of Singapore (Osaka)
14/F., Osaka Kokusai Bldg.
3-13 Azuchmachi, 2-Chome, Chuo-ku
Osaka 541, Japan
Tel: [81] (6) 261-5131/2, 262-2662
Fax: [81] (6) 261-0338 Tlx: 64596

8/F., The Imperial Tower
1-1 Uchisaiwai-cho 1-chome, Chiyoda-ku
Tokyo 100
Tel: [81] (3) 3501-6041 Fax: [81] (3) 3501-6060
Tlx: 33310

Sweden
Storgatan 42
114 55 Stockholm, Sweden
Tel: [46] (8) 6637488 Fax: [46] (8) 7823951

United Kingdom
Norfolk House
30 Charles II St.
London SW1Y 4AE, UK
Tel: [44] (71) 839-6688 Fax: [44] (71) 839-6162

United States of America
One International Place, 8/F.
Boston, MA 02110, USA
Tel: [1] (617) 261-9981 Fax: [1] (617) 261-9983

Two Prudential Plaza, Suite 970
180 Stetson Ave.
Chicago, IL 60601, USA
Tel: [1] (312) 565-1100 Fax: [1] (312) 565-1994

2049 Century Park East, Suite 400
Los Angeles, CA 90067, USA
Tel: [1] (310) 553-0199 Fax: [1] (310) 557-1044

55 East 59th St.
New York, NY 10022, USA
Tel: [1] (212) 421-2200
Fax: [1] (212) 421-2206 Tlx: 421848 TDBNY

210 Twin Dolphin Dr.
Redwood City, CA 94065, USA
Tel: [1] (415) 591-9102 Fax: [1] (415) 591-1328

1350 Connecticut Ave. NW, Suite 504
Washington, DC 20036, USA
Tel: [1] (202) 223-2570/1 Fax: [1] (202) 223-2572
Tlx: 493928

SINGAPORE TRADE DEVELOPMENT BOARD

Singapore
Singapore Trade Development Board
1 Maritime Square #10-40 (Lobby D)
World Trade Centre, Telok Blangah Rd.
Singapore 0409
Tel: 2719388 Fax: 2740770, 2782518
Tlx: 28617 TRADEV
Export Institute Tel: 3342188
Imports & Exports Help Desk Tel: 2790350
Fax: 2724720
Library Tel: 2790433
Market Research Tel: 2790434

Singapore Trade Development Board
Changi Airport Office
115 Airport Cargo Rd.
#04-18 Cargo Agents Building C
Singapore 0781
Tel: 5427179 Fax: 5425385

Australia
Singapore Trade Development Board
Suite 5503, MLC Centre, 19-29 Martin Place
Sydney, NSW 2000, Australia
Tel: [61] (2) 233-7015, 233-3391
Fax: [61] (2) 233-4361

China
Singapore Trade Development Board
c/o Embassy of the Republic of Singapore
1 Xiu Shui Bei Jie
Jianguomenwai
Beijing 100600., PRC
Tel: [86] (1) 5323926, 5323143
Fax: [86] (1) 5322215 Tlx: 22578

Singapore Trade Development Board
c/o Consulate of Singapore
400 Wulumuqi Zhong Lu
Shanghai 200031, PRC
Tel: [86] (21) 4370776, 4331362
Fax: [86] (21) 4334150 Tlx: 33540

France
Singapore Trade Development Board
20 bis av. Bourgain
92130 Issy-les-Moulineaux, France
Tel: [33] (1) 46-38-24-24 Fax: [33] (1) 41-08-94-66

Germany
Singapore Trade Development Board
c/o Internationales Schiffahrtskontor (ISKON)
Kaiserstrasse 42
Postfach 320526
D-4000 Düsseldorf 30, Germany
Tel: [49] (211) 499261/5 Fax: [49] (211) 490573
 Tlx: 8584738 NOLD

Singapore Trade Development Board
Goethestr 5
6000 Frankfurt am Main 1, Germany
Tel: [49] (69) 281743 Fax: [49] (69) 285039
Tlx: 4189605 STCFD

All addresses and telephone numbers are in the Republic of Singapore unless otherwise noted. The country code for Singapore is [65].

Hong Kong
Singapore Trade Development Board
c/o Singapore Commission, Hong Kong
901 Admiralty Centre, Tower I
18 Harcourt Rd.
Hong Kong
Tel: [852] 5286185 Fax: [852] 8610048
Tlx: 66791 TDBHK HX

Hungary
Singapore Trade Development Board
Hegyalja ut. 7/13, Alag Buda Center
1016 Budapest, Hungary
Postal Address: c/o Budabox, PO Box 700/11
1539 Budapest, Hungary
Tel: [36] (1) 2025133, 2025527
Fax: [36] (1) 2023652 Tlx: 222155 TDB

India
Singapore Trade Development Board
88A Jolly Maker Chambers 11, Nariman Point
Bombay 400-021, India
Tel: [91] (22) 2040732 Fax: [91](22) 2045051
Tlx: 11-86165

Singapore Trade Development Board
c/o Singapore High Commission
E-6 Chandragupta Marg
Chanakyapuri
New Delhi 110 021, India
Tel: [91] (11) 604162, 608149 Fax: [91] (11) 677798
Tlx: 3172169

Indonesia
Singapore Trade Development Board
c/o Embassy of Singapore
Blk X/4, Kav No 2
Jln H. R. Rasuna Said, Kuningan
Jakarta 12950, Indonesia
Tel: [62] (21) 5201489, 5201490/1
Fax: [62] (21) 5201488 Tlx: 62213 SINGA IA

Japan
Singapore Trade Development Board
c/o Consulate-General of Singapore
14/F., Osaka Kokusai Bldg.
3-13 Azuchi-machi, 2-chome, Chuo-ku
Osaka 541, Japan
Tel: [81] (6) 2622662 Fax: [81] (6) 2622664

Singapore Trade Development Board
Embassy of Singapore
12-3 Roppongi 5-chome, Minato-ku
Tokyo 106 , Japan
Tel: [81] (3) 35846032 Fax: [81] (3) 35846135

Korea
Singapore Trade Development Board
Suite 1306, Sam Ku Bldg.
70 Sokong-Dong, Chung-ku
Seoul, Korea
Tel: [82] (2) 7548404, 7548405 Fax: [82] (2) 7547017

Singapore Trade Development Board
c/o Ashin Shipping Co. Ltd.
10/F., Marine Center
51 Sokong-dong, Chung-ku
Seoul, Korea
Tel: [82] (2) 757-6944, 753-1211
Fax: [82] (2) 757-4919 Tlx: 24722

Netherlands
Singapore Trade Development Board
c/o Consulate of Singapore
Grindweg 88
3055 VD Rotterdam, Netherlands
Postal address: PO Box 4402
3006 AK Rotterdam
Tel: [31] (10) 461-5899 Fax: [31] (10) 461-5895
Tlx: 26083

Saudi Arabia
Singapore Trade Development Board
c/o Consulate of Singapore
Suite 1021, Corniche Commercial Centre
PO Box 18294
Jeddah 21415, Kingdom of Saudi Arabia
Tel: [966] (2) 6435677, 6437267
Fax: [966] (2) 6430750 Tlx: 605794

Sweden
Singapore Trade Development Board
c/o Consulate-General of Singapore
Storgatan 42
11455 Stockholm, Sweden
Tel: [46] (8) 660-0135 Fax: [46] (8) 662-2035

Switzerland
Singapore Trade Development Board
c/o Perm. Mission of Singapore to the U.N.
6 bis rue Antoine-Carteret
1202 Geneva, Switzerland
Tel: [41] (22) 3447300, 3447339
Fax: [41] (22) 3457910 Tlx: 415909 SINGH CH

United Arab Emirates
Singapore Trade Development Board
c/o Al-Futtaim Industries Pte. Ltd.
PO Box 152
Dubai, U.A.E.
Tel: [971] (4) 233961 Fax: [971] (4) 212933
Tlx: 45462 FUTAIM EM

United Kingdom
Singapore Trade Development Board
Suite 30, Westminster Palace Gardens
1-7 Artillery Row
London SW1P 1RJ, UK
Tel: [44] (71) 976-8601, 222-0770
Fax: [44] (71) 976-8598

United States of America
Singapore Trade Development Board
c/o Los Angeles World Trade Center
350 S. Figueroa St., Suite 909
Los Angeles, CA 90071, USA
Tel: [1] (213) 617-7358/9, 617-7397/8
Fax: [1] (213) 617-7367

Singapore Trade Development Board
745 Fifth Ave., Suite 1601
New York, NY 10151, USA
Tel: [1] (212) 421-2207 Fax: [1] (212) 888-2897
Tlx: 421848 TDB NY

Singapore Trade Development Board
Embassy of the Republic of Singapore
3501 International Place NW
Washington DC 20008, USA
Tel: [1] (202) 537-3100 Fax: [1] (202) 537-0876

INDUSTRY-SPECIFIC TRADE ORGANIZATIONS IN SINGAPORE

Air Cargo Agents Association [Singapore]
115 Airport Cargo Rd. #04-09
Cargo Agents Bldg. C
Singapore 1781
Tel: 5454620, 5459597 Fax: 5426820 Tlx: 38963

Air-conditioning & Refrigeration Association
58 Kensington Park Rd.
Singapore 1955
Tel: 2885491

Air Transport Association [International]
331 North Bridge Rd. #20-00
Singapore
Tel: 3399978 Fax: 3390855

Bankers' Association [Singapore Merchant]
24 Raffles Pl. #16-02, Clifford Centre
Singapore 0104
Tel: 5327565 Fax: 5323390

Banks in Singapore [The Association of]
10 Shenton Way #12-08, MAS Bldg.
Singapore 0207
Tel: 2244300, Fax: 2241785 Tlx: 29291

Book Exporters & Importers Council
c/o Chopmen Publishers
865 Mountbatten Rd. #05-28/29
Katong Shopping Centre
Singapore
Tel: 3441495 Fax: 3440180

Building Ltd. [Singapore Institute of]
108 Murray St.
Singapore
Tel: 2230258 Fax: 2244249

Building Materials Suppliers' Association [Singapore]
426 Race Course Rd.
Singapore 0821
Tel: 2984660

Computer Society [Singapore]
71 Science Park Dr.
Singapore
Tel: 7783901 Fax: 7788221

Contractors Association Ltd. [Singapore]
1 Bt. Merah Lane 2, Construction House
Singapore 0315
Tel: 2789577 Fax: 2733977 Tlx: 22406

Corrugated Box Manufacturers Association
c/o Low Kok Kim & Co.
315 Outram Rd. #15-07/08
Tan Boon Liat Bldg.
Singapore 0316
Tel: 2222068

Dental Association [Singapore]
2 College Rd.
Singapore 0316
Tel: 2202588 Fax: 2237967

Direct Marketing Association of Singapore
100 Beach Rd. #27-05, Shaw Towers
Singapore 0718
Tel: 2970438 Fax: 2998308

Electrical Contractors Association [Singapore]
315 Outram Rd. #10-09A, Tan Boon Liat Bldg.
Singapore 0316
Tel: 2263216 Fax: 2237568

Electrical Traders Association [Singapore]
35A/B Truro Rd.
Singapore 0821
Tel: 2990355 Fax: 2995495

Electronic Industries in Singapore [Association of]
470 North Bridge Rd. #03-09
Singapore
Tel: 3374643 Fax: 3399341

Freight Forwarders Association [Singapore]
7500A Beach Rd. #13-314, The Plaza
Singapore 0719
Tel: 2964645 Fax: 2922504

Furniture Manufacturers' & Traders' Association [Singapore]
16C 4/F., Geylang Lor 37, NTWU Bldg.
Singapore 1438
Tel: 7441600, 7441421 Fax: 7452917 Tlx: 20855

Granite Quarry Owners & Employers Association
141 Cecil St. #05-00, Tung Ann Association Bldg.
Singapore 0106
Tel: 2211560

Grocer's Association [Singapore]
33A Lor 15 Geylang
Singapore
Tel: 7451821

Industrial Automation Association [Singapore]
151 Chin Swee Rd. #03-13, Manhattan House
Singapore 0316
Tel: 7346911 Fax: 2355721

Leasing Association of Singapore
2 Finlayson Green #15-04
Asia Insurance Bldg.
Singapore 0104
Tel: 2217379, 2217509 Fax: 2219674 Tlx: 26739

Marine Industries [Association of Singapore]
1 Maritime Sq. #09-50, World Trade Centre
Singapore 0409
Tel: 2707883 Fax: 2731867 Tlx: 37706

Medical Practitioners of Singapore Pte. Ltd. [The Association of]
Alumni Medical Centre
2 College Rd.
Singapore 0316
Tel: 2230901

Microcomputer Trade Association
211 Henderson Rd. #01-01
Singapore
Tel: 2782855

All addresses and telephone numbers are in the Republic of Singapore unless otherwise noted. The country code for Singapore is [65].

Optometrists [Singapore Society of]
Blk. 531A Up Cross St. #04-85
Singapore
Tel: 5341195 Fax: 5382027

Pharmaceutical Industries [Singapore Association of]
30 Shaw Rd., 5/F., Roche Bldg.
Singapore 1336
Tel: 2868277 Fax: 2802167

Pineapple Industry Board [Malayan]
10 Collyer Quay, #19-06, Ocean Bldg.
Singapore 0104
Tel: 5338827

Quality Institute [Singapore]
Blk. 18, Clementi Rd. #03-08
Ngee Ann Polytechnic
Singapore 2159
Tel: 4674225 Fax: 4674226

Radio & Electrical Traders Association of Singapore
68 Lor 16 Geylang #04-01
Singapore
Tel: 7470971, 7457568

Rattan Industry Association [Singapore]
12 Lor 24A Geylang
Singapore
Tel: 7441853

Ready-Mixed Concrete Association of Singapore
Blk. 1 Thomson Rd. #03-332E
Singapore
Tel: 2568359

Real Estate Developers' Association of Singapore (REDAS)
190 Clemenceau Ave. #07-01, Singapore Shopping Centre
Singapore 0923
Tel: 3366655 Fax: 3372217

Realtors [Association of Singapore]
1 Colombo Court #07-30
Singapore 0617
Postal address: 50 Jln Sultan #24-07
Jln Sultan Centre
Singapore 0719
Tel: 2947000 Fax: 2990374

Restaurant Association of Singapore
11 Dhoby Ghaut #04-03, Cathay Bldg.
Singapore 0922
Tel: 3383774 Fax: 3390903

Rubber Association of Singapore
79 Robinson Rd. #14-01, CPF Bldg.
Singapore 0106
Tel: 2219022 Fax: 2241641, 2215316
Tlx: 20554

Shipping Association [Singapore National]
456 Alexandra Rd. #02-02, NOL Bldg.
Singapore 0511
Tel: 2733574, 2783464 Fax: 2745079 Tlx: 24021

Ship Suppliers [Singapore Association of]
1 Colombo Court #07-19
Singapore 0617
Tel: 3367755 Fax: 3390329 Tlx: 22860

Textile & Garment Manufacturers' Association of Singapore
47 Beach Rd. #06-01/02
Singapore
Tel: 3372022 Fax: 3389179

Timber Trade United Friendly Association [Singapore]
1A Lor 30 Geylang
Tel: 7435821

Training and Development Association [Singapore]
150 Orchard Rd. #06-05, Orchard Plaza
Singapore 0923
Postal address: Tanglin Rd. PO Box 323, Singapore 9124
Tel: 7376211 Tlx: 24554

UNIX Association [Singapore] (SINEX)
190 Clemenceau Ave #05-33/34
Singapore
Tel: 3343512

FINANCIAL INSTITUTIONS

BANKS

The Association of Banks in Singapore
10 Shenton Way #12-08, MAS Bldg.
Singapore 0207
Tel: 2244300, Fax: 2241785 Tlx: 29291

Singapore Merchant Bankers' Association
24 Raffles Pl. #16-02, Clifford Centre
Singapore 0104
Tel: 5327565 Fax: 5323390

Government Banks

Board of Commissioners of Currency Singapore
79 Robinson Rd. #01-01, CPF Bldg.
Singapore 0106
Tel: 2222211 Fax: 2257671 Tlx: 24722
Issues currency

Government of Singapore Investment Corporation Pte. Ltd. (GSIC)
250 North Bridge Rd. #33-00
Raffles City Tower
Singapore 0617
Tel: 3363366 Fax: 3308722 Tlx: 20484

Monetary Authority of Singapore (MAS)
MAS Bldg., 10 Shenton Way
Singapore 0207
Tel: 2255577 Fax: 2299229 Tlx: 28174
Performs functions of a central bank, except for issuance of currency

Domestic Banks

Bank of Singapore Ltd.
101 Cecil St. #01-02, Tong Eng Bldg.
Singapore 0106
Tel: 2239266 Fax: 2247407 Tlx: 27149

Chung Khiaw Bank Ltd.
10 Anson Rd. #01-01, International Plaza
Singapore 0207
Tel: 2228622 Tlx: 22027

DBS Bank (Development Bank of Singapore)
6 Shenton Way, DBS Bldg.
Singapore 0106
Tel: 2201111 Fax: 2211306 Tlx: 24455

Industrial and Commercial Bank Ltd.
2 Shenton Way #01-01, ICB Bldg.
Singapore 0106
Tel: 2211711 Fax: 2259777 Tlx: 21112

Oversea-Chinese Banking Corp. Ltd.
65 Chulia St. #08-00, OCBC Centre
Singapore 0104
Tel: 5357222 Fax: 5337955 Tlx: 21209

Overseas Union Bank Ltd.
1 Raffles Place, OUB Centre
Singapore 0104
Tel: 5338686 Fax: 5332293 Tlx: 24475

United Overseas Bank Ltd.
1 Bonham St. #01-00, UOB Bldg.
Singapore 0104
Tel: 5339898 Fax: 5342334 Tlx: 21539

Foreign Banks

Banca Commerciale Italiana
36 Robinson Rd., 03-01 City House
Singapore 0106
Tel: 2201333 Fax: 2252004 Tlx: 24545

Bank of America NT & SA (USA)
78 Shenton Way #19-00
Singapore 0207
Tel: 2236688 Fax: 2233310 Tlx: 24570

Bank of Tokyo Ltd. (Japan)
16 Raffles Quay #01-06, Hong Leong Bldg.
Singapore 0104
Tel: 2208111 Fax: 2244965 Tlx: 24363

Banque Indosuez (France)
3 Shenton Way #01-05, Shenton House
Singapore 0106
Tel: 2207111 Fax: 2297924 Tlx: 24435

Chase Manhattan Bank NA (USA)
50 Raffles Place, Shell Tower
Singapore 0104
Tel: 5304111 Fax: 2247950

Citibank, N.A. (USA)
5 Shenton Way, UIC Bldg.
Singapore 0106
Tel: 2242611 Fax: 2249844 Tlx: 24584 CITIBANK

Credit Suisse (Switzerland)
6 Battery Rd. #37-01
Standard Chartered Bank Bldg.
Singapore 0104
Tel: 2252055 Fax: 2291203 Tlx: 24650

Deutsche Bank (Asia) (Germany)
8 Shenton Way #01-01, Treasury Bldg.
Singapore 0106
Tel: 2244677 Fax: 2259442 Tlx: 21189

Hongkong and Shanghai Banking Corporation Ltd. (Hong Kong)
21 Collyer Quay #19-00, Hongkong Bank Bldg.
Singapore 0104
Tel: 5305000 Fax: 2250663 Tlx: 21259

National Australia Bank Ltd.
10 Collyer Quay #26-02/07, Ocean Bldg.
Singapore 0104
Tel: 5357655 Fax: 5344264 Tlx: 21583

Royal Bank of Canada Ltd.
140 Cecil St., #01-00, PIL Bldg.
Singapore 0106
Tel: 2247311 Fax: 2245635 Tlx: 23451

Standard Chartered Bank
6 Battery Rd.
Singapore 0104
Tel: 2258888 Fax: 2259136 Tlx: 24290

All addresses and telephone numbers are in the Republic of Singapore unless otherwise noted. The country code for Singapore is [65].

INSURANCE COMPANIES

American International Assurance Co. Ltd.
152 Beach Rd. #32-00 to #35-00, Gateway East
Singapore 0718
Tel: 2918000 Fax: 2986000

ANDA Insurance Agencies Pte. Ltd.
8 Lor 1 Gaylang
Singapore1436
Tel: 7412288 Fax: 7415155

British American General Insurance Co. Pte. Ltd.
57 Robinson Rd. #03-00
British American Insurance Bldg.
Singapore 0106
Tel: 2222311 Fax: 2223547

Export Credit Insurance Corporation of Singapore, Ltd.
460 Alexandra Rd. #18-00, PSA Bldg.
Singapore 0511
Tel: 2728866 Tlx: 21524

General Insurance Association of Singapore
1 Shenton Way #13-07, Robina House
Singapore 0106
Tel: 2218788 Fax: 2248910 Tlx: 20814

Insurance Corporation of Singapore Ltd.
137 Cecil St. #08-00, ICS Bldg.
Singapore 0106
Tel: 2218686 Fax: 2247242 Tlx: 37770

International Insurance Pte. Ltd.
64 Cecil St. #04-00/06-00, IOB Bldg.
Singapore 0104
Tel: 2238122 Fax: 2257743 Tlx: 34894

Kemper International Insurance Co. (Pte) Ltd.
143 Cecil St. #06-01, GB Bldg.
Singapore 0106
Tel: 2249477 Fax: 2250241 Tlx: 28327

Overseas Assurance Corporation Ltd.
138 Cecil St. #05-00, Cecil Court
Singapore 0106
Tel: 2251122 Fax: 2240672 Tlx: 21443

People's Insurance Co. Ltd.
10 Collyer Quay #04-01, Ocean Bldg.
Singapore 0104
Tel: 5326022 Fax: 5333871 Tlx: 26714

Singapore Aviation and General Insurance Co.
77 Robinson Rd. #13-00, SIA Bldg.
Singapore 0106
Tel: 2241111 Fax: 2257597 Tlx: 28226

Taisho Marine and Fire Insurance Pte. Ltd.
16 Raffles Quay #24-01, Hong Leong Bldg.
Singapore 0104
Tel: 2209644 Fax: 2256371 Tlx: 21742

The New Zealand Insurance Co. Ltd.
65 Chulia St. #15-00, OCBC Centre
Singapore 0104
Tel: 5335700 Fax: 5333476 Tlx: 21439

United Overseas Insurance Ltd.
156 Cecil St. #09-01, Far Eastern Bank Bldg.
Singapore 0106
Tel: 2227733 Fax: 2242718 Tlx: 25095

Yasuda Fire and Marine Insurance Co.
50 Raffles Place #03-03, Shell Tower
Singapore 0104
Tel: 2235293 Fax: 2257947 Tlx: 26329

STOCK & COMMODITY EXCHANGES

Central Depository Pte. Ltd.
1 Raffles Place #04-07/09, OUB Centre
Singapore 0104
Tel: 5357511

Rubber Exchange of Singapore
12/F., Singapore Rubber House
14 Collyer Quay
Singapore 0104
Tel: 913333

Singapore International Monetary Exchange
1 Raffles Place #07-00, OUB Centre
Singapore 0104
Tel: 5357382 Fax: 5357282 Tlx: 38000

Stock Exchange of Singapore (SES)
1 Raffles Place #24-00 OUB Centre
Singapore 0104
Tel: 5353788 Fax: 5350985 Tlx: 21853

SERVICES

ACCOUNTING FIRMS

Arthur Andersen & Co.
10 Hoe Chiang Rd.
#18-00 Keppel Towers
Singapore 0208
Tel: 2204377 Fax: 2234795

Coopers & Lybrand
9 Penang Rd. #12-00, Park Mall
Singapore 0923
Postal address: Orchard PO Box 285, Singapore 9123
Tel: 3362344, 3360877 Fax: 3362539, 3390048
Tlx: 22137, 28121

Deloitte & Touche
95 South Bridge Rd. #09-00, Pidemco Centre
Singapore 0105
Tel: 2248288 Fax: 5386166

Ernst & Young
10 Collyer Quay #21-04, Ocean Bldg.
Singapore 9007
Postal address: PO Box 384, Singapore 9007
Tel: 5357777 Fax: 5327662 Tlx: 22172

Ernst & Young
35 Robinson Rd. #17-00, City House
Singapore 0106
Postal address: PO Box 3257, Singapore 9052
Tel: 2201136 Fax: 2250465

KMPG Peat Marwick
16 Raffles Quay #22-00, Hong Leong Bldg.
Singapore 0104
Postal address: Robinson Rd. PO Box 448
Singapore 9008
Tel: 2207411 Fax: 2250984

Lo Hock Ling & Co.
(A member of Horwath International)
101A Up Cross St. #11-22
People's Park Centre
Singapore 0105
Tel: 5356111 Fax: 5336960

Moore Iyer & Co.
(A member of Moores Rowland International)
70 Shenton Way #16-01
Marina House
Singapore 0207
Tel: 2244022

Ng, Lee & Associates
(A member of DFK International)
116 Middle Rd. #08-02/04
ICB Enterprise House
Singapore 0718
Tel: 3386755 Fax: 3386129

Price Waterhouse
6 Battery Rd. #32-00
Singapore 0104
Tel: 2256066 Fax: 2252366 Tlx: 23039

Teo Foong & Wong
(A member of Nexia International)
15 Beach Rd. #03-10, Beach Centre
Singapore 0718
Tel: 3362828 Fax: 3390438

ADVERTISING AGENCIES

Ball Partnership EURO RSCG
49 Beach Rd. #05-00, Hexagon House
Singapore 0719
Tel: 3398066 Fax: 3392231

Bozell Advertising (s) Pte. Ltd.
107-108 Amoy St.
Singapore 0106
Tel: 2276818 Fax: 2271433 Tlx: 23382 BJKNE

BSB Singapore
100 Beach Rd. #30-06 Shaw Towers
Singapore 0718
Tel: 2993301 Fax: 2967525 Tlx: 786-26190

Chiat/Day/Mojo Pte. Ltd.
111 North Bridge Rd. #24-04, Peninsula Plaza
Singapore 0617
Tel: 3376503

DDB Needham Worldwide DIK Pte. Ltd.
133 New Bridge Rd. #23-09/10
Chinatown Point
Singapore 0105
Tel: 5380988 Fax: 5380588 Tlx: 22247 EXCITE

Dentsu, Young & Rubicam/Singapore
6/F., 20 Kallang Ave.
Singapore 1233
Tel: 2950025 Fax: 2962016

DMB&B Tokyu
19 Keppel Rd. #09-08, Jit Poh Bldg.
Singapore 0208
Tel: 2255522 Fax: 2257186 Tlx: DMB&B28009

Export Consultants (Pte) Ltd.
1 Scotts Rd. #18-06, Shaw Centre
Singapore 0922
Tel: 7349061

Graphic Direction Pte. Ltd.
Blk. 1 Pasir Panjang Rd. #10-35
Alexandra Distripark
Singapore 0511
Tel: 2748171, 2742060 Fax: 2731637

Interad Advertising Consultants
154 Clemenceau Ave. #01-01/07
Haw Par Centre
Singapore 0923
Tel: 3388181 Fax: 3368282

J. Walter Thompson Pte. Ltd.
190 Middle Rd., #20-02/05, Fortune Centre
Singapore 0718
Tel: 3389336 Fax: 3366087 Tlx: 33321

All addresses and telephone numbers are in the Republic of Singapore unless otherwise noted. The country code for Singapore is [65].

Leo Burnett Pte. Ltd.
30 Robinson Rd. #03-01/04-01
Robinson Towers
Singapore 0104
Tel: 2207022 Fax: 2207011 Tlx: 5139

Lintas Singapore
133 Cecil St. #14-01, Keck Seng Tower
Singapore 0106
Tel: 2224233 Fax: 2245645 Tlx: 25693

McCann-Erickson (Singapore) Private Limited
360 Orchard Rd. #03-00, International Bldg.
Singapore 0923
Tel: 7379911 Fax: 7371455 Tlx: 21663

Medicus Intercon
30 Prinsep St. #09-02, LKN Princep House
Singapore 0718
Tel: 3393800 Fax: 3392455

Ogilvy & Mather Pte. Ltd.
1 Maritime Sq. #11-10, World Trade Centre
Singapore 0409
Tel: 2738011 Fax: 2742156 Tlx: 21402

Saatchi & Saatchi Advertising Pte. Ltd.
#04-11 Beach Centre, 15 Beach Rd.
Singapore 0718
Tel: 3394733 Fax: 3393916 Tlx: 25404

Spenser/BBDO
65 Chulia St. #46-06, OCBC Centre
Singapore 0104
Tel: 5332200 Fax: 5335511 Tlx: 28635 BBDO

Starlight Advertising (Pte.) Ltd.
Blk. 261 Waterloo St. #03-36, Waterloo Centre
Singapore 0718
Tel: 3362733 Fax: 3396736

Tokai Agency Singapore Pte. Ltd.
2 Jurong East St. 21 #03-09, IMM Bldg.
Singapore 2260
Tel: 5682999 Fax: 5682993

Tropical/Alliance Communications Pte. Ltd.
52 Chin Swee Rd. #02-00
Singapore 0316
Tel: 7342133, 7343633 Fax: 7342212 Tlx: 35693

LAW FIRMS

Law Society of Singapore
1 Colombo Court #08-29/30
Singapore
Tel: 3383165 Fax: 3397358 Tlx: 42097

Allen & Gledhill
36 Robinson Rd. #18-01, City House
Singapore 0106
Tel: 2251611, 3204319 Fax: 2248210
Tlx: 21600 GLEDHIL

Arthur Loke & Partners
21 Collyer Quay #16-00, Hongkong Bank Bldg.
Singapore 0104
Tel: 2247166 Fax: 2226842 Tlx: 20740

Baker & McKenzie (USA)
21 Collyer Quay #16-00, Hongkong Bank Bldg.
Singapore 0104
Tel: 2248066 Fax: 2243872, 2241038 Tlx: 20852

Chor Pee & Company
11 Collyer Quay #18-02, The Arcade
Singapore, 0104
Tel: 2201911 Fax: 2240183 Tlx: 21570 CPHH

Clifford Chance (UK)
6 Battery Rd. #11-07
Singapore 0104
Tel: 2241588 Fax: 2247553, 2241029
Tlx: 20974 LEX S

Coudert Brothers (USA)
Tung Centre, 20 Collyer Quay
Singapore 0104
Tel: 2229973 Fax: 2241756 Tlx: 21466

David Chong & Co.
8 Robinson Rd. #08-00
Singapore 0104
Tel: 2240955

Donaldson & Burkinshaw
#15-00, Clifford Centre, 24 Raffles Place
Singapore, 0104
Postal address: PO Box 3667, Singapore 9056
Tel: 5339422 Fax: 5337806, 5333590 Tlx: 21556

Drew & Napier
24 Raffles Place, Hex 27-01, Clifford Centre
Singapore 0104
Tel: 5350733 Fax: 5354864

Ella Cheong & G. Mirandah
111 North Bridge Rd. #22-01, Peninsula Plaza
Singapore 0617
Tel: 3394040 Fax: 3370031 Tlx: 42619 ELLACO

Khattar Wong & Partners
80 Raffles Place #25-01, UOB Plaza
Singapore 0104
Tel: 5356844

Koh & Yang
10 Collyer Quay #15-08, Ocean Bldg.
Singapore 0104
Tel: 5323277 Fax: 5342441 Tlx: 28391

Lee & Lee Advocates & Solicitors
5 Shenton Way #19-00, UIC Bldg.
Singapore 0106
Tel: 2200666

Lee & Partners
90 Cecil St. #05-00, Carlton Bldg.
Singapore 0106
Tel: 2222777

Mallesons Stephen Jaques (Australia)
Level 27, OCBC Centre, 65 Chulia St.
Singapore 0104
Tel: 7372894 Fax: 7381273

May Oh & Wee
21 Collyer Quay #14-02/03, Hongkong Bank Bldg.
Singapore 0104
Tel: 2230309

IMPORTANT ADDRESSES

Murphy & Dunbar
585 North Bridge Rd. #10-03, Blanco Court
Singapore 0718
Tel: 2969996

Norton Rose (UK)
5 Shenton Way #33-08, UIC Bldg., 33/F.
Singapore 0106
Tel: 2237311, 2231022 Fax: 2245758
Tlx: NOPURA 28880

Robert W.H. Wang & Woo
50 Raffles Place #12-05, Shell Tower
Singapore 0104
Tel: 2250166

Rodyk & Davidson
6 Battery Rd. #38-01
Singapore 0104
Tel: 2252626

Shook Lin & Bok
2 Battery Rd. #06-00
Malayan Bank Chambers
Singapore 0104
Tel: 5351944

Tan, Rajah & Cheah
9 Battery Rd. #15-00, Straits Trading Bldg.
Singapore 0104
Tel: 5322271

Interlingua Language Services (S) Pte. Ltd.
141 Cecil St. #06-01
Tung Ann Association Bldg.
Singapore 0106
Tel: 2223755 Fax: 2249188

Khaw and Sons Pte. Ltd.
336 River Valley Rd. #12-02, AA Centre
Singapore 0923
Tel: 7382498 Fax: 7343410

Ong & Shi Bi-Lingual Consultant
63, Toh Tuck Place
Singapore 2159
Tel: 4667748

Worldwide Translation Services
64 Lloyd Rd.
Singapore 0923
Tel 7377672 Fax: 2355480 Tlx: 24200 TMSR

TRANSLATORS & INTERPRETERS

And Vice Versa Translating & Interpreting Services
120 King's Ave.
Singapore 2775
Tel: 7582500 Fax: 7587133

Arlette Interpreting & Translating Services
400 Orchard Rd. #06-07, Orchard Towers
Singapore 0923
Tel: 7340651 Fax: 7345197

Coleman Commercial & Language Centre
111 North Bridge Rd. #05-33/35
Peninsula Plaza
Singapore 0617
Tel: 3363462 Fax: 3360929

DavidARTS Publishing Pte. Ltd.
50 Jalan Sultan #19-06 Jalan Sultan Centre
Singapore 0719
Tel: 2930083 Fax: 2949736

Europhone Language Institute Pte. Ltd.
3 Coleman St. #04-33
Peninsula Shopping Centre
Singapore 0617
Tel: 3373617, 3363992 Fax: 3374506

FLP Singapore Pte. Ltd.
1 Selegie Rd. #08-26, Paradiz Centre
Singapore 0718
Tel: 3340723 Fax: 3340024

Frank Tan Research Associates
Translation Services Division
420 North Bridge Rd. #05-27
North Bridge Centre
Singapore 0718
Tel: 3343951/2 Fax: 3343953

All addresses and telephone numbers are in the Republic of Singapore unless otherwise noted. The country code for Singapore is [65].

TRANSPORTATION

AIRLINES

Aeroflot
15 Queen St. #01-02, Tan Chong Tower
Singapore 0718
Tel: 3361757

Air China
51 Anson Rd. #01-53, Anson Centre
Singapore 0207
Tel: 2252177

Air India
5 Shenton Way #17-01, UIC Bldg.
Singapore 0106
Tel: 2205277 Fax: 2257636

Airlanka
140 Cecil St. #02-00/B, PIL Bldg.
Singapore 0106
Tel: 2236026, 2257233

Air Mauritius
135 Cecil St. #04-02, LKN Bldg.
Singapore 0106.
Tel: 2223033 Fax: 2259726

Air New Zealand
10 Collyer Quay #24-08, Ocean Bldg.
Singapore 0104
Tel: 5358266, 5323846

Alitalia
Tel: 7373166

All Nippon
Tel: 2283222

Asiana Airlines
135 Cecil St. #01-00, LKN Bldg.
Singapore 0106
Tel: 2253866 Fax: 2270250

Biman Bangladesh
15 McCallum St. #01-02, Natwest Centre
Singapore 0106
Tel: 2217155

British Airways
101 Thomson Rd. #01-56, United Sq.
Singapore 1130
Tel: 2538444

Cathay Pacific
10 Collyer Quay #16-01, Ocean Bldg.
Singapore 0104
Tel: 5331333 Fax: 5341161

China Airlines
Tel: 7372211

Emirates
435 Orchard Rd. #19-06, Wisma Atria
Singapore 0923
Tel: 2351911, 2353966

EVA Airways
Tel: 2261533

Finnair
541 Orchard Rd. #18-01, Liat Towers
Singapore 0923
Tel: 7333377

Garuda Indonesia
101 Thomson Rd. #13-03, United Sq.
Singapore 1130
Tel: 5423013, 2502888 Fax: 2536196

Indian Airlines
70 Shenton Way #01-03, Marina House
Singapore 0207
Tel: 2254949

Japan Air Lines
16 Raffles Quay #01-01, Hong Leong Bldg.
Singapore 0104
Tel: 2210522, 2202013 Fax: 2248870

Japan Air System
137 Cecil St. #01-02, ICS Bldg.
Singapore 0106
Tel: 2210522, 2261515 Fax: 2232167

JES Air
GSA: Region Air Pte. Ltd.
50 Cuscaden Rd. #07-01, HPL House
Singapore 1024
Tel: 2356277

KLM
333 Orchard Rd. #01-02
Singapore 0923
Tel: 7377622

Korean Air
10 Collyer Quay #07-08, Ocean Bldg.
Singapore 0104
Tel: 5342111, 5343222

LOT
Tel: 2210116

Lufthansa
390 Orchard Rd. #05-01, Palais Renaissance
Singapore 0923
Tel: 7379222

Malaysia Airlines
190 Clemenceau Ave. #02-09/11
Singapore Shopping Centre
Singapore 0923
Tel: 3366777 Fax: 3362782

Northwest
435 Orchard Rd. #11-03, Wisma Atria
Singapore 0923
Tel: 2357166

Pakistan International Airlines
Tel: 2512322

Philippine Airlines
Tel: 3371103

Qantas
Tel: 7373744

IMPORTANT ADDRESSES

Royal Brunei
25 Scotts Rd. #01-04A/04B/05
Royal Holiday Inn Crowne Plaza
Singapore 0922
Tel: 2354672

Royal Jordanian
Tel: 3388188

SAS
152 Beach Rd. #23-01/04, Gateway East
Singapore 0719
Tel: 2941611

Saudi Arabian Airlines
7500A Beach Rd. #10-318
Singapore 0719
Tel: 2355660, 2917322

Sempati Air
GSA: Region Air Pte. Ltd.
50 Cuscaden Rd. #07-01, HPL House
Singapore 1024
Tel: 7345077, 2356277 Fax: 7361662

Silkair
55 Airport Blvd., 3/F. SATS Bldg.
Singapore 1781
Tel: 2212221, 5428111

Singapore Airlines
Airline Rd., Airline House
Singapore 1781
Tel: 2238888, 5423333 Fax: 5455034, 5454231

Swissair
435 Orchard Rd. #18-01, Wisma Atria
Singapore 0923
Tel: 7378133, 2351708

Thai Airways
Tel: 2249977

Turk Hava Yollari (Turkish Airlines)
545 Orchard Rd. #02-18 & #02-21
Far East Shopping Centre
Singapore 0923
Tel: 7324556

United Airlines
Tel: 2200711

UTA
Tel: 7376355

TRANSPORTATION & CUSTOMS BROKERAGE FIRMS

Companies may offer more services in addition to those listed here. Service information is provided as a guideline and is not intended to be comprehensive.

International Air Transport Association
331 North Bridge Rd. #20-00
Singapore
Tel: 3399978 Fax: 3390855

Singapore Air Cargo Agents Association
115 Airport Cargo Rd. #04-09
Cargo Agents Bldg. C
Singapore 1781
Tel: 5454620, 5459597 Fax: 5426820 Tlx: 38963

Singapore Freight Forwarders Association
7500A Beach Rd. #13-314, The Plaza
Singapore 0719
Tel: 2964645 Fax: 2922504

Singapore National Shipping Association
456 Alexandra Rd. #02-02, NOL Bldg.
Singapore 0511
Tel: 2733574, 2783464 Fax: 2745079 Tlx: 24021

American President Lines Ltd.
19 Keppel Rd., 02-01 Jit Poh Bldg.
Singapore 0208
Tel: 2259966 Fax: 2214922 Tlx: 21337
Shipping

Barwil Agencies Pte. Ltd.
200 Cantonment Rd., Southport
Singapore 0208
Tel: 2252577 Fax: 2252538 Tlx: 23057
Shipping agents, freight forwarders, ship chandlers

Calberson (SEA) Pte. Ltd.
8 Pandan Ave.
Singapore 2260
Tel: 2658383 Fax: 2660716
Airport office Tel: 5425322 Fax: 5425473
International freight forwarder, air & sea, warehousing and distribution

Chartered Materials & Services
3 Lim Teck Kim Rd. #09-01
Singapore Technologies Bldg.
Singapore 0208
Tel: 2221433 Fax: 223766
Airport office Tel: 5428811
International freight forwarding, cargo consolidation, customs clearance, trucking, door-to-door delivery, warehousing, packing and crating, aircraft/ship charter, insurance, project cargo management, materials management procurement and trading.

CTE Shipping (S)
133 New Bridge Rd. #09-06
Chinatown Point
Singapore 0105
Tel: 5380155 Fax: 5387592
Sea & air freight forwarding, transportation, import/export documentation, custom clearance, warehousing and distribution.

Curio Pack Pte. Ltd.
No. 1 Maritime Square #09-12
World Trade Center
Singapore 0409
Tel: 2701909 Fax: 2732977
International freight forwarding (including exhibition forwarding), transshipment services, packing services, warehousing and storage.

Danzas Singapore Pte. Ltd.
115 Airport Cargo Rd. #04-01
Cargo Agents Bldg. C
Singapore 1781
Tel: 5430244 (Main office), 2278783 (Sea freight), 5459500 (Air freight) Fax: 5458980
International sea and air freight forwarders

All addresses and telephone numbers are in the Republic of Singapore unless otherwise noted. The country code for Singapore is [65].

EAS Express Aircargo System
115 Airport Cargo Rd. #04-05
Cargo Agents Bldg. C
Singapore Changi Airport
Singapore 1781
Tel: 5429036 Fax: 5428764
Air freight/Express parcel/courier/sea freight.
Specialize in service to PRC, N. Korea and Vietnam

Everett Steamship Corporation SA
24 Raffles Place, 17-03 Clifford Centre
Singapore 0104
Tel: 5325481 Fax: 5325486 Tlx: 21306
Shipping

Federal Express (S) Pte. Ltd.
3 Kaki Bukit Rd. 2, Block A, Unit 3E
Eunos Warehouse Complex
Singapore 1441
Tel: 7432626 Fax: 7414225 Tlx: 22454
Air cargo, courier service

Freight Express International Ltd.
Airport Cargo Rd. #41-06-71
Cargo Agents Bldg. A
Singapore 1781
Tel: 5427711 Fax: 5420556
International air freight, forwarders and consolidators, customs brokers, sea/air service and charter brokers

Guan Guan Shipping Pte. Ltd.
23 Telok Ayer St.
Singapore 0104
Tel: 2219790 Fax: 5343504 Tlx: 21395
Shipping

Hai Sun Hup Co
200 Cantonment Rd. #09-01, Southpoint
Singapore 0208
Tel: 2204906 Fax: 2251107
Shipowners, shipping agents, general marine services, freight forwarding, warehousing, terminal operators.

Hapag-Lloyd (Asia)
200 Cantonment Rd. #08-03, Southpoint
Singapore 0208
Tel: 2244792 Fax: 2278378
Container line shipping services

Hill & Delamain (S)
115 Airport Cargo Rd. #06-21/22
Cargo Agents Bldg. C
Singapore Changi Airport
Singapore 1781
Tel: 5427622 Fax: 5427584
International freight forwarder, air cargo

Jasinta Impex Pte. Ltd.
5 Kaki Bukit Rd. 2 #03-13
City Warehouse
Singapore 1440
Tel: 7468766
International freight forwarder, air/sea freight services, transportation, packing, warehousing, customs clearance

Kawasaki Kisen Kaisha Ltd.
460 Alexandra Rd. #19-01, PSA Bldg.
Singapore 0511
Tel: 2218977
Shipping

Kin Yuen Group Singapore
158 Cecil St. #09-00
Dapenso Bldg.
Singapore 0106
Tel: 2258555 Fax: 2244427
Shipbrokers, chartering, freight forwarding

KWE-Kintetsu World Express
115 Airport Cargo Rd. #04-12
Cargo Agents Bldg. C
Singapore 1781
Tel: 5427777 Fax: 5456008
International freight forwarder, air freight consolidator, cover transportation by air, sea, road and intermodal services.

Logistics and Procurement Pte. Ltd.
Block 9002 #02-02
Sampines St. 93
Singapore 1852
Tel: 7866192 Fax: 7866507
Project cargoes specialist, NVOCC consolidator, general transportation, air & sea handling, forwarding agents, custom clearance and documentation, packing and removal, labor supplies, warehousing, container trucking.

Maersk Singapore Pte. Ltd.
Robinson Rd., PO Box 1631
Singapore 9032
Tel: 2240505 Fax: 2247649
Shipping

Ming Hoe Shipping & Transportation Agency
100 Jalan Sultan #08-06, Sultan Plaza
Singapore 0719
Tel: 2972816 Fax: 2971948
International freight forwarding, container haulage & transportation, shipping & insurance agents, customs clearance, warehousing, air cargo.

Mutiara Line Shipping
101 Cecil St. #18-06, Tong Eng Bldg.
Singapore 0106
Tel: 2201126 Fax: 2276563
Shipping agents, freight forwarders, cargo logistics

Nedlloyd Air Cargo
Singapore Changi Airport
PO Box 543
Singapore
Tel: 5425566 Fax: 5423019
Air cargo

Nedlloyd EAC Agencies Pte. Ltd.
138 Robinson Rd. #01-00, Hong Leong Centre
Singapore 0106
Tel: 2218989 Fax: 2249106 Tlx: 21261
Shipping

Neptune Orient Lines
456 Alexandra Rd., NOL Bldg.
Singapore 0511
Tel: 2789000 Fax: 2784900
Shipping

New Straits Shipping Co. Pte. Ltd.
51 Anson Rd. #09-53, Anson Centre
Singapore 0207
Tel: 2201007 Fax: 2240785 Tlx: 23150
Shipping

Nigai Nitto (S) Pte. Ltd.
Blk 4, Pasir Panjang Rd. #07-31/39
Alexandra Distripark
Singapore 0511
Tel: 2730311
Freight Forwarding (air & sea), warehousing/ packaging, import & export clearance, documentation, transportation, exhibition handling.

NYK Agencies
150 Beach Rd. #25-00
Gateway West, Singapore 0718
Tel: 2950123 Fax: 2937354 Tlx: 28599
Liner shipping agent, freight consolidators and forwarding, warehousing, container depot

Pacific International Lines Pte. Ltd.
140 Cecil St. #03-00, PIL Bldg.
Singapore 0601
Tel: 2218133 Fax: 2258741 Tlx: 24190
Shipping

Satsaco Express Transportation
Unit 112, 2/F.
Changi Airport Cargo Complex
CIAS Cargo Terminal Bldg.
Singapore 1781
Tel: 5424025 Fax: 5424350
Freight forwarding, transportation, import & export documentation

Sea Consortium Pte. Ltd.
11-12 Duxton Hill
Singapore 0208
Tel: 2239033 Fax: 2257496
Shipping agents, container leasing agents, shipbroker, chartering broker

Sea-Land Service, Inc.
200 Cantonment Rd. #05-03, Southpoint
Singapore 0208
Tel: 2225222, 3219333 Fax: 2243226
Tlx: 21442
International containerized shipping services

TNT Express Worldwide
140 Paya Lebar Rd. #02-10, A-Z Bldg.
Singapore 1440
City Office Tel: 7429000, 7453122 Fax: 7422214, 7476349
Airport Office Tel: 5423321 Fax: 5424534
Tlx: 38102
Air cargo, courier service

United Parcel Service Singapore Pte. Ltd.
3 Killiney Rd., Winsland House #06-01
Singapore 0923
Tel: 7383388 Fax: 7382683, 7382883
Air cargo, courier service

Wallem Shipping
1 Maritime Square #13-02/04
World Trade Centre
Singapore 0409
Tel: 2712611 Fax: 2719569 Tlx: 24505, 24605
Cargo handling; forwarding, liner, tramp agents, cargo brokers

Yamato Transport (S) Pte. Ltd.
1 Maritime Square #09-23
World Trade Center
Singapore 0409
Tel: 2783058 Fax: 2733288
International freight forwarders, air and sea and small parcels.

Yick Fung Shipping & Enterprises
General Agent for China Ocean Shipping Company
62 Cecil St. #02-00, TPI Bldg.
Singapore 0104
Tel: 2214466 Fax: 2255215, 2255087
Shipping

All addresses and telephone numbers are in the Republic of Singapore unless otherwise noted. The country code for Singapore is [65].

PUBLICATIONS, MEDIA & INFORMATION SOURCES

All publications are in English unless otherwise noted.

DIRECTORIES & YEARBOOKS

Annual Report—Monetary Authority of Singapore
(Annual)
10 Shenton Way, MAS Bldg.
Singapore 0207
Tel: 2255577 Fax: 2299491 Tlx: 28174 ORCHID

Asian Computer Directory
(Monthly)
Washington Plaza
1/F., 230 Wanchai Rd.
Wanchai, Hong Kong
Tel: [852] 8327123 Fax: [852] 8329208

Asian Printing Directory
(English/Chinese, Annual)
Travel & Trade Publishing (Asia)
16/F., Capitol Centre
5-19 Jardines Bazaar
Causeway Bay, Hong Kong
Tel: [852] 8903067 Fax: [852] 8952378

Asia-Pacific Journal of Operational Research
(Annual)
Operational Research Society of Singapore
National University Singapore
Department of Math
Singapore 0511
Tel: 7722737 Fax: 7795452

Asia Pacific Leather Directory
(Annual)
Asia Pacific Leather Yearbook
(Annual)
Asia Pacific Directories, Ltd.
6/F. Wah Hen Commercial Centre
381 Hennessy Rd.
Hong Kong
Tel: [852] 8936377 Fax: [852] 8935752

Asia-Pacific Travel Index
(Annual)
Asian Business Press Pte. Ltd.
100 Beach Rd., 2600 Shaw Tower
Singapore 0718

ASTRAD: ASEAN Trade Directory
(Annual)
Peter Isaacson Publications Pty. Ltd.
46-50 Porter St.
PO Box 172
Prahran, Vic. 3181, Australia
Tel: [61] (3) 520-5555 Fax: [61] (3) 525-2983

Bankers Handbook for Asia
(Annual)
Dataline Asia Pacific Inc.
3rd Fl., Hollywood Center
233 Hollywood Road
Hong Kong
Tel: [852] 8155221 Fax: [852] 8542794

Directory—Singapore Indian Chamber of Commerce
Singapore Indian Chamber of Commerce
101 Cecil St. #23-01, Tong Eng Bldg.
Singapore 0106
Tel: 2222505 Fax: 2231707 Fax: 22336

Directory of Financial Institutions
Monetary Authority of Singapore
10 Shenton Way, MAS Bldg.
Singapore 0207
Tel: 2255577 Fax: 2299491 Tlx: 28174 ORCHID

Economic Survey of Singapore
(Annual)
Singapore National Printers Ltd.
Publication Division
8 Shenton Way #B1-07
Singapore
Tel: 2230834

Fact Book
(Annual)
Stock Exchange of Singapore Ltd.
1 Raffles Pl. #24-00, OUB Centre
Singapore 0104
Tel: 5353788

Hui Yuan Ming Lu
Singapore Chinese Chamber of Commerce & Industry
47 Hill St. #09-00
Singapore 0617
Tel: 3378381 Fax: 3390605 Tlx: 33714

International Tax and Duty Free Buyers Index
(Annual)
Pearl & Dean Publishing, Ltd.
9/F. Chung Nam Bldg., 1 Lockhart Rd.
Hong Kong
Tel: [852] 8660395 Fax: [852] 2999810

Investor's Guide
Singapore International Chamber of Commerce
6 Raffles Quay #05-00, Denmark House
Singapore 0104
Tel: 2241255 Fax: 2242785 Tlx: 25235 INTCHAM

Jurong Industries Directory
Business Media
39 Stamford Rd., Unit 136
Stamford House
Singapore 0617

Kompass Directory of American Business in Singapore
(Annual)
Kompass South East Asia Ltd.
326-C King George's Ave.
Singapore 0820
Tel: 2969684 Fax: 2972561

Kompass Directory of Australian Business in
Singapore
(Annual)
Kompass South East Asia Ltd.
326-C King George's Ave.
Singapore 0820
Tel: 2969684 Fax: 2972561

Kompass Directory of European Business in
Singapore
(Annual)
Kompass South East Asia Ltd.
326-C King George's Ave.
Singapore 0820
Tel: 2969684 Fax: 2972561

Kompass Directory of French Business in
Singapore
(Annual)
Kompass South East Asia Ltd.
326-C King George's Ave.
Singapore 0820
Tel: 2969684 Fax: 2972561

Kompass Directory of Japanese Business in
Singapore
(Annual)
Kompass South East Asia Ltd.
326-C King George's Ave.
Singapore 0820
Tel: 2969684 Fax: 2972561

Kompass Singapore
(Annual)
Kompass South East Asia Ltd.
326-C King George's Ave.
Singapore 0820
Tel: 2969684 Fax: 2972561

Lloyd's Singapore Port Services Index
(Annual)
LLP-Times Maritime & Business Publishing Co.
Times Centre, 1 New Industrial Rd.
Singapore 1953
Tel: 2848844 Fax: 2881186 Tlx: 25713

Report on the Labour Force Survey of Singapore
(Annual)
Singapore National Printers Ltd.
Publication Division
8 Shenton Way #B1-07
Singapore
Tel: 2230834

The Singapore Accountant
(Annual)
Institute of Certified Public Accountants of
Singapore
116 Middle Rd.
1CB Enterprise House
Singapore
Tel: 3367020

Singapore Book World
(Annual; English, Chinese, and Malay)
Chopmen Publishers
865 Mountbatten Rd., 05-28/29, Katong Shopping
Centre
Singapore
Tel: 3441495 Fax: 3440180

Singapore Builders Directory
(Annual)
Far East Media Representatives Bldg.
320 Serangoon Rd.
Singapore

Singapore Business Yearbook
(Annual)
Times Periodicals Pte. Ltd.
1 New Industrial Rd.
Times Centre
Singapore 1953
Tel: 2848844 Fax: 2874720 Tlx: 25713 TIMESS

Singapore Manufacturers and Products Directory
Straits Times Press
82 Genting Lane
Singapore 1334
Tel: 7444875 Fax: 7449949 Tlx: 55959

Singapore Shipping & Air Transportation
Industries Directory
(Bimonthly)
Victor Kamkin Inc
4956 Boiling Brook Parkway
Rockville MD 20852
Tel: [1] (301) 8815973

Singapore Standards Yearbook
(Biennial)
Singapore Institute of Standards & Industrial
Research (SISIR)
1 Science Park Dr.
Singapore 0511
Tel: 7787777 Fax: 7780086 Tlx: 28499 SISIR

Singapore's Judicial & Legal Directory
Legal Publications Pte. Ltd.
206 Colombo Court
Singapore 0617

Singapore Telex & Telefax Directory / Telecoms
(Annual)
Telecommunication Authority of Singapore
31 Exeter Rd. #05-00 Comcentre
Singapore 0923
Tel: 7343344, 7387788 Fax: 7328428, 7330073
Tlx: 33311 TELECOM

The SMA Directory
Singapore Manufacturers' Association
20 Orchard Rd., SMA House
Singapore 0923
Tel: 3388787 Fax: 3385385 Tlx: 24992

Times Business Directory of Singapore
(Annual)
Times Trade Directories Pte. Ltd.
Times Centre, 1 New Industrial Rd.
Singapore 1953
Tel: 2848844 Fax: 2881186 Tlx: 25713

Times Guide to Computers
(Annual)
Times Periodicals Pte. Ltd.
1 New Industrial Rd.
Times Centre
Singapore 1953 Singapore
Tel: 2848844 Fax: 2874720 Tlx: 25713 TIMESS

All addresses and telephone numbers are in the Republic of Singapore unless otherwise noted. The country code for Singapore is [65].

Who's Who in Singapore
(Annual)
City Who's Who Pte. Ltd.
100 Beach Rd. Suite 1306, Shaw Towers
Singapore 0718

World Jewelogue
(Annual)
Headway International Publications Co.
907 Great Eagle Center
23 Harbour Rd.
Hong Kong
Tel: [852] 8275121 Fax: [852] 8277064

Yearbook of Statistics: Singapore
(Annual)
Singapore National Printers Ltd.
Publication Division
8 Shenton Way #B1-07
Singapore
Tel: 2230834

NEWSPAPERS

Asian Wall Street Journal
Dow Jones Publishing Co. (Asia)
2/F. AIA Bldg., 1 Stubbs Rd.
GPO Box 9825
Hong Kong
Tel: [852] 5737121 Fax: [852] 8345291

Business Times
Times House
390 Kim Seng Rd.
Singapore 0923
Tel: 7370011 Fax: 7335271

International Herald Tribune
7/F. Malaysia Bldg., 50 Gloucester Rd.
Wanchai, Hong Kong
Tel: [852] 8610616 Fax: [852] 8613073

Lianhe Wanbao
(Chinese)
News Centre
82 Genting Lane
Singapore 1334 Singapore
Tel: 7444875 Fax: 7449949 Tlx: 55959

Lianhe ZaoBao
(Chinese)
News Centre
82 Genting Lane
Singapore 1334 Singapore
Tel: 7444875 Fax: 7449949 Tlx: 55959

New Paper
News Centre
82 Genting Lane
Singapore 1334 Singapore
Tel: 7444875 Fax: 7449949 Tlx: 55959

Shin Min Daily News
(Chinese)
News Centre
82 Genting Lane
Singapore 1334 Singapore
Tel: 7444875 Fax: 7449949 Tlx: 55959

The Straits Times
Singapore Newspaper Services Pte. Ltd.
Circulation Department
82 Genting Lane
Singapore 1334 Singapore
Tel: 7444875 Tlx: 55959

GENERAL BUSINESS & TRADE PERIODICALS

ASEAN Business Quarterly
Asia Research Pte. Ltd.
PO Box 91, Alexandra Post Office
Singapore

ASEAN Economic Bulletin
(Three times a year)
Institute of Southeast Asian Studies
Heng Mui Keng Terrace
Singapore 0511
Tel: 7780955 Fax: 7781735 Tlx: 37068 ISEAS

Asia Labour Monitor
(Bimonthly)
Asia Monitor Resource Center
444-446 Nathan Road, 8th Fl., Flat B
Kowloon, Hong Kong
Tel: [852] 3321346

Asia Pacific Journal of Management: APJM
(Semiannual)
National University of Singapore
Faculty of Business Administration
10 Kent Ridge Crescent
Singapore 0511
Tel: 7765641 Tlx: UNISPO33943

Asian Business
(Monthly)
Far East Trade Press, Ltd.
2/F Kai Tak Commercial Bldg.
317 Des Voeux Rd.
Central, Hong Kong
Tel: [852] 5457200 Fax: [852] 5446979

Asian Finance
(Monthly)
3rd Fl., Hollywood Center
233 Hollywood Road
Hong Kong
Tel: [852] 8155221 Fax: [852] 8504437

Asian Monetary Monitor
(Bimonthly)
GPO Box 12964
Hong Kong
Tel: [852] 8427200

Asiaweek
(Weekly)
Asiaweek Ltd.
199 Des Voeux Road
Central, Hong Kong
Tel: [852] 8155662 Fax: [852] 8155903

Business Week, Asia Edition
(Weekly)
2405 Dominion Centre
43-59 Queens Rd. East
Hong Kong
Tel: [852] 3361160 Fax: [852] 5294046

All publications are in English unless otherwise noted.

Economic Bulletin
(Monthly)
Singapore International Chamber of Commerce
6 Raffles Quay #05-00, Denmark House
Singapore 0104
Tel: 2241255 Fax: 2242785 Tlx: 25235 INTCHAM

The Economist, Asia Edition
(Weekly)
The Economist Newspaper, Ltd.
1329 Chater Rd.
Hong Kong
Tel: [852] 8681425

Far Eastern Economic Review
(Weekly)
Review Publishing Company Ltd.
6-7th Fl., 181-185 Gloucester Road
Hong Kong
Tel: [852] 8328381 Fax: [852] 8345571

Journal of Business
(Three times a year)
Asia Pacific-Journal of Management
School of Management
National University of Singapore
Kent Ridge
0511 Singapore
Tel: 7723022

Newsweek International, Asia Edition
(Weekly)
Newsweek, Inc.
47/F., Bank of China Tower
1 Garden Rd.
Central, Hong Kong
Tel: [852] 8104555

Showcase
Singapore International Chamber of Commerce
6 Raffles Quay #05-00, Denmark House
Singapore 0104
Tel: 2241255 Fax: 2242785 Tlx: 25235 INTCHAM

Singapore Business
(Monthly)
Times Trade Directories Pte. Ltd.
Times Centre, 1 New Industrial Rd.
Singapore 1953
Tel: 2848844 Fax: 2881186 Tlx: 25713

Singapore Economic Review
(Semiannual)
Department of Economics and Statistics
University of Singapore
Singapore 0511
Tel: 7723941 Fax: 7752646

Singapore Journal of Primary Industries
(Semiannual)
Director of Primary Production Bldg., 2-3/F.
Maxwell Rd.
Singapore 0106
Tel: 3226634 Fax: 2206068 Tlx: 28851

Singapore Management Review
(Semiannual)
Singapore Institute of Management
3/F., Thong Teck Bldg.
Scotts Rd.
Singapore 0922 Singapore
Tel: 7378866 Tlx: SIM 50259

Time, Asia Edition
(Weekly)
Time, Inc.
31/F., East Tower, Bond Centre
89 Queensway
Hong Kong
Tel: [852] 8446660 Fax: [852] 5108799

World Executives Digest
(Monthly)
3/F. Garden Square Bldg., Greenbelt Drive Cor.
Legaspi Makati
Metro Manila, Philippines
Tel: [63] (2) 8179126

INDUSTRY-SPECIFIC PERIODICALS

Asia Computer Weekly
(Bimonthly)
Asian Business Press Pte., Ltd.
100 Beach Rd. #26-00 Shaw Towers
Singapore 0718
Tel: 2943366 Fax: 2985534

Asiamac Journal: The Machine-Building and Metal Working Journal for the Asia Pacific Region
(Quarterly; English, Chinese)
Adsale Publishing Company
21st Fl., Tung Wai Commercial Building
109-111 Gloucester Road
Hong Kong
Tel: [852] 8920511 Fax: [852] 8384119, 8345014
Tlx: 63109 ADSAP HX

Asian Architect and Contractor
(Monthly)
Thompson Press Hong Kong Ltd.
Tai Sang Commercial Building, 19th Fl.
24-34 Hennessy Road
Hong Kong

Asian Aviation
(Monthly)
Asian Aviation Publications
2 Leng Kee Rd. #04-01, Thye Hong Centre
Singapore 0315
Tel: 4747088 Fax: 4796668

Asian Computer Monthly
(Monthly)
Computer Publications Ltd.
Washington Plaza, 1st Fl.
230 Wanchai Road
Wanchai, Hong Kong
Tel: [852] 9327123 Fax: [852] 8329208

All addresses and telephone numbers are in the Republic of Singapore unless otherwise noted. The country code for Singapore is [65].

Asian Defence Journal
(Monthly)
Syed Hussain Publications (Sdn)
61 A&B Jelan Dato, Haji Eusoff
Damai Complex
PO Box 10836
50726 Kuala Lumpur, Malaysia
Tel: [60] (3) 4420852 Fax: [60] (3) 4427840

Asian Electricity
(11 per year)
Reed Business Publishing Ltd.
5001 Beach Rd. #06-12, Golden Mile Complex
Singapore 0719
Tel: 2913188 Fax: 2913180

Asian Electronics Engineer
(Monthly)
Trade Media Ltd.
29 Wong Chuck Hang Rd.
Hong Kong
Tel: [852] 5554777 Fax: [852] 8700816

Asian Hospital
(Quarterly)
Techni-Press Asia Ltd.
PO Box 20494
Hennessy Road
Hong Kong
Tel: [852] 5278682 Fax: [852] 5278399

Asian Hotel & Catering Times
(Bimonthly)
Thomson Press (HK)
19/F., 23-34 Hennessy Rd.
Tai Sang Commercial Bldg.
Hong Kong
Tel: [852] 5283351 Fax: [852] 8650825

Asian Manufacturing
Far East Trade Press Ltd.
2nd Fl., Kai Tak Commercial Building
317 Des Voeux Road
Central, Hong Kong
Tel: [852] 5453028 Fax: [852] 5446979

Asian Medical News
(Bimonthly)
MediMedia Pacific Ltd.
Unit 1216, Seaview Estate
2-8 Watson Rd.
North Point, Hong Kong
Tel: [852] 5700708 Fax: [852] 5705076

Asian Meetings & Incentives
(Monthly)
Travel & Trade Publishing (Asia)
16/F., Capitol Centre
5-19 Jardines Bazaar
Causeway Bay, Hong Kong
Tel: [852] 8903067 Fax: [852] 8952378

Asian Oil & Gas
(Monthly)
Intercontinental Marketing Corp.
PO Box 5056
Tokyo 100-31, Japan
Fax: [81] (3) 3667-9646

Asian Plastic News
(Quarterly)
Reed Asian Publishing Pte., Ltd.
5001 Beach Rd. #06-12, Golden Mile Complex
Singapore 0719
Tel: 2913188 Fax: 2913180

Asian Printing: The Magazine for the Graphic Arts Industry
(Monthly)
Travel & Trade Publishing (Asia)
16/F., Capitol Centre
5-19 Jardines Bazaar
Causeway Bay, Hong Kong
Tel: [852] 8903067 Fax: [852] 8952378

Asian Security & Safety Journal
(Bimonthly)
Elgin Consultants, Ltd.
Tungnam Bldg.
Suite 5D, 475 Hennessy Rd.
Causeway Bay, Hong Kong
Tel: [852] 5724427 Fax: [852] 5725731

Asian Shipping
(Monthly)
Asia Trade Journals Ltd.
7th Fl., Sincere Insurance Building
4 Hennessy Road
Wanchai, Hong Kong
Tel: [852] 5278532 Fax: [852] 5278753

Asian Sources: Computer Products
Asian Sources: Electronic Components
Asian Sources: Gifts & Home Products
Asian Sources: Hardware
Asian Sources: Timepieces
(All publications are monthly)
Asian Sources Media Group
22nd Fl., Vita Tower
29 Wong Chuk Hang Road
Wong Chuk Hang, Hong Kong
Tel: [852] 5554777 Fax: [852] 8730488

Asian Water & Sewage
(Quarterly)
Techni-Press Asia, Ltd.
PO Box 20494, Hennessy Rd.
Hong Kong
Fax: [852] 5278399

Asia Pacific Broadcasting & Telecommunications
(Monthly)
Asian Business Press Pte., Ltd.
100 Beach Rd. #26-00, Shaw Towers
Singapore 0718
Tel: 2943366 Fax: 2985534

Asia Pacific Dental News
(Quarterly)
Adrienne Yo Publishing Ltd.
4th Fl., Vogue Building
67 Wyndham Street
Central, Hong Kong
Tel: [852] 5253133 Fax: [852] 8106512

All publications are in English unless otherwise noted.

IMPORTANT ADDRESSES

Asia Pacific Food Industry Business Report
(Monthly)
Asia Pacific Food Industry Publications
24 Peck Sea St., #03-00 Nehsons Building
Singapore 0207
Tel: [65] 2223422   Fax: [65] 2225587

Asiatechnology
(Monthly)
Review Publishing Company Ltd.
6-7th Fl., 181-185 Gloucester Road
GPO Box 160
Hong Kong
Tel: [852] 8328381   Fax: [852] 8345571

Asia Travel Guide
(Monthly)
Interasia Publications, Ltd.
190 Middle Rd. #11-01, Fortune Center
Singapore 0718
Tel: 3397622   Fax: 3398521

Asia Travel Guide
Asia Pacific Food Industry
(Monthly)
Asia Pacific Food Industry Publications
24 Peck Sea St., #03-00 Nehsons Building
Singapore 0207
Tel: [65] 2223422   Fax: [65] 2225587

ATA Journal: Journal for Asia on Textile & Apparel
(Bimonthly)
Adsale Publishing Company
Tung Wai Commercial Building, 21st Fl.
109-111 Gloucester Road
Wanchai, Hong Kong
Tel: [852] 8920511   Fax: [852] 8384119

Building & Construction News
(Weekly)
Al Hilal Publishing (FE) Ltd.
50 Jalan Sultan #20-06, Jalan Sultan Centre
Singapore 0719
Tel: 2939233   Fax: 2970862

Business Traveller Asia-Pacific
(Monthly)
Interasia Publications
200 Lockhart Rd., 13/F
Wanchai, Hong Kong
Tel: 5749317   Fax: [852] 5726846

Butterworths Law Digest Malaysia,
Singapore and Brunei
(Monthly)
Malayan Law Journal Pte. Ltd.
3 Shenton Way #14-03
Shenton House
Singapore 0104
Tel: 2203684   Tlx: 42890 BGASIA

Cargo Clan
(Quarterly)
Emphais (HK), Ltd.
10/F. Wilson House
19-27 Wyndam St.
Central, Hong Kong
Tel: [852] 5215392   Fax: [852] 8106738

Cargonews Asia
(Bimonthly)
Far East Trade Press, Ltd.
2/F Kai Tak Commercial Bldg.
317 Des Voeux Rd.
Central, Hong Kong
Tel: [852] 5453028   Fax: [852] 5446979

Catering & Hotel News, International
(Biweekly)
Al Hilal Publishing (FE) Ltd.
50 Jalan Sultan #20-26, Jalan Sultan Centre
Singapore 0719
Tel: [852] 2939233   Fax: [852] 2970862

CB Magazine: Training Technologists for Tomorrow
Singapore Polytechnic Department of Civil
Engineering and Building
Civil Engineering and Building Club
Dover Rd.
Singapore 0513

Electronic Business Asia
(Monthly)
Cahners Publishing Company
275 Washington St.
Newton, MA 02158, USA
Tel: [1] (617) 964-3030   Fax: [1] (617) 558-4506

Energy Asia
(Monthly)
Petroleum News Southeast Asia Ltd.
6th Fl., 146 Prince Edward Road W
Kowloon, Hong Kong
Tel: [852] 3805294   Fax: [852] 3970959

Far East Health
(10 per year)
Update-Siebert Publications
Reed Asian Publishing Pte.
5001 Beach Rd. #06-12, Golden Mile Complex
Singapore 0719
Tel: 2913188   Fax: 2913180

Fashion Accessories
(Monthly)
Asian Sources Media Group
22nd Fl., Vita Tower
29 Wong Chuk Hang Road
Wong Chuk Hang, Hong Kong
Tel: [852] 5554777   Fax: [852] 8730488

Industrial News and Research
Singapore Institute of Standards & Industrial
Research (SISIR)
1 Science Park Dr.
Singapore 0511
Tel: 7787777   Fax: 7780086   Tlx: 28499 SISIR

International Construction
(Monthly)
Reed Business Publishing, Ltd.
Reed Asian Publishing Pte.
5001 Beach Rd. #06-12, Golden Mile Complex
Singapore 0719
Tel: 2913188   Fax: 2913180

All addresses and telephone numbers are in the Republic of Singapore unless otherwise noted. The country code for Singapore is [65].

International Journal of High Speed Electronics
(Quarterly)
World Scientific Publ. Company
Farrer Rd., PO Box 128
Singapore 9128
Tel: 3825663   Fax: 3825919

Journal—Singapore Computer Society
(Three times a year)
Singapore Computer Society
71 Science Park Dr.
Singapore
Tel: 7783901   Fax: 7788221

Lloyd's Maritime Asia
(Monthly)
Lloyd's of London Press (FE)
Rm. 1101 Hollywood Centre
233 Hollywood Rd.
Hong Kong
Tel: [852] 8543222   Fax: [852] 8541538

Media Asia
(Quarterly)
Asian Mass Communication Research and Information Center
39 Newton Rd.
Singapore
Tel: 2515105   Fax: 2534535   Tlx: AMICSI 55524

Media: Asia's Media and Marketing Newspaper
(Biweekly)
Media & Marketing Ltd.
1002 McDonald's Bldg., 46-54 Yee Wo St.
Causeway Bay, Hong Kong
Tel: [852] 5772628   Fax: [852] 5769171

Medicine Digest Asia
(Monthly)
Rm. 1903, Tung Sun Commercial Centre
194-200 Lockhart Rd.
Wanchai, Hong Kong
Tel: [852] 8939303   Fax: [852] 8912591

Monthly Digest of Statistics
(Monthly)
Government of Singapore
Chief Statistician
Maxwell Rd.
PO Box 3010
Singapore 9050
Tel: 3209689   Tlx: 20826 STATS

Oil & Gas News
(Weekly)
Al Hilal Publishing (FE) Ltd.
50 Jalan Sultan #20-06, Jalan Sultan Centre
Singapore 0719
Tel: 2939233   Fax: 2970862

Petrochemicals & Refining
(Monthly)
Petroleum News Publishing Pte. Ltd.
41 Middle Rd. #01-00
Singapore 0718
Tel: 3361728   Fax: 3367919

Petroleum News
(Monthly)
Petroleum News Publishing Pte. Ltd.
41 Middle Rd. #01-00
0718 Singapore
Tel: 3361728   Fax: 3367919

Polymers & Rubber Asia
(Bimonthly)
Upper West St.
Reigate Surrey RH2 9HX, UK
Tel: [44] (737) 242599   Fax: [44] (737) 223235   Tlx: 932699

Shippers' Times
(Bimonthly)
Singapore National Shippers Council
47 Hill St., SCCI Bldg.
Singapore 0617
Tel: 3372441   Tlx: 24473 FRETER

Shipping & Transport News
(Monthly)
Al Hilal Publishing (FE) Ltd.
50 Jalan Sultan #20-06, Jalan Sultan Centre
Singapore 0719
Tel: 2939233   Fax: 2970862

SIAJ (Journal of the Singapore Institute of Architects)
(Bimonthly)
Singapore Institute of Architects
20 Outram Park #02-393
Singapore 0316
Tel: 2203456

Singapore Air Cargo
(Monthly)
119 Tong Xing Complex, Ubi Ave. 4
Singapore 1440
Tel: 7478088   Fax: 7479119

Singapore Medical Journal
(Bimonthly)
Singapore Medical Association
4 A College Rd.
Singapore 0316 Singapore
Tel: 2231264

Singapore Shipping 'n' Shipbuilder
(Monthly)
Cosmic Media
PO Box 3163
Singapore

Southeast Asia Building Magazine
(Monthly)
Safan Publishing Pte.
510 Thomson Rd.
Block A #08-01, SLF Complex
Singapore 1129
Tel: 2586988   Fax: 2589945

Travel News Asia
(Bimonthly)
Far East Trade Press, Ltd.
2/F Kai Tak Commercial Bldg.
317 Des Voeux Rd.
Central, Hong Kong
Tel: [852] 5453028   Fax: [852] 5446979

Travel Trade Gazette Asia
(Weekly)
Asian Business Press Pte., Ltd.
100 Beach Rd. #26-00 Shaw Towers
Singapore 0718
Tel: 2943366 Fax: 2985534

What's New in Computing
(Monthly)
Asian Business Press Pte. Ltd.
100 Beach Rd. #26-00, Shaw Towers
Singapore 0718
Tel: 2943366 Fax: 2985534 Tlx: 25280 ABPSIN

RADIO & TELEVISION

Singapore Broadcasting Corporation (SBC)
Caldecott Hill, Andrew Rd.
Singapore
Tel: 2560401 Fax: 2538119 (Publicity), 3551503
(Radio programs)
General inquiries on SBC programming
Tel: (800) 2520331 (Office hours)
Tel: (800) 2520089 (Eves, weekends)
Government operated SBC consists of three television stations and several radio stations, with programming in four languages.

Rediffusion
1 Jln Selanting
Singapore 2159
Tel: 4671144 Fax: 4663888
Operates commercial radio services.

LIBRARIES

Development Bank of Singapore (DBS)
Resource Centre
DBS Bldg., 6 Shenton Way
Singapore 0106
Tel: 2201111 Fax: 5342886 Tlx: 2211306

Ministry of Trade and Industry Library
8 Shenton Way, Treasury Bldg.
Singapore 0106
Tel: 3209258 Tlx: 24702 MTI

National Library of Singapore
Stamford Rd.
Singapore 0617
Tel: 3377355 Tlx: NATLIB 26620

National University of Singapore Library
10 Kent Ridge Crescent
Singapore 0511
Tel: 7722069 Fax: 7773571 Tlx: 33943 UNISPO

Ngee Ann Polytechnic Library
535 Clementi Rd.
Singapore 2159
Tel: 4666555 Tlx: 39206

Singapore Trade Development Board Library
1 Maritime Square #10-40 (Lobby D)
World Trade Centre, Telok Blangah Rd.
Singapore 0409
Tel: 2790433

All addresses and telephone numbers are in the Republic of Singapore unless otherwise noted. The country code for Singapore is [65].

Index

A

Accelerated Depreciation Allowances, 43
acceptance
   legal definition of, 192
accord and satisfaction
   legal definition of, 192
accounting firms
   addresses, 283
accounting industry
   foreign investment in, 32
Acts of Parliament, 189
ACU. *See* Asian Currency Unit
addresses and telephone numbers, 267–297
advantages of doing business in Singapore, 163
advertising, 166
   addresses of agencies, 283
advice
   definition of, 234
advising bank
   definition of, 234
aerospace industry, 9, 65–66
   foreign investment in, 35
   trade fairs, 84–85
agency
   legal definition of, 192
agents, 175
   legal definition of, 192
   legal discussion of, 200
   tips on helping, 167
agriculture, 9
   trade fairs, 85
agrotechnology parks, 9
air cargo facilities, 247. *See also* Changi International Airport
air travel time to/from Singapore, 136
aircraft equipment industry
   market, 27. *See also* aerospace industry
airlines, 136
   addresses, 286–287
airport. *See* Changi International Airport
aliens. *See also* immigration; visa requirements
   legal discussion of, 197
amendment
   definition of, 234

annual fixed investment, 9
Anti-Dumping Duties, 59
apparel industry, 9, 25, 76–77
   trade fairs, 131–132
appliance industry, 9
art (commercial)
   trade fairs, 108–110
ASEAN. *See* Association of Southeast Asian Nations
Asian Currency Unit (ACU), 207, 210, 220
   licenses, 208, 209, 217
Asian dollar market, 10, 220
assignments
   legal discussion of, 197
Association of Banks in Singapore, 208
Association of Southeast Asian Nations (ASEAN), 16, 51, 54
   Free Trade Area, 16
ATA carnets, 55
attorneys. *See also* law firms
   role of, 190
audiovisual equipment industry
   trade fairs, 125–127
audiovisual rentals, 143
authentication
   legal definition of, 192
authoritarianism, 17–19. *See also* government control
automatic teller machines (ATMs), 208, 209, 211, 220
automation equipment industry, 28–29, 72
   trade fairs, 103–104
automobile and auto parts industry
   trade fairs, 87
avionics equipment industry, 65. *See also* aerospace industry
   market, 27

B

back-to-back letter of credit. *See* letter of credit
bailment
   legal definition of, 192
bank deposits
   legal discussion of, 197

Bank for International Settlements (BIS), 210
Bank of Credit and Commerce International (BCCI), 208
Banking Act, 207
banking industry, 10, 207–211
   computerization of, 208
   requirements and restrictions, 210
   trends and prospects, 211
banks
   addresses, 281
   commercial, 208
      services offered by, 209
   domestic, 209
   foreign, 208–209
   full license, 208
   merchant, 210–211
   offshore license, 208
   restricted license, 208
   retail services offered by, 209
banquet etiquette, 151–152
BCCI. *See* Bank of Credit and Commerce International
BCCS. *See* Board of Commissioners of Currency of Singapore
beneficiary
   definition of, 234
bill of exchange
   definition of, 234
   legal definition of, 192
bill of lading
   definition of, 233
   for exports, 64
   for imports, 58
biotechnology
   foreign investment in, 35
birth rate, 160
BIS. *See* Bank for International Settlements
Board of Commissioners of Currency of Singapore (BCCS), 207
   responsibility of currency issue by, 219
body language, 150
   Malay, 155
bona fide, 192
bonus pay, 8, 184, 185–186
British East India Company, 3
British Sale of Goods Act, 190, 199, 200–204
budi (Malay behavior code), 153
bus travel, 140–141
business cards
   exchange of, 140, 149
business centers, 143
business culture and etiquette, 139–140, 145–158
   further reading, 158
Business Development Scheme, 44
business directories
   addresses, 290
business entities and formation, 171–180
   further reading, 180
   glossary, 172
   reminders and recommendations, 179
   selected useful addresses, 179–180

business hours, 141–142
business licenses, 177
business negotiations, 157–158
   etiquette of, 156–157
business organization
   forms of, 171–176
business parks, 40, 45–46
business registration, 176–179
   authorizations needed, 177
   fees and expenses, 177
Business Registration Act, 175, 176, 177, 179
business service industries, 10
   trade fairs, 114–115
business services
   addresses, 143
business travel, 135–144
business yearbooks
   addresses, 290

C

C&F (cost and freight)
   definition of, 234
capacity to contract
   legal definition of, 192
   legal discussion of, 198
capital requirements
   for companies, 173
carrier
   legal definition of, 192
cash
   personal customs limits, 137
cash in advance
   terms of, 223
censorship, 17, 20
censorship fees, 55
Central Procurement Office
   procedures for registering with, 36
Central Provident Fund (CPF), 5, 17, 185, 186–187, 212–213
   employee contributions to, 245–246
   employer contributions to, 8, 240
central registration number (CR No.), 61
   application process for, 56, 61–62
certificate of manufacture
   definition of, 233
certificate of origin
   definition of, 233
   for exports, 64
   for imports, 58
chambers of commerce. *See also* International Chamber of Commerce; Singapore International Chamber of Commerce
   addresses, 275–276
Changi International Airport, 136, 137, 163, 247
   foreign investment in, 35
   ground travel to/from, 137
chattel
   legal definition of, 192
chemical industry, 9, 24–25, 70–71
   trade fairs, 112–114

INDEX

Chicago Mercantile Exchange
  SIMEX agreement with, 216
Chinese businesspeople, 146
Chinese culture, 146–153
Chinese export trade fairs, 101–103
Chinese language
  use of, 146. *See also* languages in use
chopsticks
  etiquette of use, 151
CIDB. *See* Construction Industry Development Board
CIF (cost, insurance, and freight) terms
  definition of, 234
Civil Aviation Authority, 35
Civil Law Act, 189, 190, 200
civil service
  efficiency of, 20
classification
  advance ruling for imports, 55
climate, 136
CLOB (Central Limit Order Book) system, 214
CLOB International, 215
clothing appropriate for business, 136
collecting bank
  definition of, 235
commerce sector of economy, 10
commercial and industrial space
  leasing of, 45–46, 143
commercial invoice
  definition of, 233
  for exports, 63–64
commercial register
  legal discussion of, 197–198
commodities exchanges, 217
  addresses, 282
common law, 198
Communists
  establishment of PAP by, 4
companies, 171–174
  governance of, 173–174
  limited by guarantee, 171
  limited by shares, 171
  limited by shares and guarantee, 171–172
  listing requirements for stock exchange, 215
  public vs. private, 172–173
  registration procedures for, 178
  unlimited, 172
Companies Act, 171
  dissolution of companies under, 174
  formation of business entities under, 176
  regulation of commercial register by, 197
  regulation of companies by, 174
  regulation of company registration by, 179, 190
  regulation of foreign business entities, 177
computer industry, 9, 24, 66–68
  trade fairs, 87–91
computer rental firm addresses, 143
computer software industry, 66
  market, 28
Confucianism, 139–140, 145, 146–147

construction industry, 10, 34, 72–73
  trade fairs, 91–94
Construction Industry Development Board (CIDB), 34
  procedures for registering with, 36–37
consular invoice
  definition of, 233
consumer prices
  by category, 161
  consumer price index (CPI)
    (1982-1992), 7, 160
  increases
    (1975-1992), 160
consumption patterns, 161
contracts
  for sale of goods
    legal discussion of, 200–201
  international sales provisions, 194–196
  legal discussion of, 198–199. *See also* sales: legal discussion of
  practical application of, 190
Controller of Work Permits, 182
Copyright Act, 190, 199
Copyright International Protection Regulations, 199
copyrights
  legal discussion of, 199–200
corporate debt
  trading of, 216
corruption, 6
counteroffer
  legal definition of, 192
countertrade, 55, 61
countervailing duties, 59
courier services, 143
CPF. *See* Central Provident Fund
credit availability, 45
credit cards, 220
  issuance of, 209
crime, 6
  punishment of, 145
cuisine (Singaporean), 138–139
currency, 137, 219
  internationalization of, 45
  limits on international transactions, 219
  restrictions on, 210
current issues, 17–21
Customs and Excise Department, 53, 54, 57, 63
customs brokerage firms
  addresses, 287–289
customs clearance
  personal, 137, 144
customs declaration
  inward (for imports), 56
  outward (for exports), 62
customs duties. *See also* tariffs
  basis of assessment, 54
  on samples and advertising matter, 55
customs fines and penalties, 55

D

DBS. *See* Development Bank of Singapore
death rate, 160
debt markets, 216
demographics, 159–162
dental care industry
  market, 29–30
dentists, 144
design service industries, 33
Design Venture Program, 33
designs
  registered, 199–200
Development Bank of Singapore (DBS), 209, 211
dictionary (Chinese/English), 251–266
dining etiquette
  Chinese, 151–152
  Malay and Indian, 154–155
diplomatic mission addresses
  foreign in Singapore, 271–274
  of Singapore, 269–271
direct marketing
  advantages/disadvantages of, 165
discrepancy
  definition of, 235
distributorships, 175
  advantages/disadvantages of opening, 165
dock receipt
  definition of, 233
document against acceptance (D/A), 223, 224
  definition of, 235
document against payment (D/P), 223, 224
  definition of, 235
documentary collection, 223, 224
  definition of, 235
  procedures, 224
  tips for buyers and suppliers, 225
  types of, 224
dollar (Singapore). *See* currency; foreign exchange
dollar (US)
  acceptance of, 219
Double Tax Deduction Scheme, 44
draft
  definition of, 234
drinking etiquette, 151, 152
Dutch influence, 3

E

Economic Development Board (EDB), 15, 52, 163, 179
  administration of tax incentive programs, 42–43
  approval of foreign workers, 182
  assistance schemes, 213
  concessions for companies limited by shares, 171
  description of, 172
  functions of, 46
  incentives for foreign investment, 179
  worldwide office addresses, 276–277
economy, 3–16
  context of, 5–6
  government control of, 5
  government development strategy, 14–15
  history of, 3–4
  labor, 181–183
  political outlook for, 15–16
  sectors of, 8–9
  size of, 4–5
  structure of, 8
  underground, 6
EDB. *See* Economic Development Board
education industry
  trade fairs, 94
educational system, 160, 183
electric current standards, 55
electronics industry, 9, 23–24, 68–69
  market, 28
  production and test equipment market, 28
  trade fairs, 95–97
emergency information, 143–144
emigration from Hong Kong, 183
employer organizations, 187
employment
  conditions of, 184–185
  termination of, 184–185
Employment Act, 184
employment benefits, 185–187
energy consumption, 162
energy industries
  trade fairs, 98–100
English language, 163
  use of, 146, 183. *See also* languages in use
environmental protection, 55
environmental service industries
  foreign investment in, 33, 35
  trade fairs, 98–100
equitable assignment
  legal definition of, 192
equities markets, 213–216
ethnic composition of Singapore, 140, 145, 146, 155, 160, 181
ex dock
  definition of, 234
ex parte
  legal definition of, 192
ex point of origin
  definition of, 234
expenses (typical daily business travel), 144
export controls, 61
export licenses, 61
  definition of, 233
export marketing
  five tips for, 166–168
export procedures, 61–64
exporters
  leading, 49
exporting opportunities, 27–31
exports
  by country, 51

leading, 12, 48–49
top 10 by commodity, 49
top 10 by percentage increase, 48
total (1982-1992), 11
external debt, 9

F

face (concept), 140, 147
factory buildings
   prices of, 45
family
   importance of, 140, 146, 147–148, 153
FAS (free alongside ship) terms
   definition of, 234
fax service, 142
FCPA. *See* Foreign Corrupt Practices Act
Federation of Malaya, 4
fertility rate, 160
film and video industries
   foreign investment in, 33
finance companies, 212
financial institutions, 207–218
   further reading, 218
financial markets, 213–218
financial service industries, 10, 41
   foreign investment in, 32
   trade fairs, 114–115
financing
   specialized, 213
   underground, 213
fish (ornamental) industry, 26
   trade fairs, 86
fishing industry, 9
floating rate notes (FRNs), 45
FOB (free on board) terms
   definition of, 234
food. *See also* cuisine (Singaporean)
   Hindu restrictions on, 155
   Muslim restrictions on, 153
food industry, 69–70
   trade fairs, 105–106
food processing industry, 9, 72
   market, 30
   trade fairs, 105–106
Foreign Corrupt Practices Act (FCPA), 191
foreign exchange, 219–221
   controls, 14, 179, 219–220, 240
   further reading, 221
   legal discussion of, 200
   market, 10, 210, 217, 220
   operations, 220
   personal, 137
   rates, 221
      detailed (1991-1993), 221
      year-end (1981-1993), 220
   use of ACUs in, 210
foreign investment, 14, 41–46
   assistance, 46
   by Japan, 14, 42
   by the European Community, 42
   by the United States, 14, 42
   climate and trends, 41
   ease of, 163
   incentive schemes, 43–44
   incentives, 14
   opportunities, 32–33
   outward, 15
   policy, 42
   restrictions, 14
   size of, 14
   tax incentives, 42–43
foreign investors
   leading, 41–42
foreign reserves. *See* international reserves
foreign trade, 11–12, 47–52
   balance of, 11–12, 49–50
   deficit, 12
   early growth of, 3–4
   government strategy and development for, 51
   growth and development of, 47
   growth of, 11
   partners, 13–14, 50–51
      leading, 50
   total, 48
      (1982-1992), 11
   with emerging economies, 51
   with Hong Kong, 13–14
   with Japan, 13–14, 50
   with Malaysia, 13–14, 50
   with Saudi Arabia, 13–14
   with the European Community, 13–14, 50
   with the United States, 13–14, 50
forestry industry
   trade fairs, 86–87
forward foreign future
   definition, 235
franchising
   foreign investment in, 33
   trade fairs, 115
free trade zones, 38
Frustrated Contract Act, 190, 198, 199
furniture industry, 9
   trade fairs, 106–108
futures exchange, 32, 216–217

G

General Agreement on Tariffs and Trade (GATT), 52
   Convention Covering Anti-Dumping Duties, 59
   impact on banking industry, 211
geography of Singapore, 3
gifts
   etiquette of giving and receiving, 148–149
giftware industry
   trade fairs, 108–110
Goh Chok Tong (Prime Minister), 16, 21, 145
   appointment of, 18–19
Gold Exchange of Singapore, 216, 217
gold markets, 217–218

government
  agency addresses, 267
  securities issued by, 216
  structure of, 189
government control, 5
  examples of, 11
  of economy, 14–15, 15
  of foreign trade, 51
government run corporations
  addresses, 274
graphic arts equipment industry
  market, 30–31
greeting etiquette
  Chinese, 149–150
  Malay, 154
gross domestic product (GDP), 4
gross national product (GNP)
  (1982-1992), 5
  growth rate, 161
  per capita, 4–5, 161, 181
Growth Triangle, 15, 21
guidebooks on Singapore, 144

H

handshaking etiquette, 149, 154
Harmonized System, 53, 61
health care industry
  market, 29
health expenditures, 162
health precautions for travelers, 143
health products industry, 73–74
high-technology industries, 9–10, 41
  effects of Information Technology program on, 19–20
  R&D efforts, 38
Hinduism, 155
history
  of culture, 145
  of economy, 3–4
holidays, 141
hospitals and clinics, 143
hotels, 138
housewares industry
  trade fairs, 106–108
human resources, 183–184

I

ICC. *See* International Chamber of Commerce
IMF. *See* International Monetary Fund
immigration
  relaxation of restrictions on, 20
immunization requirements, 136
import declaration, 58
import licenses, 58
  definition of, 233
import policy, 53–56
import procedures, 53, 56–59
import regulations
  food, health and safety, 54

importing
  opportunities, 23–26
  regulation of, 53
imports
  by country, 51
  leading, 13, 48–49
  top 10 by commodity, 48
  top 10 by percentage increase, 48
  total (1982-1992), 11
income. *See* gross national product (GNP): per capita
Indian culture, 155–156
industrial estates, 40, 45
  map of, 39
  Suzhou (China), 15, 41
industrial materials industry, 70–71
  trade fairs, 112–114
Industrial Relations Act, 187
industrial standards, 54–55
industries
  reviews of, 65–77
industry sector of economy, 9–10
inflation, 7, 161
  (1982-1992), 6
information technology industries
  trade fairs, 87–91
Information Technology program, 19–20
infrastructure, 47, 163
  development of, 51
  financial, 213, 217
inspection
  certificates
    definition of, 233
  of imports, 55
Institute of Technical Education, 184
instrument industry, 9, 77
  market, 29
  trade fairs, 116
insurance
  accident, 186
  documents
    definition of, 233
insurance certificates
  for imports, 58
insurance companies
  addresses, 282
insurance industry, 212
  foreign investment in, 32
intellectual freedom, 18
intellectual property rights, 6, 190
interactive databases, 19
interest rates, 44, 209, 217, 221
Internal Security Act, 4
International Chamber of Commerce (ICC), 224, 226, 234, 236
  arbitration of trade disputes by, 46
international funds management, 216
International Monetary Fund (IMF), 219
international organization membership, 16, 51–52
international payments, 219, 223–236
  further reading, 236

glossary, 234
international reserves, 11–12, 18, 50, 221
international trade documents
    glossary, 233
Internet, 20
INTRACO (Import Export Company), 48
    address, 274
investment guarantee agreements, 213
irrevocable confirmed credit. *See* letter of credit
irrevocable credit. *See* letter of credit
issuance
    definition of, 235
issuing bank
    definition of, 235
IT 2000 Plan, 19, 20, 28

J

Japan
    occupation by, 3
jewelry industry, 9, 26
    trade fairs, 108–110
joint ventures, 176
    advantages/disadvantages of, 165–166
    registration procedures for, 178
Jurong Environmental Engineering (JEE), 34
Jurong Town Corporation (JTC), 249
    industrial land development by, 34
    management of commercial property by, 45
    management of industrial estates, 40

L

labeling requirements
    for imports, 58–59
labor, 7–8, 181–187
    availability, 181–182
    costs, 8
    policy, 185
    shortage, 7, 20–21, 181, 182
labor force, 7
    composition of, 181
    foreign, 7, 20, 182–183
        fees for, 182
    women in, 7, 21, 184
labor relations, 187
    arbitration, 187
labor unions, 8, 187
languages in use, 146, 163, 169, 181
law, 189–206
    geographical scope of, 190
    governing business, 189–190
    practical application of, 190
law digest, 197–206
law firms
    addresses, 284–285
Lee Kuan Yew (former Prime Minister), 4, 16
    legacy of, 17–19, 145
legal disputes
    resolution of, 190
legal glossary, 192–194

legal system
    basis of, 189
letter of credit (L/C), 223, 226, 235
    amendment of, 227
    application, 230, 231
    back-to-back, 229, 234
    common problems, 228
    confirmed, 235
    deferred payment, 229, 235
    irrevocable, 235
    irrevocable confirmed credit, 229
    irrevocable credit, 229
    issuance of, 227
    legal obligations of banks toward, 228
    opening, 230
    parties to, 226
    red clause, 229
    revocable, 229, 235
    revolving, 229, 235
    special, 229
    standby, 229, 235
    steps in using, 226
    tips for buyers and sellers, 232
    transferable, 229
    types of, 229
    unconfirmed, 229
    utilization of, 228
letters of introduction, 149–150
libraries
    addresses, 297
licensing agreements, 175
    registration procedures for, 178
life expectancy, 160
literacy rate, 183
loans
    availability of, 45
Local Enterprise Finance Scheme, 43–44
Local Enterprise Technical Assistance Scheme, 44
Loco-London market, 218
Loco-Singapore market, 218
London Metals Exchange, 218

M

machine industry, 9, 25, 72–73
    trade fairs, 116
magazines. *See also* periodicals
    English-language, 143
Malay culture, 153–155
Malay language
    use of, 146. *See also* languages in use
manufacturing, 9–10
marketing, 163–169
marking requirements
    for imports, 58–59
MAS. *See* Monetary Authority of Singapore
Mass Rapid Transport System (MRT), 35, 140
maternity leave, 184
medical product industry, 73–74
    market, 29
    trade fairs, 117–123

messenger services, 143
metal industry, 25, 70
    trade fairs, 124–125
Ministry of Finance (MOF), 172, 176
    description of, 172
Ministry of Trade and Industry (MTI)
    description of, 172
MOF. *See* Ministry of Finance
Monetary Authority of Singapore (MAS), 45, 177, 179, 207, 207–208
    description of, 172
    monitoring of stock exchange by, 214
    requirements for bank operations, 210
money markets, 216
motor vehicles in use, 162
MRT. *See* Mass Rapid Transport System
MTI. *See* Ministry of Trade and Industry
multimedia equipment industry
    trade fairs, 125–127
Muslims, 153, 155

N

names
    Chinese, 151
    Indian, 156
    Malay, 154
Nanyang Technological University, 183
National Association of Securities Dealers (NASD)
    links to, 216
National Computer Board, 19
National Design Center, 33
National Information Technology Plan, 28
National Science and Technology Board
    administration of R&D Assistance Scheme, 43
National Trade Union Congress (NTUC), 8, 187
National University of Singapore, 183
National Wages Council (NWC), 8, 185
negotiable certificates of deposit (NCDs), 45
negotiable instrument, 192
negotiation
    definition of, 235
newspaper advertising, 166
newspapers
    addresses, 292
    circulation of, 162
    English-language, 142–143
nexus
    legal definition of, 193
Ngee Ann Polytechnic, 33
note issuance facilities (NIFs), 45
NTUC. *See* National Trade Union Congress
nutrition, 162
NWC. *See* National Wages Council

O

OCBC. *See* Overseas-Chinese Banking Corporation
offer
    legal definition of, 193
office space
    leasing of, 46, 143
offices
    branch, 174
        operating requirements, 174
        registration procedures for, 178
    liaison, 175
    representative, 174–175, 179
        advantages/disadvantages of, 163
        registration procedures for, 178
oil refining industry, 9, 23. *See also* petrochemical industry
oil rig fabrication industry, 9
Ong Teng Cheong (President), 16
    1993 election of, 18
open account (O/A) terms, 223
    definition of, 235
orchid-growing industry, 26
    trade fairs, 86
Overseas Union Bank (OUB), 209
Overseas-Chinese Banking Corporation (OCBC), 209

P

packaging industry
    trade fairs, 127–129
packing list
    definition of, 233
    for exports, 64
PAP. *See* People's Action Party
partnerships, 175–176
    registration procedures for, 178, 178–179
    structure and operation of, 175–176
patents, 6
    legal discussion of, 200
PCS. *See* Petrochemical Corporation of Singapore
pension funds. *See* Central Provident Fund
People's Action Party (PAP), 4, 16, 17
    opposition to, 145
periodicals
    general business and trade
        addresses, 292
    industry-specific
        addresses, 293–297
personal connections
    cultivating, 148
    tips for building, 164–165
    value of, 147–148
Petrochemical Corporation of Singapore (PCS), 35
petrochemical industry, 70
    foreign investment in, 35
    trade fairs, 129–130
petroleum refining industry, 23
Pharmaceutical Department
    procedures for registering with, 37
pharmaceutical industry, 9, 73–74
    foreign investment in, 32
    trade fairs, 117–123
pin-yin. *See* transliteration systems
political outlook

for economy, 15–16
population, 3, 181
   (1980-2000), 159
   age structure, 159
   by age and sex, 159
   density, 3
   growth of, 3
   growth rate and projections, 159
port facilities, 247, 249
   foreign investment in, 35
Port of Singapore Authority, 35
   regulation of transit trade by, 57, 62
Portuguese
   influence of, 3
Postal Savings Bank (POSBank), 211–212
   postal savings system, 208
postal service, 142
power of attorney
   legal definition of, 193
   legal discussion of, 200
presidential election, 16
prima facie
   legal definition of, 193
prime rate, 7, 209
principal
   legal definition of, 193
   legal discussion of, 200
printing industry, 9, 25–26
   market, 30–31
   trade fairs, 127–129
private enterprises
   restrictions on, 177
Product Development Assistance Scheme, 43
promissory note
   definition of, 234
public procurement
   entering bids on
      advantages/disadvantages of, 166
   invitations to bid on, 37
   opportunities, 34–35
   process, 36–37
   regulatory authorities, 36
Public Utilities Board (PUB), 34
public works projects, 15–16, 34–35
publishing industry, 25–26
   trade fairs, 94–95

Q

quantum meruit
   legal definition of, 193

R

radio
   addresses, 297
   English-language, 143
   sets in use, 162
radio advertising, 166
Raffles, Sir Thomas Stamford, 3
RAS Commodities Exchange, 217

RCB. *See* Registry of Companies and Businesses
red clause letter of credit. *See* letter of credit
reexports, 12–13, 48, 57–58, 62–63
refrigeration equipment industry
   market, 30
Registry of Companies and Businesses (RCB), 174, 176, 179, 197
   description of, 172
Registry of Manufactures, 177
religions, 160
remitter
   definition of, 235
remitting bank
   definition of, 235
repatriation of profits, 179, 219–220
   from interest, 44
rescind
   legal definition of, 193
Research and Development Assistance Scheme, 43
restaurants, 138–139
retail industry, 10
retail stores
   advantages/disadvantages of opening, 165
retirement age, 21, 186
revocable credit. *See* letter of credit
revolving underwriting facilities (RUFs), 45
road construction
   foreign investment in, 34–35
robotics equipment industry, 72
   market, 28–29
   trade fairs, 103–104
rubber industry. *See also* RAS Commodities Exchange
   establishment of, 3

S

salaam (Malay greeting), 154
salaries. *See* wages
sales. *See also* British Sale of Goods Act
   glossary of terms, 234
   legal discussion of, 200–204
   seven rules for, 168–169
savings rate, 9, 12, 50, 213
seal
   legal definition of, 193
secretarial services, 143
Securities Industry Act and Regulations, 214
security and safety industries
   market, 31
service industries, 10–11, 41
   foreign investment in, 32
SES. *See* Stock Exchange of Singapore
SESDAQ. *See* Stock Exchange of Singapore Dealing and Automated Quotation System (SESDAQ)
shares
   issuance of, 173
ship repair and shipbuilding industries, 9, 74–75, 249
   foreign investment in, 32
   trade fairs, 84–85

shipper's export declaration
  definition, 233
SICC. *See* Singapore International Chamber of Commerce
sick leave, 184
SIMEX. *See* Singapore International Monetary Exchange
Singapore Broadcasting Corporation, 166
Singapore Convention Bureau
  trade fair listings by, 79–80
Singapore Customs Ordinance, 59
Singapore Federation of Chambers of Commerce and I, 52
Singapore Government Gazette, 37
Singapore Hotel Association, 138
Singapore Institute of Standards and Industrial Re, 54–55
Singapore International Chamber of Commerce (SICC), 179
  functions of, 46
Singapore International Monetary Exchange (SIMEX), 32, 216, 217
Singapore market
  seven ways to enter, 163–166
Singapore National Employers Federation, 187
Singapore Petroleum Company
  address, 274
Singapore Science Park, 38
  tenant application process, 38
Singapore Telecom, 34
  sale of shares in, 10
Singapore Tourist Promotion Board (STPB)
  office addresses, 135
Singapore Trade Development Board (STDB), 19, 52, 80, 179
  administration of Design Venture Program, 33
  administration of Double Tax Deduction Scheme, 44
  approval of representative offices, 175
  description of, 172
  functions of, 46
  regulation of central registration numbers by, 56
  regulation of exports, 61
  regulation of importing, 53
  worldwide office addresses, 277–278
Singlish, 146
Skills Development Fund, 183
skin care products industry
  market, 31
small and medium enterprises (SMEs), 9
social planning by government, 5
social security system. *See* Central Provident Fund
sole proprietorships, 176
  registration procedures for, 178, 178–179
special trade zones, 38–40
sporting goods industry
  trade fairs, 130–131
sports facilities
  foreign investment in, 33
standby letter of credit. *See* letter of credit

stationery industry
  trade fairs, 108–109
Statute of Frauds
  legal definition of, 193
  legal discussion of, 204
STDB. *See* Singapore Trade Development Board
Stock Exchange of Singapore (SES), 10, 213–214
  Dealing and Automated Quotation System (SESDAQ), 215
  investment by CPF in, 213
  regulation and operation of, 214–215
stock exchanges
  addresses, 282
  foreign investment in, 32
  secondary, 215–216
STPB. *See* Singapore Tourist Promotion Board
Straits Settlements, 3
Straits Times Index, 214
Strategic Economic Plan, 15
subway travel, 140
Supreme Court, 189

T

table manners
  Chinese, 152
  Malay and Indian, 154–155
tariffs, 14, 53–54
  ASEAN Tariff Amendments, 54
  classification of, 61
tax codes
  revision of, 14
tax concessions
  on ACU operation income, 210
tax credits, 239
tax incentives, 237–238
tax treaties, 239, 246
  withholding rates under, 240–241
taxes, 38, 163
  administration of, 238, 246
  capital gains, 238, 245
  concessions for nonresidents, 44–45
  corporate, 237–241
  estate, 44–45, 245
  excise, 55
  exemptions, 237–238
  income
    corporate, 237–239
    personal, 243–244
    personal nonresident, 245
  motor vehicle registration, 55–56
  on film admission tickets, 55
  personal, 243–246
  sales, 163
  skills development levy, 240
  value-added tax (VAT), 14
taxis, 141
technical assistance agreements, 175
  registration procedures for, 178
Technology Parks Pte Ltd., 38

Telecommunications Authority of Singapore, 34
telecommunications equipment industry, 24, 75–76
　market, 27
　trade fairs, 87–89
telecommunications system, 10, 19, 34, 142–143
telegram service, 142
telephone
　international dialing codes, 142
　sets in use, 162
　system, 142
Teletech Park, 40
television
　addresses, 297
　English-language, 143
　sets in use, 162
television advertising, 166
telex service, 142
textile industry, 9, 25, 76–77
　trade fairs, 131–132
thing in action
　legal definition of, 193
time zones, 137
timepiece industry
　trade fairs, 108–110
tipping, 137
tobacco consumption, 162
tool industry, 77
　trade fairs, 132–133
tourism industry
　foreign investment in, 35
　trade fairs, 111–113
trade association addresses
　general, 275
　industry-specific, 279
trade delegations
　etiquette of, 156–157
trade fairs, 79–134
　advantages/disadvantages of, 163–165
　tips for attending, 81–83
Trade Marks Act, 190, 204–206
trademarks, 6, 190
　legal discussion of, 204–206
　registration process, 205
TradeNet, 19, 56–57, 61, 63
trading firms
　advantages/disadvantages of hiring, 165
translators and interpreters, 143
　addresses, 285
transliteration systems, 251
transportation, 10. *See also* Changi International Airport; port facilities
　domestic, 140–141
　map of, 248
transportation firms, 287–289
transshipment, 57–58
　for exports, 62–63
travel. *See* business travel
traveler's checks, 137, 220

U

ultra vires
　legal definition of, 193
unconfirmed credit. *See* letter of credit
unemployment, 7–8, 20, 183
　comparative, 182
Unfair Contract Terms Act, 190, 198
Uniform Customs and Practices (UCP), 226
Uniform Rules for Collections (URC), 224
United Kingdom
　colonization of Singapore by, 3
　granting of independence by, 3–4
　influence on legal system, 189, 189–190
United Nations, 16, 52
United Overseas Bank (UOB), 209
UOB. *See* United Overseas Bank
utilities
　foreign investment in, 34

V

vacation benefits, 184
validity
　definition of, 235
VAT. *See* taxes: value-added tax
venture capital financing, 45
visa requirements, 135–136
Vision 1999, 15
vocational training, 183–184

W

Wade-Giles. *See* transliteration systems
wages, 185–187
　average weekly
　　by industry, 186
　　by occupation, 186
　　comparative, 185
　minimum, 186
　monthly manufacturing (1986-1991), 161
　overtime, 184
　policy, 185
warehousing, 57–58
Westernization
　fears of, 18, 147
women in business and society, 150, 153, 154. *See also* labor force: women in
Workmen's Compensation Act, 186
workweek, 8, 184
world trade center
　address, 275